The Wisdom of the Hive

THE WISDOM OF THE HIVE

The Social Physiology of Honey Bee Colonies

THOMAS D. SEELEY

HARVARD UNIVERSITY PRESS

Cambridge, Massachusetts

London, England 1995

Printed in the United States of America

Library of Congress Cataloging-in-Publication Data

Seeley, Thomas D.
The wisdom of the hive : the social physiology of honey bee colonies / Thomas D. Seeley.
p. cm.
Includes bibliographical references and index.
ISBN 0-674-95376-2 (acid-free)
1. Honeybee—Food. 2. Honeybee—Behavior. I. Title.
QL568.A6S445 1995
595.79'9—dc20 95-3645

Designed by Gwen Frankfeldt

To Saren and Maira, who waited patiently,
and to Robin, who helped in all ways

Contents

Preface *xi*

PART I. INTRODUCTION 1

1. The Issues 3
    - 1.1. The Evolution of Biological Organization *3*
    - 1.2. The Honey Bee Colony as a Unit of Function *7*
    - 1.3. Analytic Scheme *16*
2. The Honey Bee Colony 22
    - 2.1. Worker Anatomy and Physiology *23*
    - 2.2. Worker Life History *28*
    - 2.3. Nest Architecture *31*
    - 2.4. The Annual Cycle of a Colony *34*
    - 2.5. Communication about Food Sources *36*
    - 2.6. Food Collection and Honey Production *39*
3. The Foraging Abilities of a Colony 46
    - 3.1. Exploiting Food Sources over a Vast Region around the Hive *47*
    - 3.2. Surveying the Countryside for Rich Food Sources *50*
    - 3.3. Responding Quickly to Valuable Discoveries *52*
    - 3.4. Choosing among Food Sources *54*
    - 3.5. Adjusting Selectivity in Relation to Forage Abundance *59*

3.6. Regulating Comb Construction *61*
3.7. Regulating Pollen Collection *63*
3.8. Regulating Water Collection *65*

Summary 66

PART II. EXPERIMENTAL ANALYSIS 69

4. Methods and Equipment 71

4.1. The Observation Hive *71*
4.2. The Hut for the Observation Hive *74*
4.3. The Bees *75*
4.4. Sugar Water Feeders *77*
4.5. Labeling Bees *79*
4.6. Measuring the Total Number of Bees Visiting a Feeder *81*
4.7. Observing Bees of Known Age *81*
4.8. Recording the Behavior of Bees in the Hive *81*
4.9. The Scale Hive *82*
4.10. Censusing a Colony *83*

5. Allocation of Labor among Forage Sites 84

How a Colony Acquires Information about Food Sources 85

5.1. Which Bees Gather the Information? *85*
5.2. Which Information Is Shared? *88*
5.3. Where Information Is Shared inside the Hive *88*
5.4. The Coding of Information about Profitability *90*
5.5. The Bees' Criterion of Profitability *94*
5.6. The Relationship between Nectar-Source Profitability and Waggle Dance Duration *98*
5.7. The Adaptive Tuning of Dance Thresholds *102*
5.8. How a Forager Determines the Profitability of a Nectar Source *113*

Summary 119

How a Colony Acts on Information about Food Sources 122

5.9. Employed Foragers versus Unemployed Foragers *122*
5.10. How Unemployed Foragers Read the Information on the Dance Floor *124*

5.11. How Employed Foragers Respond to Information about Food-Source Profitability *132*
5.12. The Correct Distribution of Foragers among Nectar Sources *134*
5.13. Cross Inhibition between Forager Groups *142*
5.14. The Pattern and Effectiveness of Forager Allocation among Nectar Sources *145*

Summary 151

6. Coordination of Nectar Collecting and Nectar Processing 155

How a Colony Adjusts Its Collecting Rate with Respect to the External Nectar Supply 156

6.1. Rapid Increase in the Number of Nectar Foragers via the Waggle Dance *156*
6.2. Increase in the Number of Bees Committed to Foraging via the Shaking Signal *158*

How a Colony Adjusts Its Processing Rate with Respect to Its Collecting Rate 162

6.3. Rapid Increase in the Number of Nectar Processors via the Tremble Dance *162*
6.4. Which Bees Become Additional Food Storers? *173*

Summary 174

7. Regulation of Comb Construction 177

7.1. Which Bees Build Comb? *177*
7.2. How Comb Builders Know When to Build Comb *181*
7.3. How the Quantity of Empty Comb Affects Nectar Foraging *187*

Summary 191

8. Regulation of Pollen Collection 193

8.1. The Inverse Relationship between Pollen Collection and the Pollen Reserve *194*
8.2. How Pollen Foragers Adjust Their Colony's Rate of Pollen Collection *195*
8.3. How Pollen Foragers Receive Feedback from the Pollen Reserves *198*

8.4. The Mechanism of Indirect Feedback *201*
8.5. Why the Feedback Flows Indirectly *204*
8.6 How a Colony's Foragers Are Allocated between Pollen and Nectar Collection *207*

Summary 209

9. Regulation of Water Collection 212

9.1. The Importance of Variable Demand *213*
9.2. Patterns of Water and Nectar Collection during Hive Overheating *215*
9.3. Which Bees Collect Water? *218*
9.4. What Stimulates Bees to Begin Collecting Water? *220*
9.5. What Tells Water Collectors to Continue or Stop Their Activity? *221*
9.6. Why Does a Water Collector's Unloading Experience Change When Her Colony's Need for Water Changes? *226*

Summary 234

PART III. OVERVIEW 237

10. The Main Features of Colony Organization 239

10.1. Division of Labor Based on Temporary Specializations *240*
10.2. Absence of Physical Connections between Workers *244*
10.3. Diverse Pathways of Information Flow *247*
10.4. High Economy of Communication *252*
10.5. Numerous Mechanisms of Negative Feedback *255*
10.6. Coordination without Central Planning *258*

11. Enduring Lessons from the Hive 263

Glossary *269*
Bibliography *277*
Index *291*

Preface

In the fall of 1978, having just completed a Ph.D. thesis, I wondered what to study next with the bees, my favorite animals for scientific work. One subject that greatly attracted me was the organization of the food-collection process in honey bee colonies. The recent work by Bernd Heinrich, beautifully synthesized in his book *Bumblebee Economics,* had demonstrated the success of viewing a bumble bee colony as an economic unit shaped by natural selection to be efficient in its collection and consumption of energy resources. I was intrigued by the idea of applying a similar perspective to honey bees. Because colonies of honey bees are larger than those of bumble bees and possess more sophisticated communication systems, it was obvious that they must embody an even richer story of colony design for energy economics. Of course, much was known already about the inner workings of honey bee colonies, especially the famous dance language by which bees recruit their hivemates to rich food sources. This communication system had been deciphered in the 1940s by the Nobel laureate Karl von Frisch, and its elucidation had set the stage for one of his students, Martin Lindauer, to conduct in the 1950s several pioneering studies which dealt explicitly with the puzzle of colony-level organization for food collection. Their discoveries and those of many other researchers provided a solid foundation of knowledge on which to build, but it was also clear that many mysteries remained about how the thousands of bees in a hive function as a coherent system in gathering their food.

It seemed that the best way to begin this work was to describe the foraging behavior of a whole colony living in nature, for simply observing a phenomenon broadly is generally an invaluable first step toward understanding it. So in the summer of 1979, Kirk Visscher and I teamed up to determine the spatiotemporal patterns of a colony's foraging operation. To do this, we established a colony in a glass-walled observation hive, monitored the recruitment dances of the colony's foragers, and plotted on a map the forage sites being advertised by these dances. This initial study revealed the amazing range of a colony's foraging—more than 100 square kilometers around the hive—and the surprisingly high level of dynamics in a colony's forage sites, with almost daily turnover in the recruitment targets. It also presented us with the puzzle of how a colony can wisely deploy its foragers among the kaleidoscopic array of flower patches in the surrounding countryside. From here on, the course of the research arose without a grand design as I and others simply probed whatever topic seemed most interesting in light of the previous findings. Even the central theme of this book—the building of biological organization at the group level—emerged of its own accord from these studies.

This book is not just about honey bees. These aesthetically pleasing and easily studied insects live in sophisticated colonies that vividly embody the answer to an important question in biology: What are the devices of social coordination, built by natural selection, that have enabled certain species to make the transition from independent organism to integrated society? The study of the honey bee colony, especially its food collection, has yielded what is probably the best-understood example of cooperative group functioning outside the realm of human society. This example deepens our understanding of the mechanisms of cooperation in one species in particular and, by providing a solid baseline for comparative studies, helps us understand the means of cooperation within animal societies in general. In writing this book, I have tried to summarize—in a way intelligible to all—what is currently known about how the bees in a hive work together as a harmonious whole in gathering their food. This book will have served its purpose if readers can gain from it a sense of how a honey bee colony functions as a unit of biological organization.

I owe deep thanks to many people and institutions that have helped me produce what I report here. First, there are the many summer assistants without whose help most of the experiments presented here could not have been done. In temporal succession, they are Andrea Masters, Pepper Trail, Jane Golay, Ward Wheeler, Andrew Swartz,

Roy Levien, Oliver Habicht, Mary Eickwort, Scott Kelley, Samantha Sonnak, Kim Bostwick, Steve Bryant, Tim Judd, Erica Van Etten, Barrett Klein, Cornelia König, and Anja Weidenmüller. Several graduate students at Cornell have also contributed greatly to the body of work contained in this book, through their dissertation research: Kirk Visscher, Francis Ratnieks, Scott Camazine, Stephen Pratt, and James Nieh. Susanne Kühnholz, from the University of Würzburg, also joined our group and contributed important findings. John Bartholdi, Craig Tovey, and John Vande Vate of the School of Industrial and Systems Engineering, Georgia Institute of Technology, have taught me much about the operations research approach to the analysis of group organization. I am also most grateful to the United States National Science Foundation (Animal Behavior Program) and Department of Agriculture (Hatch Program) for providing me and others with the financial assistance which was indispensable for most of the research reported here. Equally essential to the success of my own research program has been the support of Professor William Shields and his colleagues at the Cranberry Lake Biological Station (School of Environmental Science and Forestry, State University of New York), who have kindly hosted me and my assistants, and so made possible the performance of many experiments requiring a setting where the bees can find few natural sources of food.

The writing of this book began while I was on sabbatical leave with my family, living in the farmhouse at Tide Mill Farm, in Edmunds, Maine. All of the Bell family—our landlords, neighbors, and friends—were most welcoming and accommodating, and a special note of warm thanks goes to them for making our stay so enjoyable. During this time I received a Guggenheim Fellowship, which was essential to getting the book started. The completion of the writing was made possible by a fellowship at the Institute for Advanced Study in Berlin, which was kindly arranged by Professor Rüdiger Wehner of the University of Zürich. Professor Wolf Lepenies and his colleagues in Berlin were most supportive, and I and my family remember fondly our four months in Berlin. While in Germany, I benefited greatly from interactions with marvelous coworkers at the Institute: Scott Camazine, Jean-Louis Deneubourg, Nigel Franks, Sandra Mitchell, and Ana Sendova-Franks. I am very grateful to Kraig Adler, Chairman of the Section of Neurobiology and Behavior (NBB) at Cornell University, who kindly helped arrange the temporary seclusion that I needed for writing, and to my other friends and colleagues in NBB for providing over the years a delightful environment in which to study animal behavior. And I am forever indebted to Roger A. Morse, Professor of Apiculture at Cornell University, who introduced me to the wonderland of the honey bee colony more than 25 years ago.

A number of individuals have given generously of their time, reading, criticizing, and providing many insightful comments on the manuscript, including Scott Camazine, Wayne Getz, Susanne Kühnholz, Rob Page, Stephen Pratt, Tom Rinderer, Kirk Visscher, and David Sloan Wilson. I also appreciate the permissions from Scott Camazine, Kenneth Lorenzen, and William Shields to use their photographs, and from various publishers to reproduce material for which they hold the copyright: Association for the Study of Animal Behaviour *(Animal Behaviour)*; Cornell University Press; Ecological Society of America *(Ecology)*; Entomological Society of America *(Journal of Economic Entomology)*; Harvard University Press; International Bee Research Association; Macmillan Journals Ltd. *(Nature)*; Masson *(Insectes Sociaux)*; Pergamon Press (*Journal of Insect Physiology*); Princeton University Press; and Springer-Verlag *(Journal of Comparative Physiology* and *Behavioral Ecology and Sociobiology)*. Very special thanks are due to Margaret C. Nelson, who created all the illustrations for this book. Her ability to render my smudgy hand drawings on graph paper into clean computer-based artwork has been a constant source of amazement and delight. I feel extremely fortunate to have had such a talented and conscientious coworker in producing this book. Finally, Michael Fisher and Nancy Clemente of Harvard University Press expertly and enthusiastically edited the manuscript, and were sympathetic to my need to write without a deadline. To all, I give thanks.

Tom Seeley

Ithaca, New York
January 1995

INTRODUCTION

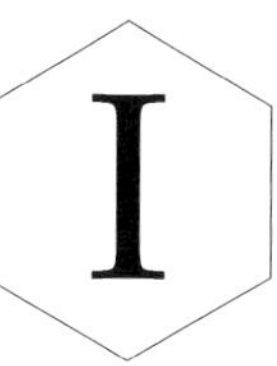

The Issues

1

This book is about how a colony of honey bees works as a unified whole. Attention will be concentrated on the mechanisms of group integration underlying a colony's food-collection process, an aspect of colony functioning which has proven particularly open to experimental analysis. Everyone knows that individual bees glean nectar from flowers and transform it into delicious honey, but it is not so widely known that a colony of bees possesses a complex, highly ordered social organization for the gathering of its food. This rich organization reflects the special fact that in the case of honey bees natural selection acts mainly at the level of the entire colony, rather than the single bee. A colony of honey bees therefore represents a group-level unit of biological organization. By exploring the inner workings of a colony's foraging process, we can begin to appreciate the elegant devices that nature has evolved for integrating thousands of insects into a higher-order entity, one whose abilities far transcend those of the individual bee.

1.1. The Evolution of Biological Organization

In a famous essay titled "The Architecture of Complexity" (1962), the economist Herbert A. Simon presented a parable about two watchmakers. Both built fine watches and both received frequent calls from customers placing orders; but one, Hora, grew richer while the other, Tempus, became poorer and eventually lost his shop. This difference in the two craftsmen's fates was traced to a fundamental difference between their methods of assembling a watch, which for both individuals consisted of 1000 parts. Tempus's procedure was such that if he had a watch partially assembled and then had to put it down—to take an order, for example—it fell apart and had to be reassembled from scratch. Hora's watches were no less complex than those of Tempus but were designed so that he could put together stable subassemblies of 10 parts each. In turn, 10 of the subassemblies would form a larger and also stable subassembly, and 10 of those subassemblies would constitute a complete watch. Thus each time Hora had to put a watch down he sacrificed only a small part of his labors and consequently was far more successful than Tempus at finishing watches.

The lesson of this story is that complex entities are most likely to arise through a sequence of stable subassemblies, with each higher-level unit being a nested hierarchy of lower-level units. Bronowski (1974) has summarized this idea as the principle of building complexity through "stratified stability." Certainly this principle applies to the evolution of life. Over the past 4 billion years, the entities that constitute functionally organized units of life have increased their range of complexity through a nested series of stable units: replicating molecules, prokaryotic cells, eukaryotic cells, multicellular organisms, and certain animal societies (Figure 1.1). To explain why natural selection has favored the formation of ever larger, ever more complex units of life, Hull (1980, 1988) and Dawkins (1982) have pointed out that all functional units above the level of replicating molecules (genes) can be viewed as "interactors" or "vehicles" built by the replicators to improve their survival and reproduction, and that in certain ecological settings larger, more sophisticated interactors propagate the genes inside them better than do smaller, simpler ones. For example, a multicellular organism is sometimes a better gene-survival machine than is a single eukaryotic cell by virtue of the organism's larger size, often greater mobility, and many other traits (Bonner 1974; Valentine 1978). Likewise, the genes inside organisms sometimes fare better when they reside in an integrated society of organisms rather than in just a single organism, because of the superior defensive, foraging, and homeostatic abilities of functionally organized groups (Alexander 1974; Wilson 1975).

What is especially puzzling about the evolution of life is how each of the transitions to a higher level of biological organization was achieved. In each case, individual units honed by natural selection to be successful, independent entities, must have begun somehow to interact cooperatively, eventually evolving into a larger, thoroughly

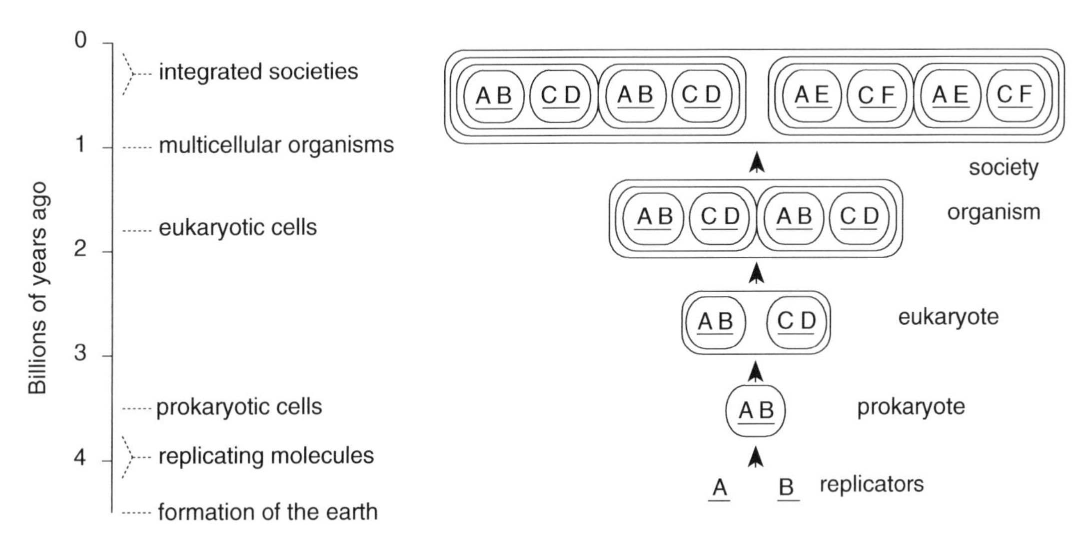

Figure 1.1 Chronology of the origins of the different levels of functionally organized units of life, from replicating molecules (the origin of life) to advanced animal societies. Each unit above the original level of replicating molecules consists of an assemblage of the previous level's units functioning as a (largely) harmonious whole. Animal societies that possess this level of functional unity include the colonies of many marine invertebrates (such as siphonophores, salps, and graptolites; Bates and Kirk 1985; Mackie 1986), some social insects (such as honey bees, fungus-growing termites, and army ants; Badertscher, Gerber, and Leuthold 1983; Franks 1989; Seeley 1989b), and a few social mammals (such as naked mole-rats and dwarf mongooses; Rood 1983; Sherman, Jarvis, and Alexander 1991).

integrated unit composed of mutually dependent parts. To fully understand each such transition, we must solve two general puzzles. The first deals with ultimate causation: *why exactly is there strong cooperation among the lower-level entities?* In particular, why doesn't natural selection among lower-level entities—genes in a chromosome, DNA-containing organelles in a cell, cells in an organism, organisms in a society—disrupt integration at a higher level? (Why is meiosis usually fair? Why are mitochondrial cancers so rare? Why do the bees in a hive mostly cooperate?) This is a fundamental problem in evolutionary biology, one which remains largely unexplored at the level of subcellular cooperation, but which recently has begun to attract increasing attention for all levels of biological organization (reviewed by Eberhard 1980, 1990; Buss 1987; Maynard Smith 1988; Werren, Nur, and Wu 1988; Wilson and Sober 1989; Leigh 1991; Williams 1992). The second puzzle lies in the realm of proximate causation: *how exactly do the lower-level entities work together to form the higher-level entity?* The challenge here is to solve the mysteries of physiology, for each level of functional organization: cell, organism, and society. Biologists have primarily investigated the intricacies of cellular and organismal physiology; hence our understanding of social physiology—the elaborate inner workings of the highly integrated animal societies—is relatively poor, and the field therefore offers rich opportunities for future study.

In this book, I aim to contribute to a better understanding of the proximate mechanisms involved in the transition from independent organism to integrated society by describing the investigations that I and others have done on the social physiology of the honey bee colony. My account will not cover all aspects of colony physiology. Rather, it will focus on just the complex process of food collection, which has been the main subject of my own research for the past 15 years. Why devote so much effort to examining this one process in this one social insect? This is a fair question; after all, every case in biology is at least partly special or even unique. Indeed, the organization of every animal society has been determined by the particular circumstances of its evolutionary history; so the precise description we give of a specific process in one society will not apply in detail to any other. I believe, however, that mechanisms analogous to those underlying a bee colony's foraging abilities are likely to underlie the functioning of many other insect societies. By establishing a detailed description for the particular case of honey bee foraging, I develop ideas that inform other studies even though no other case will look exactly like this honey bee example.

I believe too that this investigation of the food-collection process in honey bee colonies provides a paradigm of the analytic work needed to disclose the mechanisms which integrate a group of organisms into a functional whole. As we shall see, a honey bee colony operates as a thoroughly integrated unit in gathering its food. It monitors the flower patches in the countryside surrounding its hive; it distributes its foraging activity among these patches so that nectar and pollen are collected efficiently, in sufficient quantity, and in the nutritionally correct mix; and it properly apportions the food it gathers between present consumption and storage for future needs. In addition, a colony precisely controls its building of beeswax combs for honey storage, strictly limiting this costly process to times of clear need. And it adaptively adjusts its water collection in accordance with its need for water to cool the hive and feed the brood. Hence in acquiring its food, a honey bee colony presents us with many intriguing forms of precise, coherent colony behavior. What is equally important, however, is that a honey bee colony provides us with an insect society which is remarkably open to analytic studies. For instance, a colony can be laid open with minimal disturbance (by means of an observation hive; see Chapter 4) so that we can peer inside it and see the normally hidden activities of the individual bees that generate the behavior of the whole colony. Moreover, a colony's entire foraging process is amenable to experimental manipulation, which of course is critical to the incisive analysis of any complex biological system. We can precisely alter the components of a colony, the nutritional conditions inside its hive, or the foraging opportunities outside, and then monitor the individual responses of the bees or the collective response of the colony, or both. In short, the food-collection process of a honey bee colony is a model system for the study of social physiology. I should stress at the outset, however, that analysis of the bee colony's foraging process is far from complete; so the story which follows is just the best current description of a colony's sophisticated internal organization. Further research over the next few years will certainly extend and refine our present understanding.

1.2. The Honey Bee Colony as a Unit of Function

In the previous section, I asserted that "a honey bee colony operates as a thoroughly integrated unit in gathering its food." To individuals accustomed to thinking about biological phenomena in light of natural selection theory, this summary of the nature of a bee colony's foraging operation may seem simplistic. After all, the 20,000 or so worker bees in a colony (Figure 1.2) arise through sexual, not clonal, reproduction by their mother queen. Because of segregation and recombination of a queen's genes during meiosis, and because a queen typically mates with 10 or more males (Page 1986), the workers in a single hive will possess substantially different genotypes. Natural selection theory tells us that whenever there is genetic heterogeneity within a group there is great potential for conflict among the group's members. Recent theoretical and empirical studies have revealed, however, that even though the *potential* for conflict within a bee colony is indeed high, the *actual* conflict is remarkably low (see Ratnieks and Reeve 1992 for a general discussion of the distinction between potential and actual conflict in animal societies). These important studies have also generated several remarkable insights into why there is so little conflict within a beehive.

Let me begin my review of this research by noting that there is a

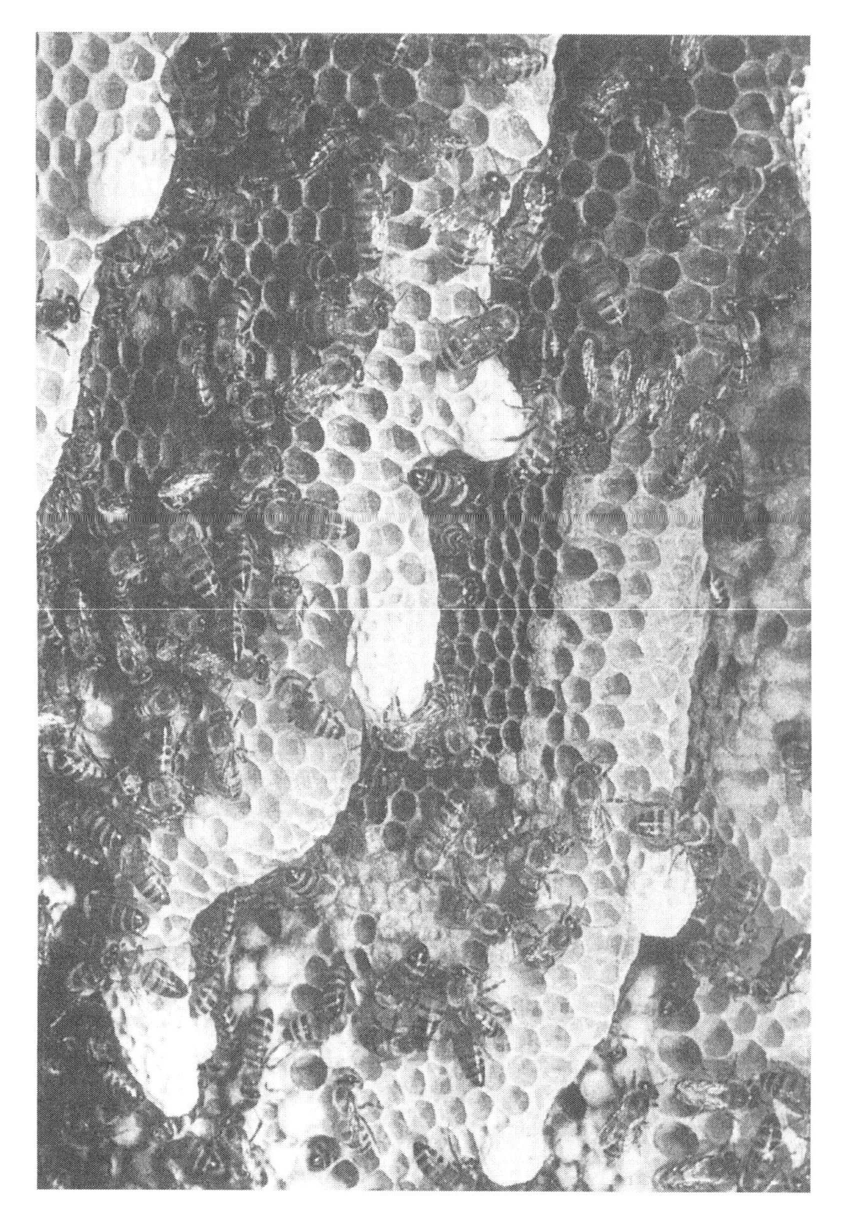

Figure 1.2 Partial view of a honey bee colony which has constructed its beeswax combs inside a tree cavity (cut open to reveal the nest). This colony consists of some 20,000 worker bees, one queen bee, and several hundred drones. Each honey bee colony is one gigantic family, for all the workers (females) and virtually all the drones (males) are the daughters and sons of the queen. The peanut-shaped structures on the margins of the combs are special cells in which queens are reared. Photograph by S. Camazine.

fundamental similarity between the somatic cells of a metazoan body and the workers in a honey bee colony with a queen: both lack direct reproduction; hence both are themselves genetic dead ends. Nevertheless, both can foster the propagation of their genes into future generations by helping other individuals that carry their genes to form genetic propagules. Somatic cells toil selflessly to enable their body's germ cells to produce gametes, and worker bees toil almost as selflessly to enable their colony's queen—their mother—to produce new queens and males. Thus the hard labor of a worker bee should be viewed as her striving to propagate her genes as they are represented in her mother's germ cells and stored sperm. This fact, coupled with the fact that usually there is just one queen in a honey bee colony, implies that the genetic interests of all of a colony's workers have a common focus, and so overlap greatly, even though these bees are far from genetically identical.

What is the evidence that worker honey bees in queenright colonies—ones containing a fully functioning queen—have essentially no personal reproduction? Although worker honey bees cannot mate, they do possess ovaries and can produce viable eggs; hence they do have the potential to have male offspring (in bees and other Hymenoptera, fertilized eggs produce females while unfertilized eggs produce males). It is now clear, however, that this potential is exceedingly rarely realized as long as a colony contains a queen (in queenless colonies, workers eventually lay large numbers of male eggs; see the review in Page and Erickson 1988). One supporting piece of evidence comes from studies of worker ovary development in queenright colonies, which have consistently revealed extremely low levels of development. All studies to date report far fewer than 1% of the workers have ovaries developed sufficiently to lay eggs (reviewed in Ratnieks 1993; see also Visscher 1995a). For example, Ratnieks dissected 10,634 worker bees from 21 colonies and found that only 7 had a moderately developed egg (half the size of a completed egg) and that just one had a fully developed egg in her body. A second, and still more powerful, indication of the virtual absence of worker reproduction in queenright honey bee colonies is a recent study by Visscher (1989) using colonies each of which was headed by a queen which carried a genetic marker (cordovan allele) that allowed easy visual discrimination of male progeny of the queen and the workers (Figure 1.3). Each summer for 2 years, Visscher trapped and inspected all the drones reared in each of his 12 study colonies.

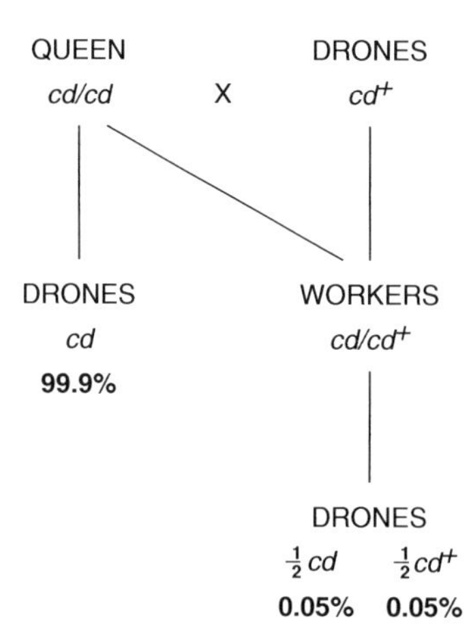

Figure 1.3 The genetic system used by Visscher (1989) to assess the frequency of worker reproduction in honey bee colonies. Although worker bees do not mate, they can lay unfertilized eggs which will develop into drones. To distinguish the drones produced by the queen from those produced by workers, he used colonies headed by queens which were homozygous for the cordovan allele (*cd/cd*) and which were mated with males hemizygous for the wild-type allele (cd^+). Therefore all the workers in each colony were heterozygous for the cordovan allele (cd/cd^+). Thus all the male offspring of the queen were *cd*, whereas the male offspring of the workers were, on average, half *cd* and half cd^+. Drones that possess the cordovan allele have a distinctive reddish-brown cuticle *(bottom left)*, whereas those with the wild-type allele have a normal, black cuticle *(bottom right)*. Photograph by T. D. Seeley.

Of the 57,959 drones captured, only 37 (approximately 0.05%) possessed a black, wild-type cuticle. This implies that only about 74, or 0.1%, were derived from worker-laid eggs. Thus it is clear that workers give rise to only a minute fraction of a queenright colony's drones. But to fully appreciate the significance of this finding, we need to calculate the probability of personal reproduction for a worker bee. Visscher measured the production of worker-derived drones for 12 colonies of bees, each of which produced approximately 150,000 worker bees each summer (Seeley 1985). Hence the probability of personal reproduction by a worker bee in one of Visscher's colonies was approximately 74 drones/(12 colonies × 150,000 worker bees/colony) = 0.00004, or essentially zero, drones per worker bee.

Why do worker bees have virtually no personal reproduction in the presence of their queen? The traditional explanation is that the mother queen prevents her daughter workers from having sons by means of "queen control" pheromones. Because 50% of the queen's genes are represented in her sons, but only 25% in her grandsons, the queen's genetic interests are better served by limiting the colony's production of males to her sons rather than allowing a mix of her sons and grandsons, all else being equal. A worker's genetic interests, however, are better served by producing sons, each of whom carries 50% of her genes, rather than by helping the queen produce males who are her (the worker's) brothers, since they carry only 25% of the worker's genes. Clearly, there is much potential for conflict between the queen and the workers over the provenance of the males. Nevertheless, I think that there is compelling evidence that the pheromones released by the honey bee queen (reviewed in Winston and Slessor 1992) function not as a drug inhibiting the development of the workers's ovaries, but instead as a signal indicating the presence of the queen (Seeley 1985; Woyciechowski and Lomnicki 1989; Keller and Nonacs 1992). One piece of the evidence is that workers are attracted to their queen and show specific behavioral adaptations to help disperse the queen's pheromones, such as licking the queen (Figure 1.4) and then crawling rapidly about the hive, all the while contacting other workers (Seeley 1979; Naumann et al. 1991). These worker adaptations can evolve and be maintained more easily if they serve the genetic interests of the workers and the queen rather than just those of the queen. A second and more telling fact is that the queen's pheromones are neither necessary nor sufficient for inhibiting workers' ovaries. Instead, they strongly inhibit the workers from rearing additional *queens*. It is now clear that the pheromones that provide the proximate stimulus for workers to refrain from laying eggs come mainly from the brood, not from the queen (reviewed in Seeley 1985; see also Willis, Winston, and Slessor 1990).

If not the queen's domination of the workers by biochemical means, what is it that ultimately prevents worker reproduction in queenright colonies? Recent theoretical considerations and experimental data strongly support the idea that this nonreproduction is a result of

Figure 1.4 A queen bee surrounded by a retinue of worker bees. The workers lick the queen and brush her with their antennae, thereby acquiring from her a blend of pheromones which they then spread throughout the colony to communicate that their mother queen is alive and well. The principal effect of this chemical signal on the workers is to inhibit them from rearing replacement queens. Photograph by K. Lorenzen.

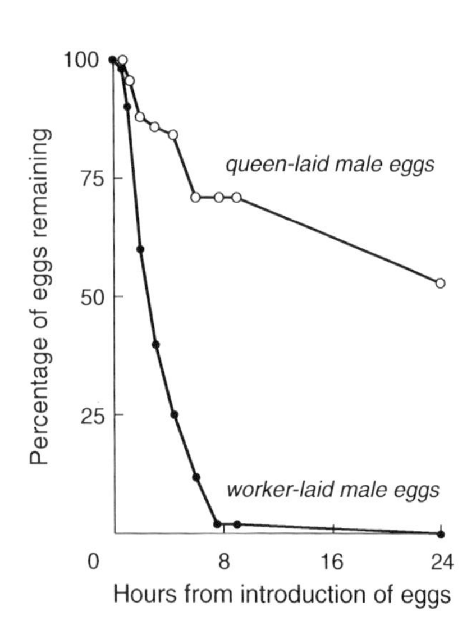

Figure 1.5 The time course of egg removal for worker- and queen-laid male eggs placed within the brood area of a populous, queenright honey bee colony. After Ratnieks and Visscher 1989.

"worker policing": the mutual prevention of reproduction by workers. The idea here—first proposed by Starr (1984) and Seeley (1985), and developed more rigorously by Woyciechowski and Lomnicki (1987) and Ratnieks (1988)—is that in honey bees and other social Hymenoptera, if the queen of a colony mates with more than 2 males (and their sperm are contributed equally and used randomly), then the workers in the colony are more closely related to the queen's sons (relatedness = 0.25) than to the sons of a randomly chosen worker (average relatedness = $0.125 + 0.25/n$, where n is the number of males with whom the queen mated). For example, in honey bees where the queen mates with 10 or more males, the average relatedness of a worker to a nephew is less than 0.15. This suggests that in species like the honey bee each worker should try to prevent other workers in her colony from reproducing, either by destroying worker-laid eggs or by showing aggression toward workers attempting to lay eggs. Ratnieks and Visscher (1989) then went on to demonstrate that worker honey bees can actually police one another by destroying worker-laid eggs. When they experimentally presented workers with queen-laid male eggs and worker-laid male eggs, they found that worker bees discriminated between them, preferentially removing the latter from cells (Figure 1.5). This discrimination is evidently mediated by a yet unidentified pheromone produced in the queen's Dufour gland and applied to queen-laid eggs (Ratnieks in press). Most recently, Ratnieks (1993), Visscher and Dukas (1995), and Visscher (1995a) have demonstrated that worker egg laying and worker policing—by means of both egg destruction and aggression toward laying workers—actually occur, though are rare, in queenright colonies. For example, Visscher established colonies with genetic markers which enabled him to distinguish queen-laid and worker-laid male eggs, assayed the freshly laid male eggs in these colonies, and arrived at an estimate of 10% for the proportion of the male eggs laid in a queenright colony that derive from workers. He also reported that the vast majority of these worker-laid eggs were destroyed by policing workers within two hours of being laid; so it is not surprising that in the end only 0.1% of the adult males produced by a colony derive from workers. It should be noted too that the rate of egg laying by workers detected by Visscher (about 5 eggs per colony per day) implies that only about one in 10,000 workers in a queenright colony lays an egg each day, a number which corroborates the repeated reports, mentioned earlier, of almost no workers in queenright colonies with ovaries sufficiently developed to lay eggs.

The virtual absence of worker reproduction implies that there is a reproductive bottleneck in a queenright honey bee colony, with virtually all the workers' gene propagation occurring through the shared channel of the reproductive offspring of their queen. This important fact does not, however, *by itself* imply that a complete congruence of the genetic interests of the workers has evolved, hence that a colony's workers should be regarded as a totally cooperative group. (Indeed, as just noted, there is a low level of active conflict among the workers over the production of males, though the negative effects of this conflict on colony functioning are probably minimal, given both the rarity of laying workers and the presumably low cost of worker policing.) As Dawkins (1982, 1989) has stated very clearly, for any group of biological entities to evolve into a coherent unit, the channel into the future for the group members' genes not only must be shared, *but also must be fair.* Only if the genes carried in the group's genetic propagules are an unbiased sample of the genes in the group, with each member of the group being guaranteed an equal chance of having its genes propagated, should we expect selection to favor strong cooperation by every individual for the common good. This situation generally prevails at the level of multicellular organisms, where typically all the cells in an organism (except the haploid gametes) have the same genes, and the rules of meiosis ensure that the gametes contain an unbiased sample of the genes in these cells (Buss 1987). But does this situation also pertain to a colony of honey bees? Do a colony's genetic propagules, its drones and virgin queens, carry an unbiased sample of the genes in the colony's workers?

This question has intrigued more investigators of honey bee sociobiology than perhaps any other in the past 10 years, and though it cannot yet be answered fully, the general form of the answer can be discerned. To begin, let me restate the question more precisely. The genes in a colony's workers come exactly half from eggs in the queen's ovaries and half from the sperm stored in her spermatheca. Thus the critical question is: Are the genes in a colony's drones and virgin queens an unbiased sample of the genes in the queen's ovaries and stored sperm? Consider first the drones, which derive virtually exclusively from unfertilized eggs of the queen. It is crystal clear that the drones must contain an impartial sample of the genes in the queen's ovaries, for this is guaranteed by the rules of meiosis. So far, so good. Now consider the virgin queens, which derive from fertilized eggs of the queen and therefore represent both the genes in the queen's ovaries and those in her stored sperm. Again, the rules of meiosis in the queen guarantee that virgin queens contain an unbiased sample of the genes in the queen's ovaries, but we cannot conclude a priori that virgin queens will embody a random sample of the genes in their mother's stored sperm. The reason is that natural selection theory indicates that in colonies headed by multiply mated queens, such as honey bee colonies, workers can potentially increase the propagation of their genes by biasing their queen-rearing efforts in favor of virgin queens sharing the same father (full-sister queen, genetic relatedness = 0.75) over ones with a different father (half-sister queen, relatedness = 0.25) (reviewed in Getz 1991). Such biasing, if done to different degrees by the different patrilines—each one the offspring of a single drone—composing a colony, could result in the genes of some workers being represented disproportionately in the virgin queens.

Do worker honey bees bias their queen-rearing efforts in favor of full-sister queens? Six separate studies have addressed this question. In each case, different experimental techniques were used to present the workers in colonies rearing queens with a choice between full-sister and half-sister female larvae, or alternatively between related and unrelated female eggs and larvae. All studies indicate either only a small bias (Page and Erickson 1984; Noonan 1986; Visscher 1986; Page, Robinson, and Fondrk 1989), or no bias, (Breed, Velthuis, and Robinson 1984; Woyciechowski 1990), in favor of more closely related queens.[1] For example, Noonan (1986) established colonies each of which was headed by a queen homozygous for the cordovan allele and mated with two males, one bearing the cordovan allele and the other bearing the wild-type allele. Thus the workers constituting each colony belonged to just two visually distinguishable (cordovan and wild-type) patrilines. After housing these colonies in observation hives and dequeening each one to induce queen rearing, Noonan painstakingly recorded the patriline membership of each worker bee seen visiting a queen cell to feed the queen larva inside (Figure 1.6). Finally Noonan reared out the queens to determine the phenotype, hence the patriline membership, of each one, and examined her records for evidence that the workers preferentially fed queen larvae of the same patriline. She found that the workers' feeding visits to the queen larvae were biased by about 5% in favor of full-sister larvae (Figure 1.7). The possibility remains, however, that this small bias in favor of closer kin, like that reported in several of the other studies, is an artifact of abnormal experimental conditions, such as the presence of a patriline carrying a mutation with strong effects on cuticle color and perhaps odor (reviewed in Page and Breed 1987; Frumhoff 1991). Nevertheless, the weight of the evidence suggests that worker bees do show a weak preference for rearing full sisters as queens, but also that in the end the distribution of virgin queens among patrilines deviates very little, if at all, from the distribution of the queen's stored sperm among these patrilines (Visscher 1995b). Most mysterious of all is why natural selection favors such minimally partisan queen rearing, given the striking difference in relatedness between full and half sisters. Both Page, Robinson, and Fondrk (1989) and Ratnieks and Reeve (1992) have stated the puzzle in theoretical terms—either the costs of more nepotistic queen rearing are high (possibly because it reduces the total number of queens reared) or its benefits are low (pos-

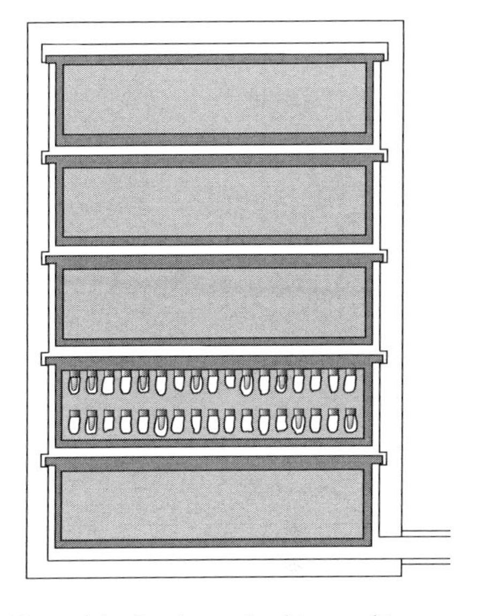

Figure 1.6 An observation hive used to explore the possibility of nepotism during queen rearing in honey bee colonies. After the colony was rendered queenless by removing the frame of comb containing the queen from the hive, 30–40 larvae from the removed frame were transferred into small beeswax cups that were mounted in a modified frame which was then placed inside the hive. A portion of the transferred larvae were reared into queens in the pendulous, peanut-shaped queen cells. After Noonan 1986.

1. Oldroyd, Rinderer, and Buco (1990) point out that the statistical analysis performed in the 1989 study by Page, Robinson, and Fondrk tends to yield false positive results, but a reanalysis of this study's data by Visscher (1995b) suggests that when correctly analyzed these data do reflect a slight bias in favor of full-sister queens.

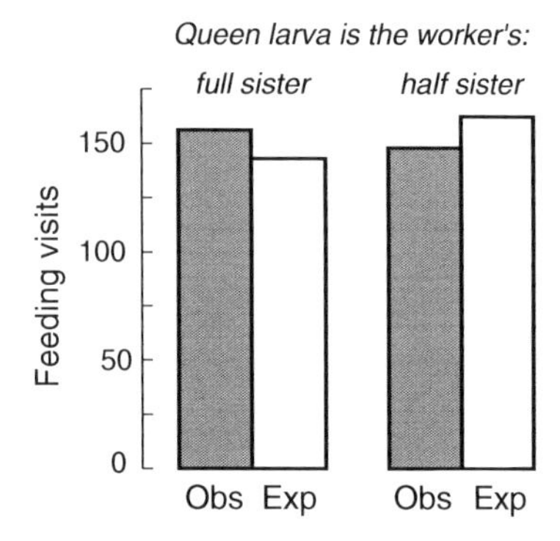

Figure 1.7 Feeding visits of workers of two genetically marked patrilines to queen cells containing developing queens. Comparisons of the observed and expected numbers of visits indicate a small (approximately 5%), but statistically significant ($P < 0.04$), bias toward feeding full-sister queen larvae. Based on data in Noonan 1986.

sibly because workers are unable to accurately discriminate full and half sisters), or both—but empirical investigations of this important subject have not yet been undertaken.

Whatever the cause of the surprisingly weak patriline bias during queen rearing, the effect is that the virgin queens produced by a honey bee colony contain a nearly unbiased sample of the genes in the mother queen's ovaries and sperm.[2] Thus we arrive at the conclusion that the genetic propagules of a honey bee colony, its virgin queens and drones, constitute an essentially impartial channel into the future for the genes of a colony's workers. Even though the workers in a colony are not genetically identical, their genetic destiny is shared in the fate of their colony, and their colony passes the workers' genes into the future with a high degree of fairness. Hence it is understandable that the workers of a honey bee colony work together strongly for the common good, and that a honey bee colony is a coherent unit of function.

1.3. Analytic Scheme

All scientific truths are rooted in the details of experimental investigation, which constitute the soil in which these truths develop. To grow such truths, then, one must use fertile methods of investigation. With such thoughts in mind, I decided to present in this book not only *what we know* about how a honey bee colony works (the truths) but also *how we know* this information (the experiments). Thus this book provides a case study of how behavioral experiments can construct a view of the biological world. This will be accomplished principally by presenting the experiments themselves in Part II, but here, at the outset, I will present a few general thoughts about effective methods for unscrambling the inner workings of honey bee colonies and other highly integrated animal societies.

The fundamental challenge of physiology, at all levels of biological organization, is to explain the abilities of units at one level in terms of the actions and interactions of lower-level units. This is always difficult because living systems are characterized by what Weaver (1948)

2. At present, the evidence supporting this statement applies only to immature virgin queens—queens in the pupal stage. This is so because the existing studies of bias in queen production have always isolated the queens from the workers before the queens emerge from their cells, and such isolation eliminates the possibility that workers introduce bias through differential care of the adult virgin queens.

has termed "organized complexity": the complexity which arises when a system consists of diverse parts bound together into an organic whole through numerous interactions, each of which has highly specific features. In such systems, the causal network for any particular property of the intact system is often staggeringly complicated. In living systems, this complexity is evidently the result of natural selection always having to build on what went before, so that even a fundamentally simple mechanism eventually becomes encumbered with subsidiary gadgets which serve, among other things, to adaptively modulate the basic mechanism. Over time, complexity is added to complexity. Moreover, each functional process within a living system is likely to evolve its own, more or less separate, set of mechanisms, so that in the end the whole system is an amazing conglomeration of devices. Thus it is that today, after some 60 million years of evolution, a honey bee colony is an astonishingly intricate web of contrivances for social life.

Given this internal complexity, it is clear that in order to understand the inner machinery of a living system we must penetrate inside it, to examine directly its innermost workings, and not simply monitor it from the outside. The interior of a biological system is the real field of action for physiological investigation. If, instead, one examines simply the exterior of a system, one is limited to measuring the inputs and outputs of the intact system and attempting to infer what goes on in between, the so-called top-down, black-box, or phenomenological approach. One danger of looking only at the outside is that it is easy to overlook things inside, especially those whose effects on the system are weak. For example, classical genetics—which used the black-box approach almost exclusively (Dawkins 1986)—provided no hint of the existence of introns in the genomes of eukaryotes. A second and greater danger of the top-down approach is that it is exceedingly easy to err in one's attempts to deduce the bits and pieces of living machinery that implement a given system-level property. Generally, the top-down approach involves building a mathematical model of the postulated mechanisms underlying a phenomenon, then seeing if the model's predictions (generated usually through computer simulation) match what is actually observed in the real world. The problem is that one's model of the inner workings may not correctly describe them, even if its predictions fit some of the facts. At least one theoretical biologist, Francis Crick (1988), has said that because the mechanisms of life have evolved by natural selection, they are usually too accidental and too intricate to be discerned by intuition alone. The human mind is attracted to elegance and simplicity, whereas evolution tends to produce rather complicated combinations of tricks; hence the top-down approach is likely to lead to a falsely simplified view of the phenomena of life.

It is fortunate for the study of social physiology, therefore, that one can easily peer inside many animal societies and closely examine the devices of social coordination that nature actually uses. When a honey bee colony is installed in a glass-walled observation hive, for example, one can observe all the behaviors of every bee inside a normal, functioning colony. Thus the observation hive makes possible a detailed yet harmless vivisection of a bee colony. Moreover, because bees are macroscopic entities, the observation and recording of a bee colony's internal processes is straightforward and minimally invasive. And with bees it is possible to apply individually identifiable labels to *all* the thousands of members of a colony (Figure 1.8), thereby enabling one to resolve the colony's inner workings at the level of single, identified bees.

The complexity of living systems also means that large problems cannot be addressed en bloc, but only after they are divided into a set of distinct, smaller problems. Thus the process of food collection by a honey bee colony is dissected into the subprocesses of nectar, pollen, and water collection, and each of these subprocesses is further broken down into still smaller topics of study. But even while trying to get inside a system, by breaking it open and isolating its different components in order to understand the hidden mechanisms of each, we must continue to consider the system an integrated whole, because the parts we examine are interdependent and mutually generative. We focus our attention on separate parts for the sake of ease in experimental analysis, not because they should be conceived of as independent entities. To understand the functional significance of any given piece of a system, we always have to refer to the whole system and see the part's effects on the whole. Indeed, many of the surprises in physiological investigation arise because the effects of a single component are unexpectedly broad. We will see, for example, that a forager in a honey bee colony can strongly influence (and be influenced by) not only other foragers, but also bees involved in operations distinct from food gathering, such as food processing, comb building, and brood rearing. The need to study biological systems at multiple levels simultaneously also arises because the most power-

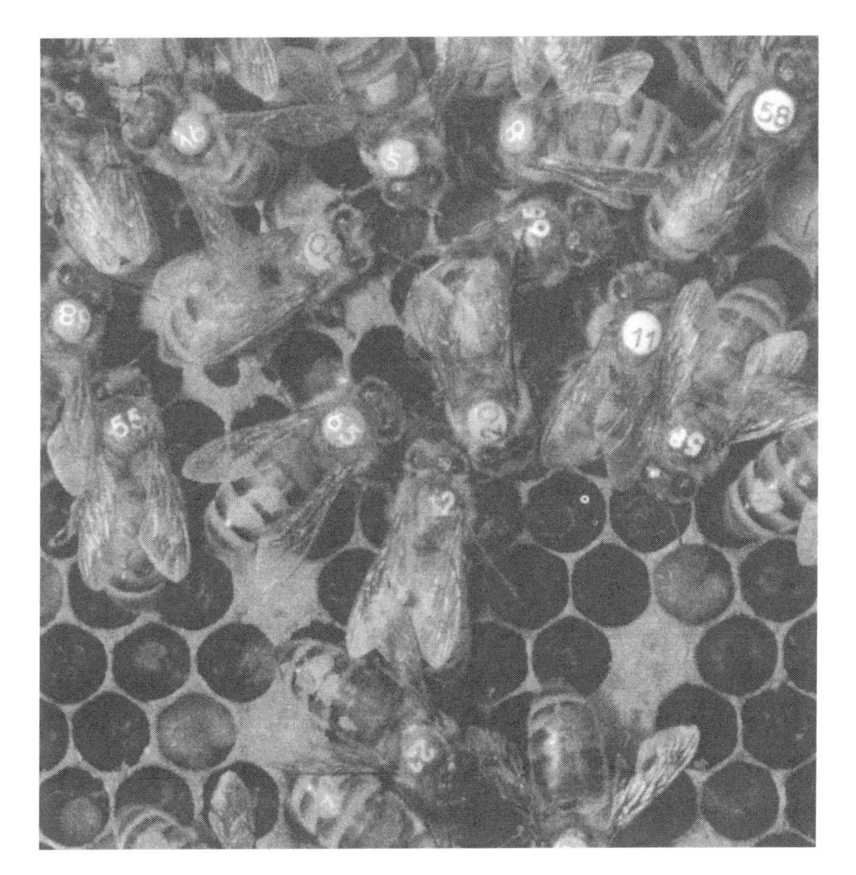

Figure 1.8 View inside an observation hive occupied by a colony of bees each of which has been labeled for individual identification. On every bee's thorax a colored (white, yellow, red, blue, or green) and numbered (0–99) plastic tag has been glued, and on the abdomen a paint mark (1 of 8 different colors) has been dotted. This labeling system makes possible the discrimination of 4000 individuals, which is a convenient population size for colonies used in experimental studies. Photograph by T. D. Seeley.

ful way to identify important physiological problems is to observe the intact system. Thus looking at a system from the top down helps us to see what the questions are, while looking from the bottom up enables us to see the answers. John Maynard Smith (1986) expressed this point succinctly when he wrote, "Most [biological] problems are best solved by starting at both ends and trying to meet in the middle." In this book, I will strive to show how one should view a honey bee colony both as a seamless whole and as a patchwork of parts.

What generalizations can be drawn about the techniques for unscrambling a complicated system such as a honey bee colony? Drawing upon the ideas of Crick (1988) and my own experience, I suggest that four main approaches are needed for a complete analysis. The first is to break the system open, identify its components, and characterize how each one works as an isolated part. For a bee colony, this entails describing the different types of workers, the rules of behavior for each, and, if appropriate, the physiological bases of the behaviors of individual bees. The second approach is to map the location of each part in the system, determine its connections to other parts, and find out how it interacts with them. Thus one needs to plot for each labor group in a beehive the spatial distribution of its members, and then analyze how its members interact with other individuals, be it through transfer of information or the flow of energy-matter, or both. The third main approach is to study the behavior of the intact system and its components while interfering delicately with one or more of its parts, to determine what effects such alterations have on performance at all levels. This is generally the most challenging part of the analysis because one must leave the system as intact as possible—so that what one observes can be related to normal system functioning—but at the same time induce specific alterations inside the system. The challenge of designing and executing experiments that fulfill both these goals makes this third phase often the most exciting in a physiological investigation, one that demands the greatest mental and manual dexterity. Success at this stage depends critically on the choice of a study system that is open to gentle experimental alterations of its inner machinery. Such is the case for the honey bee colony, where one can, to cite just a few examples, remove particular members of the colony, insert barriers to block the flow of information and matter, and manipulate the physical environment inside the hive, all with ease and high precision.

The three approaches just described are likely to yield strong suggestions about how a system works, but testing the accuracy and completeness of one's understanding requires a fourth stage in the analysis: performing a simulation of the system by means of a mathematical model which embodies one's current understanding of the system's design (Simon 1981). Here one takes a bottom-up approach to model building, using experimental results rather than intuition (the top-down approach), to give shape to the model (Figure 1.9). Usually this requires translating a verbal understanding of what happens inside the system into a mathematical form, and this is itself useful, since it imposes an exactness on the verbal postulates which is usually lacking in one's initial formulations. But the principal aim of this fourth step is to check whether the set of processes identified through experimental analyses, interacting as supposed, does indeed produce the actual performance of the intact system. The human mind is notoriously poor at predicting the performance characteris-

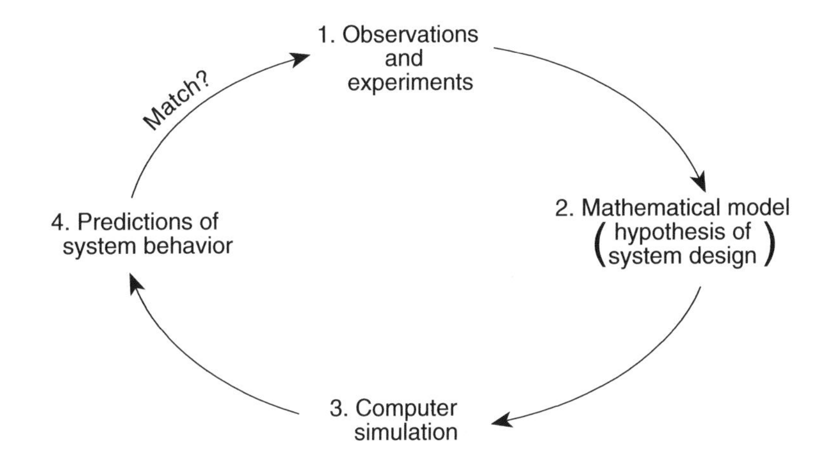

Figure 1.9 Cycle of studies needed to achieve a thorough understanding of a complex system such as a colony of honey bees.

tics of multivariable systems. Fortunately, the electronic computer is extremely good at simulating complex systems, and hence it provides a means of evaluating one's understanding of a system's overall design. If the predictions from the computer simulation fail to agree with the observations of the real system's behavior, then one knows immediately that he has a poor grasp of at least one important aspect of the system's design. In this situation, one needs to perform additional empirical investigations, which will yield improved knowledge of the system's inner workings, and at this point one can again evaluate the sufficiency of one's understanding by building and testing a refined mathematical model of the system. Each repetition of the cycle of observation, experiment, model building, and computer simulation yields a more accurate picture of the subject.

The great nineteenth-century physiologist Claude Bernard (1865) said that the science of life is "a superb and dazzlingly lighted hall which may be reached only by passing through a long and ghastly kitchen." Although for studies of honey bee colonies the image of "a long and ghastly kitchen" probably should be replaced with that of "a warm and flower-filled garden," I treasure Bernard's statement because it expresses vividly the strength of feelings associated with both the product and the process of physiological investigation. Indeed, I believe that one must appreciate both these dimensions of any scientific study if one wants to understand it accurately and with feeling. An important theme of this book is, therefore, the expression of both the ingenious methods and the exciting discoveries that characterize recent studies of the organization of honey bee colonies.

2 The Honey Bee Colony

In this chapter, I discuss the natural history of colonies of the honey bee, *Apis mellifera,* the familiar bee used for most of the world's beekeeping. This remarkable social insect is native to Europe, the Middle East, and the whole of Africa, and has been introduced by beekeepers to the Americas, Asia, Australia, and the Pacific Islands. Most of the information in this chapter applies to *Apis mellifera* across its immense range, but some ecological and sociological aspects pertain only to the cold temperate regions of the world, particularly parts of northern Europe and North America. In these seasonally cold regions, a honey bee colony must stockpile a large quantity—20 or more kg—of honey as fuel for keeping itself warm throughout the winter, and certain features of the social organization of temperate-zone colonies reflect this need to amass a huge energy reserve.

The honey bee has been the subject of scientific observations since ancient times, and today there are scores of excellent books that describe its basic biology. The most important of these are Ribbands's *Behaviour and Social Life of Honeybees* (1953), Snodgrass's *Anatomy of the Honey Bee* (1956), Lindauer's *Communication among Social Bees* (1961), von Frisch's *Dance Language and Orientation of Bees* (1967), Michener's *Social Behavior of the Bees* (1974), Seeley's *Honeybee Ecology* (1985), Erickson, Carlson, and Garment's *A Scanning Electron Microscope Atlas of the Honey Bee* (1986), Winston's *Biology of the Honey Bee* (1987), Ruttner's *Biogeography and Taxonomy of Honeybees* (1988), Crane's *Bees and Beekeeping* (1990), and Moritz and Southwick's *Bees as Superorganisms* (1992). These publications should be consulted for detailed information on the topics touched on here. In this chapter, I make no attempt to provide thorough reviews of the subjects raised; rather I aim to provide readers with selected background information that is needed for a ready understanding of the subsequent chapters.

2.1. Worker Anatomy and Physiology

Figure 2.1 shows an adult worker bee from the side. As in most other insects, the body consists of three anatomical sections: (1) the head, with mouthparts and sensory organs such as eyes and antennae; (2) the thorax, a locomotory center which is almost entirely filled with muscles that operate the membranous wings and jointed legs, and (3) the abdomen, more spacious than the other parts, which holds the organs for various functions, including digestion, circulation, and stinging.

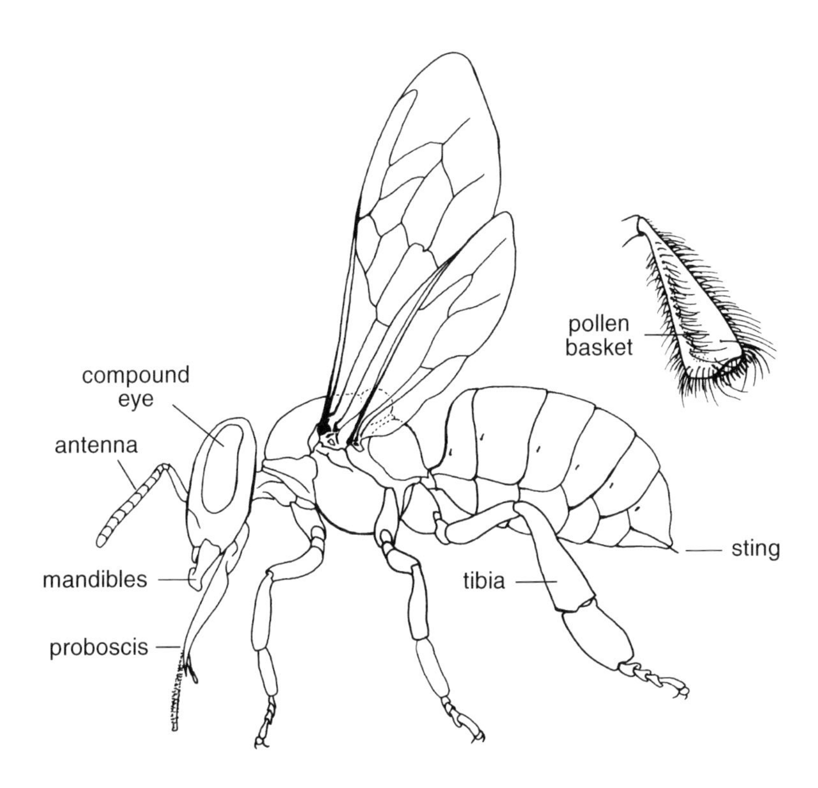

Figure 2.1 External structure of the worker honey bee with the hairy covering removed. *Upper right:* detail showing the pollen basket on the outer surface of the hind legs. After Snodgrass 1956.

2.1.1. EXTERNAL STRUCTURE

The mouthparts of a bee comprise two sets of tools, one for chewing and one for sucking. The principal chewing structures are the rigid, jawlike mandibles. They are used to manipulate wax, masticate pollen pellets, gather plant resins, groom hivemates, cut open flowers to reach otherwise inaccessible nectar, and even grip an enemy to gain a firm purchase for implanting the sting. Sucking up liquids is accomplished with the proboscis, a folding structure built of several mouthparts that form a tube around the bee's tongue. Liquids in this tube move upward toward the mouth (located at the base of the proboscis) as a result of the in-and-out movements of the bee's tongue, suction from the mouth, and perhaps also capillary action. The proboscis evolved for the function of taking in nectar, but it is also used for gathering water, exchanging food with nestmates, licking substances such as pheromones from other bees, and spreading nectar and water for rapid evaporation inside the hive. When not in use, the proboscis is folded out of the way in a large groove on the underside of the head. Solid food, mainly pollen, cannot be ingested through the proboscis, but is taken directly into the mouth after being broken up into small particles by the mandibles.

Figure 2.2 A worker honey bee foraging on buckwheat flowers. Note the proboscis, which is unfolded to probe for nectar, and the load of pollen packed on the outer surface of the hind leg. Photograph by T. D. Seeley.

The legs of a bee serve not only in locomotion, but also in food collection, for they bear special structures for transporting pollen, a dry, dustlike material. The outer side of the broad tibial segment of each hind leg is adapted to form a pollen-holding device, the so-called pollen basket. Its surface is smooth, slightly concave, and bordered by a fringe of long incurved hairs. Pollen, after being moistened with nectar, is packed into this basket and held in place by the hairs. Bees that are engaged in pollen collection are recognized instantly by the conspicuous balls of bright-colored pollen packed onto their hind legs (Figure 2.2). The pollen baskets are also used for transporting resin, which is gathered from sticky tree buds and used in nest construction.

The sting apparatus lies tucked inside a special sting chamber within the last abdominal segment. It is a modified ovipositor, or egg-laying tube. The shaft of the sting consists of two barbed lancets and a stylet which fit together to form a venom canal inside the sting's shaft. Venom is produced in a poison gland, which widens to form a sac in which venom is stored. When the bee stings, she forces venom into the venom canal and the sharp lancets are pushed into the tissue of the animal under attack. When the bee tries to retract her sting from tough skin, or the enemy tries to brush off a stinging bee, the barbed lancets ensure that the sting apparatus remains embedded.

2.1.2. INTERNAL ORGANS

The alimentary canal of a honey bee is shown in Figure 2.3. Just inside the mouth is the cibarium, or pump, which the bee can dilate and contract to draw liquid food up the proboscis and into the esophagus. Food then passess through the thorax via the esophagus and into the honey stomach (or crop), which is tremendously expandable. When a forager has gathered a full load of nectar, the honey stomach is stretched until its walls are transparent and its bulk presses the rest of the viscera to the rear of the abdomen. The contents of the honey stomach are voluntarily regurgitated when the bee applies pressure to the distended crop by contracting the telescoping abdominal segments. Pollen grains are transported to the honey stomach in solution, and then are removed from the honey stomach by a special valve, the proventriculus, which passes them and some of the liquid food into the midgut. Here is where enzymes are added and most of

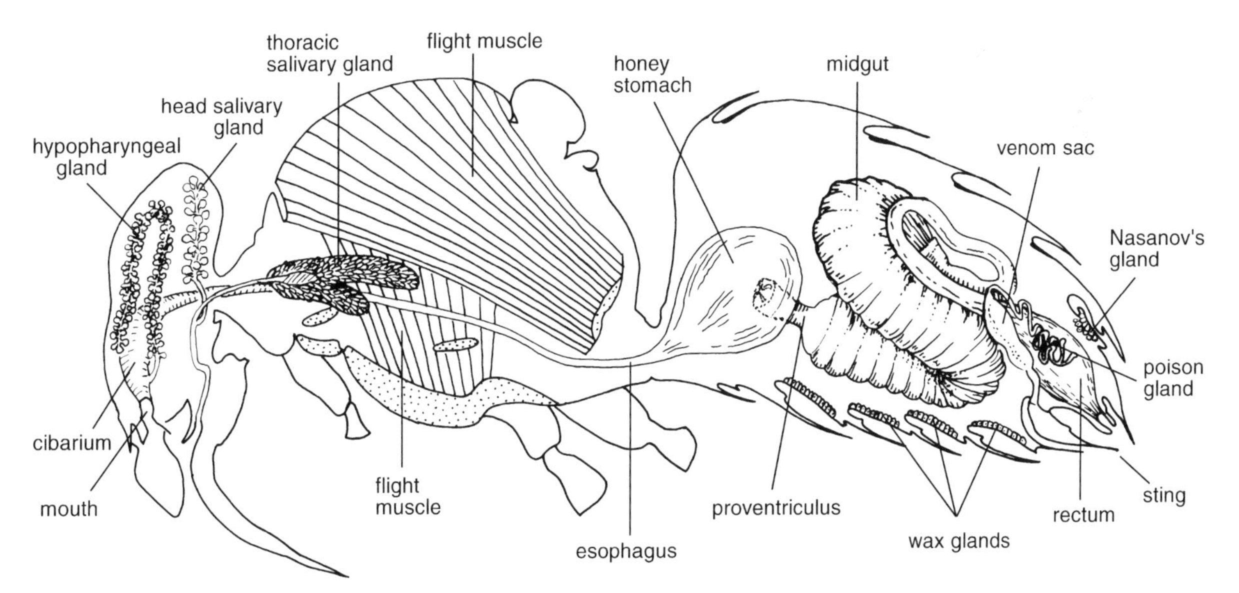

Figure 2.3 Some of the internal organs of a worker honey bee. After Michener 1974.

the digestion and absorption occur. Posterior to this is the rectum, where water and feces are stored until the bee can fly from the hive to defecate.

The large flight muscles virtually fill the thorax. These tissues, whose metabolic activity is among the highest known, not only provide the power for flights to flowers outside the hive but also serve several functions inside the hive. Their action produces fanning of the wings by bees on the combs, to produce air currents which ventilate the hive. They also serve to generate heat. By uncoupling the flight muscles from the flight mechanism, and contracting them isometrically, a bee inside the hive can produce heat but no locomotion. The flight muscles also produce small vibrations of the wings to make sounds during the communication dances of bees (see Section 2.5).

Adult bees possess numerous glands whose external secretions are either building materials, food, or communication substances (pheromones). Here I describe the locations and functions of the externally secreting glands which play a role in a colony's collection and storage of food. The *hypopharyngeal glands* are paired structures, one in each side of the head, which discharge just inside the mouth. The main ducts of these glands are massively elongated, and receive the discharge from individual cells along their entire length. They produce two quite different secretions: in young bees (nurses), proteinaceous food for both the larval brood and older adults; and in older bees (food storers and foragers), enzymes which break down the sucrose in nectar, an important step in the honey-making process. Both the head and the thorax have *salivary glands,* which discharge through a common duct near the base of the proboscis. Their secretion is used to clean the queen's body and is added to wax for softening when it is manipulated. In the abdomen, *wax glands* produce beeswax from which a colony's combs are built. The wax is secreted as a scale, whereupon it is chewed and mixed with salivary gland secretions until it is malleable. The *Nasanov's gland,* located on the upper surface of the abdomen, produces several volatile compounds that, when dispersed by fanning the wings, attract nestmates. For example, these secretions are used outside the hive to advertise the location of a rich food source.

2.1.3. SENSORY ORGANS

A worker honey bee is exquisitely endowed with the sensory equipment needed to perceive a wide range of the mechanical, visual, chemical, and temperature stimuli in its world. Many of the mechanoreceptors are sensory hairs (trichoid sensilla) which respond to specific patterns of deflection or vibration. They are distributed widely over the body and appendages. Those located at the tip of each antenna, for instance, enable foragers to detect differences in the surfaces of flower petals, which can be useful in locating the nectar in flowers. Other sensory hairs are grouped in bristle fields located at the neck and the base of the abdomen. They are deflected by any tendency of the head or abdomen to hang downward, and in this way serve in the perception of the gravitational force. Accordingly, the bee can know which way is down, and so can construct vertical hanging combs and perform communication dances—which are oriented with respect to the vertical—in an appropriate fashion. Other mechanoreceptors are spindle-shaped stretch receptors (chordotonal sensilla), located in the joints of appendages, which register vibrations and positions of the body. For example, as a recent study (Dreller and Kirchner 1993) has revealed, bees following a dancing nestmate perceive the airborne sounds produced by the dancer by sensing induced vibrations of their antennae with chordotonal sensilla (Johnston's organ) located in the second antennal segment. Similar receptors in the legs enable bees to respond to the substrate-borne vibrations used in other communication signals.

The bee's sense of smell is based on olfactory receptors located on the antennae. Each one contains some 3000 pore plates (sensilla placodea) which electrophysiological study has shown function in odor perception. There are also several other types of sensilla, some of which are no doubt involved in temperature perception, which is most acute in the antennae. Experimental studies, based mainly on testing the bee's ability to discriminate between odors to locate food, indicate that bees are often far more sensitive to certain odors than are humans, especially floral odors and bee pheromones, and are roughly equal to humans in the task of discriminating between different odors. Other chemoreceptors are located on the mouthparts, where they provide a sense of taste. The best studied are those involved in the perception of sweetness. Bees are not highly sensitive to sugars, for even in starved bees the behavioral response threshold is approximately 1/16 mol/L. This is not surprising, however, since the sugar solutions which bees deal with in nature when gathering nectar are generally quite concentrated, in the range of 0.5 to 2.5 mol/L.

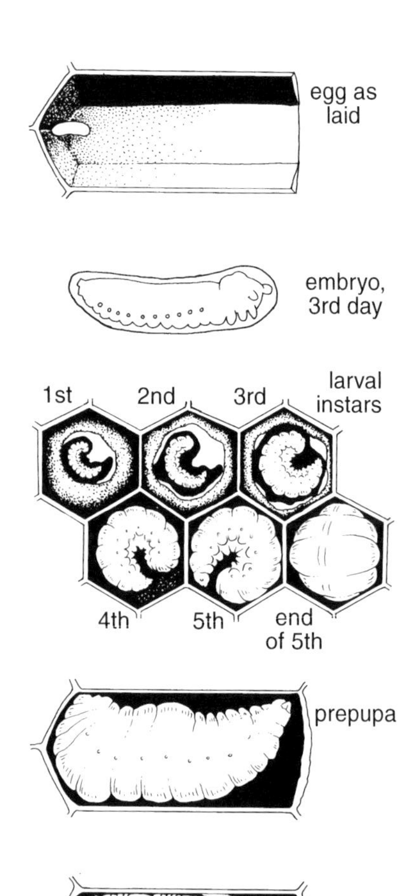

Figure 2.4 Stages in the development of a worker honey bee, from egg to pupa. Workers develop in the nearly horizontal cells that form the combs inside a beehive. There are five larval instars, or stages, each one separated by a molt in which the larva sheds its old skin and begins a new phase of growth. The so-called prepupa is merely the pupa in its early developmental stages within the skin of the fifth instar larva. When this larval skin is finally cast off, the insect appears in the form of an adult bee, and is called a pupa. After Dade 1977.

The bee's principal visual receptors consist of two compound eyes which cover large parts of the sides of the head. Each eye is made up of a sheet of some 6900 visual units called ommatidia, spread over a convex surface so that each covers a different portion of the bee's visual field. The divergence between the visual fields of adjacent ommatidia is 1–4°, which is at least 100 times greater than the angular divergence between adjacent cones in the foveal region of the human eye. However, as Wehner and Srinivasan (1984) recently pointed out, the reduction in the bee's visual acuity that results from the high angular divergence of its visual units is largely compensated for by the small interaction distances between a bee and its subjects of visual scrutiny, such as a flower on which it is about to land. The bee presumably perceives its light environment by integrating in the nervous system the information received by the photoreceptor mosaic of the thousands of ommatidia in each compound eye. Each ommatidium registers both color and intensity, so that the bee is endowed with color vision as well as form vision. The bee's visible spectrum differs, however, from our own, for a bee is highly sensitive to ultraviolet radiation (with wavelengths as short as 300 nm) but basically insensitive to red light (wavelengths greater than about 650 nm).

2.2. Worker Life History

2.2.1. DEVELOPMENT

The development of worker bees is typical for an insect that undergoes complete metamorphosis. Each individual passes through four stages: egg, larva, pupa, and adult. The changes in appearance of a worker as it develops from an egg to an adult are shown in Figure 2.4 (see also Figure 2.6). The embryo grows inside the egg for 3 days, consuming the protein-rich egg yolk. The larva, a whitish grub, then hatches from the egg and begins an intensive feeding stage, with its food supplied by the adult bees—a mixture of honey, pollen, and brood food secreted from the hypopharyngeal glands of the adult nurse bees. Larvae grow enormously, undergoing four molts (sheddings of old cuticle or skin) and multiplying their weight by a factor of more than 2000 during the 6-day-long larval stage. A larval bee's feeding ceases when she has lived about 8 days (5 of which have been spent as a larva), at which time the adult workers construct a wax capping that seals the larva in its cell. The fully grown larva then spins a cocoon of silk inside its cell, and orients itself with the head outward. A few days after cocoon spinning, the insect sheds its skin once again, now appearing as a fully formed pupa. The intricate process of pupal development actually starts before the last larval skin is shed, while the bee is a "prepupa." Pupal development is a reconstruction process in which a second set of cells, which had remained inactive in the larva, suddenly starts to divide rapidly. Their nourishment comes from the large larval cells, which are digested. This group of newly active cells forms the adult tissue, eventually replacing all the larval tissue, to give rise to the pupa, with its appearance of an adult bee. Finally the newly formed adult worker gnaws through the wax cap of the cell with her mandibles and emerges as a soft, young bee.

2.2.2. ADULT ACTIVITIES

When a worker emerges from her cell in the comb, her anatomical features are fixed, but the full development of her glandular system takes places only afterward, in a complex pattern which mirrors the changes in the bee's behavior over her life. Typically, brood food is secreted by young workers, beeswax by middle-aged bees, and enzymes for converting nectar into honey by older workers. Figure 2.5 portrays the sequence of activities that unfolds over the lives of bees, as determined by monitoring the activities of one cohort of bees living in an observation hive. During the first few days of adult life, a worker functions primarily as a *cell cleaner,* cleaning and polishing recently vacated brood cells. She also devotes time to eating some of the pollen that is stored nearby, which favors the rapid activation of her hypopharyngeal glands. The worker also spends some 20% of her time resting—standing motionless on the combs or in a cell—and another 20% patrolling—walking about the combs, as if searching for work. By the time she reaches 3 days of age, she functions as a *nurse,* for her hypopharyngeal glands have begun secreting brood food and she has started spending much time feeding the brood. She also performs the other tasks that arise within the broodnest, including tending the queen, capping brood, and grooming and feeding nestmates. This pattern continues for the next 10 days or so, or until she is about 12 days old. At this point she leaves the central broodnest to work primarily in the peripheral, food-storage region of the hive. Here she functions mainly as a *food storer.* Her

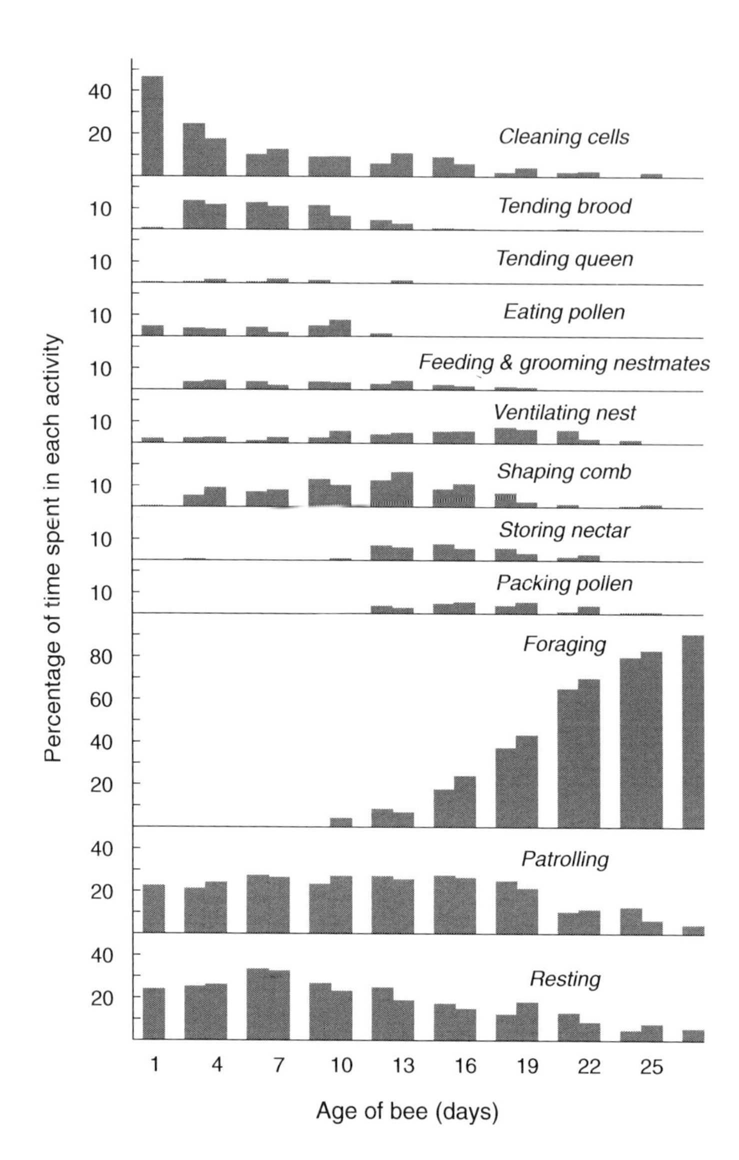

Figure 2.5 The behavioral changes of worker bees as they grow older. At each age, individuals specialize on a subset of the tasks needed to maintain the colony's well-being. Typically, young workers concentrate on the jobs occurring in the central broodnest, such as cleaning cells, feeding brood, and tending the queen. Middle-aged bees work mainly on the periphery of the combs, receiving and storing nectar, packing pollen, and ventilating. The old workers function almost entirely outside the hive as foragers. Based on the data in figure 1 of Seeley 1982.

hypopharyngeal glands are secreting the enzymes needed for producing honey, and her poison gland has filled the venom sac. Shuttling between the hive entrance and the upper honeycombs, she receives nectar from the returning nectar foragers, converts it to honey, and deposits this in the storage cells. She also packs pollen in cells, ventilates the hive by fanning her wings, helps guard the hive entrance, and continues grooming and feeding her hivemates. Also, if additional comb is needed for honey storage, these middle-aged bees will activate their wax glands and build comb. Finally, from the age of about 20 days until the end of life, a worker toils outside the hive as a forager, gathering nectar, pollen, water, resin, or some combination of these substances.

The general sequence of activities depicted in Figure 2.5—from cell cleaner to nurse bee to food storer and finally to forager—is more or less fixed for worker bees, but there is tremendous variation among individuals in the effort expended on the different activities within each of the four sets of tasks. One worker may never undertake a certain activity, while another may specialize in it for several days. For example, some food-storer bees never guard the hive entrance, while others spend a week or more specializing as guards. Likewise, some forager bees concentrate on pollen foraging while others devote their entire foraging careers to the collection of water. Much of this behavioral variation traces to the underlying genetic variability of worker bees, a product of the queen's curious habit of mating with a dozen or more males (discussed in detail in Page and Robinson 1991). It should also be noted that the age ranges for the different activities will differ dramatically under different sets of conditions. The timings for activities shown in Figure 2.5 are representative only for a colony experiencing plentiful nectar; if the colony studied had been experiencing a nectar dearth, there would have been less need for labor devoted to foraging and the transition from food storer to forager would have been delayed, even by a week or more. It must be stressed that the activities of workers are adjusted in accordance with the needs of their colony, and that these needs can vary greatly depending on the conditions both inside and outside the hive. Indeed, it is probably possible for bees of almost any age to perform a particular task if the occasion demands it, as has been recently discussed in detail by Robinson (1992).

2.3. Nest Architecture

As is well known, the combs inside a beehive hang vertically and each is made of two layers of horizontal cells, with openings on opposite sides of the comb (Figure 2.6). These cells serve both as containers for stored food (honey and pollen) and as cradles for the developing immature bees. They are precisely hexagonal and form a beautiful, regular array, except around the edges of the comb, where they are

Figure 2.6 Sectional view of a part of a honey bee comb, showing pupae in cells. Photograph by K. Lorenzen.

attached to the substrate, or where there is a transition between the smaller cells used for rearing workers ("worker cells") and the largers ones used for rearing drones ("drone cells") (see Figure 7.1). Both cell types are also used for food storage. The principal material used in comb construction is the wax secreted from the glands on the underside of each worker's abdomen. A typical nest in nature is an impressive edifice, consisting of some 100,000 cells in a half dozen or so combs whose total surface area is approximately 2.5 m^2. Building this requires over 1200 g of wax. Such a large structure is needed to provide storage space for the 20 or more kilograms of honey that must be stockpiled for winter food, as well as to provide nursery space for the 20,000 or more immature bees that a colony contains in the spring, its time of most rapid growth.

Figure 2.7 depicts the general layout of a honey bee nest inside a

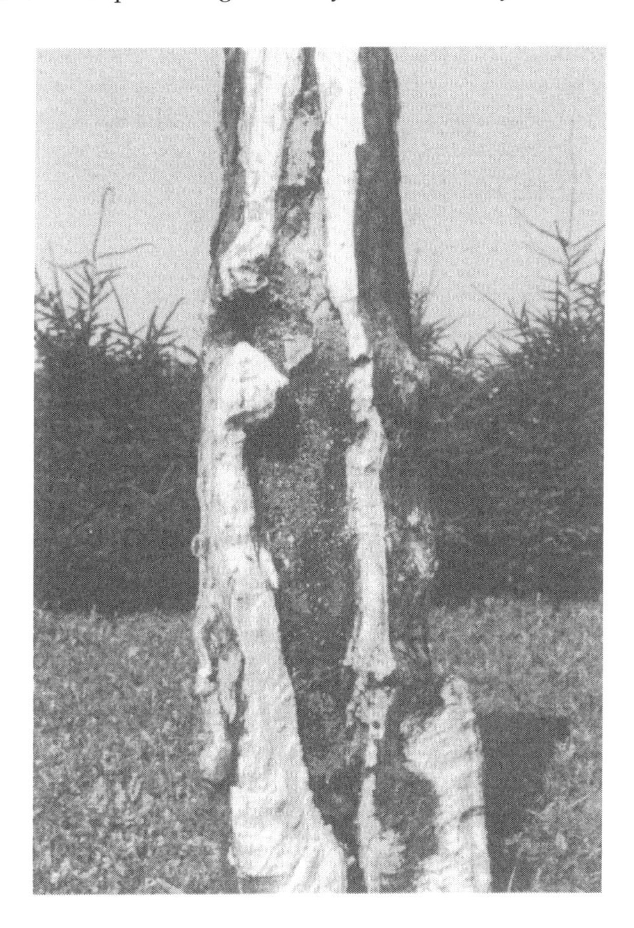

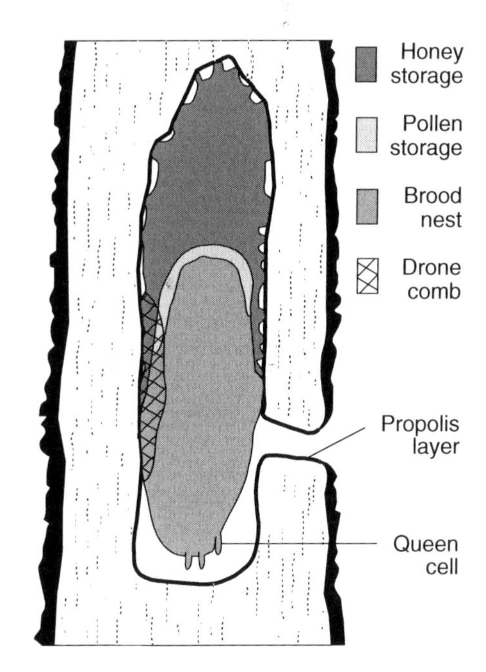

Figure 2.7 Cross-section of a typical nest in a hollow tree. *Left:* actual nest; the entrance was through the transected knothole in the left side of the tree. *Right:* schematic diagram. After Seeley and Morse 1976. Photograph by T. D. Seeley.

hollow tree. The tree cavity is carefully checked by the bees before it is occupied, to be sure it offers sufficient space (no less than about 25 L), has an entrance opening which is not too large (maximum diameter about 4 cm), and ideally has an entrance which is near the floor of the cavity for easy removal of debris, faces south for warmth, and is high off the ground for safety from predators. Inside the hollow, the bees scrape off any loose, decayed wood from the inner surfaces and coat them with propolis (dried tree resin) to produce a clean, firm interior surface. They then build their beeswax combs downward, attaching them to the cavity's roof and walls and leaving small passageways along the edges. The combs consist mainly of the smaller, worker cells, with about 15% of the comb area devoted to the larger, drone cells. These are normally produced in discrete patches on the edge of the combs, as shown in Figure 2.7. Special peanut-shaped queen cells are constructed on the margins of the combs when new queens are reared in the spring (see Section 2.4). Food storage and brood production are spatially segregated, with honey typically stored in the upper region of the combs, brood reared in the lowermost areas, and the pollen reserve forming a narrow band between the honey and the brood.

2.4. The Annual Cycle of a Colony

The honey bee colony's annual cycle in cold temperate regions can be thought of as beginning shortly after the winter solstice, when the days start to grow longer but snow still blankets the countryside. At this time a colony, which is living as a tight ball of bees inside the hive, raises the core temperature of its cluster to about 34°C and starts to rear brood. At first, only 100 or so young bees are produced, but by early spring, when the first flowers blossom, several thousand cells hold developing bees, and the pace of colony growth quickens daily. Come late spring, honey bee colonies will already have expanded to full size, 30,000 or so individuals, and will begin to reproduce. Reproduction involves not only the standard process of rearing males, which simply fly from the hive and mate, but also an intricate process of colony fission in which the colony rears several new queens and, when these queens are nearly mature, divides itself with about half the workers plus the old queen leaving in a swarm (Figure 2.8). After flying a short distance from the parent hive, the swarm condenses into a beardlike cluster on a tree branch. From here the swarm's scout

Figure 2.8 A swarm of honey bees, consisting of one queen and approximately 12,000 workers. These bees have recently left their old nest and have settled on these branches to rest quietly until the swarm's scouts have located a new home site. Photograph by T. D. Seeley.

bees explore for nest cavities, select the one which is most suitable, and finally direct the other bees in the swarm to the new home site.

For a week or so following the departure of the mother queen, the workers in the parental hive are queenless; then the first virgin queen emerges. If the first ("prime") swarm's departure has greatly weakened the parental colony, the remaining workers allow the virgin queen that emerges first to search through the nest for her rival sister queens and to kill them while still in their cells. Frequently, however, by the time the first virgin queen has appeared, sufficient worker brood has also emerged to restore the parent colony's strength. In this situation, the workers guard the remaining queen cells against destruction by the first virgin queen, start shaking her to prepare her for flight, and eventually push her out of the nest in an afterswarm. This process is repeated with later emerging queens until the colony is weakened to the point where it cannot support further fissioning. If more than one queen remains in the parental nest, the workers allow these queens to fight each other until just one remains. The reproductive process is completed when the surviving virgin queens fly from their hives to mate with males from the surrounding colonies. During the remainder of the summer and on into the autumn, the colonies in new nest sites strive to build combs, and all the colonies intensively rear brood and gather food to rebuild both their populations and their food reserves before the arrival of the cold, flowerless days of winter.

2.5. Communication about Food Sources

When a worker bee discovers a rich source of pollen or nectar, she is able to recruit nestmates to it and thereby strengthen her colony's exploitation of this desirable feeding site. The principal mechanism of this recruitment communication is the waggle dance, a unique behavior in which a bee, deep inside her colony's hive, performs a miniaturized reenactment of her recent journey to a patch of flowers. Bees following the dance learn the distance to the patch, the direction it lies in, and the odor of the flowers, and can translate this information into a flight to the specified patch. Thus a waggle dance is a truly symbolic message, one which is separated in time and space from both the actions on which it is based and the behaviors it will guide.

To examine how bees communicate using waggle dances, let us follow the behavior of a bee upon her return from a rich new food source.

Her find is a large cluster of flowers located a moderate distance from her nest, say 1500 m, and along a line 40° to the right of the line running between the sun and her nest (Figures 2.9 and 2.10). Excited by her discovery, she scrambles inside her colony's hive and immediately crawls onto one of the vertical combs. Here, amidst a massed throng of her sisters, she performs her recruitment dance. This involves running through a small figure-eight pattern: a waggle run followed by a turn to the right to circle back to the starting point, another waggle run, followed by a turn and circle to the left, and so on in a regular alternation between right and left turns after waggle runs. The waggle run portion of the dance is the most striking and informative part of the bee's performance, and is given special emphasis both by the vigorous waggling—the lateral vibrating of the body, with sideways deflections greatest at the tip of the abdomen and least at the head—and by the dorso-ventral vibrating of the wings at approximately 260 Hz. Usually several bees will trip along behind a dancer, their antennae always extended toward her. These followers detect the dance sounds with their antennae. The flagellum (outermost portion) of a worker bee's antennae has a resonant frequency of about 260–280 Hz, matching the vibration frequency of the wing vibrations. Moreover, the Johnston's organ, the vibration detector at the base of the flagellum, is maximally sensitive to vibrations in the 200 to 350 Hz band.

The direction and duration of each waggle run is closely correlated with the direction of and the distance from the flower patch being advertised by the dancing bee. Flowers located directly in line with the sun are represented by waggle runs in an upward direction on the vertical comb, and any angle to the right or left of the sun is coded by a corresponding angle to the right or left of the upward direction. In the example illustrated in Figures 2.9 and 2.10, the flowers lie 40° to the right of the sun, and the waggle run is correspondingly directed at an angle of 40° to the right of vertical. The distance beween the nest and the recruitment target is evidently encoded in the duration of the

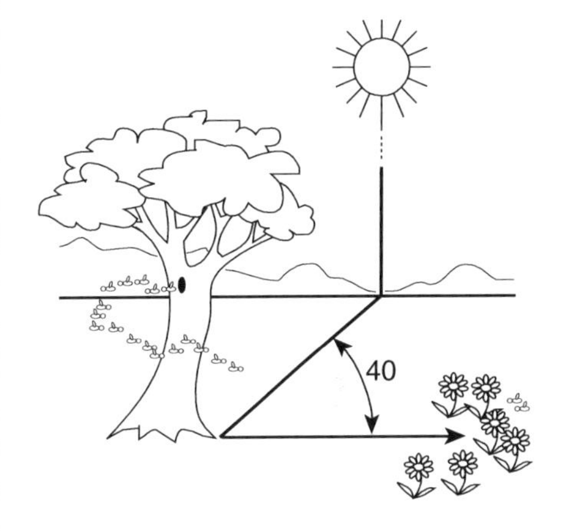

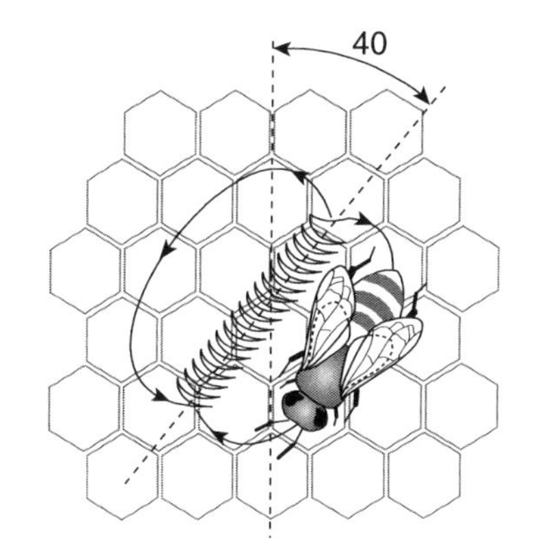

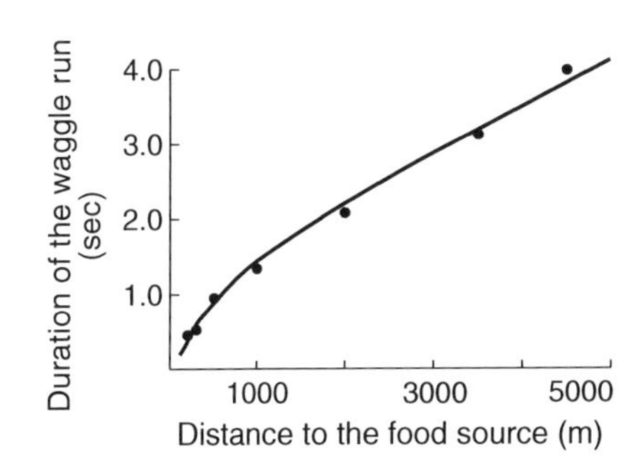

Figure 2.9 The waggle dance of the honey bee. *Top:* The patch of flowers lies along a line 40° to the right of the sun as a bee leaves her colony's nest inside a hollow tree. *Middle:* To report this food source when inside the nest, the bee runs through a figure-eight pattern, vibrating her body laterally as she passes through the central portion of the dance, the waggle run. *Bottom:* the relationship between the distance to the flowers and the duration of the waggle run (based on data in table 13 of von Frisch 1967). After Seeley 1985.

Figure 2.10 A worker bee performing a waggle dance and several bees following her dance. Photograph by K. Lorenzen.

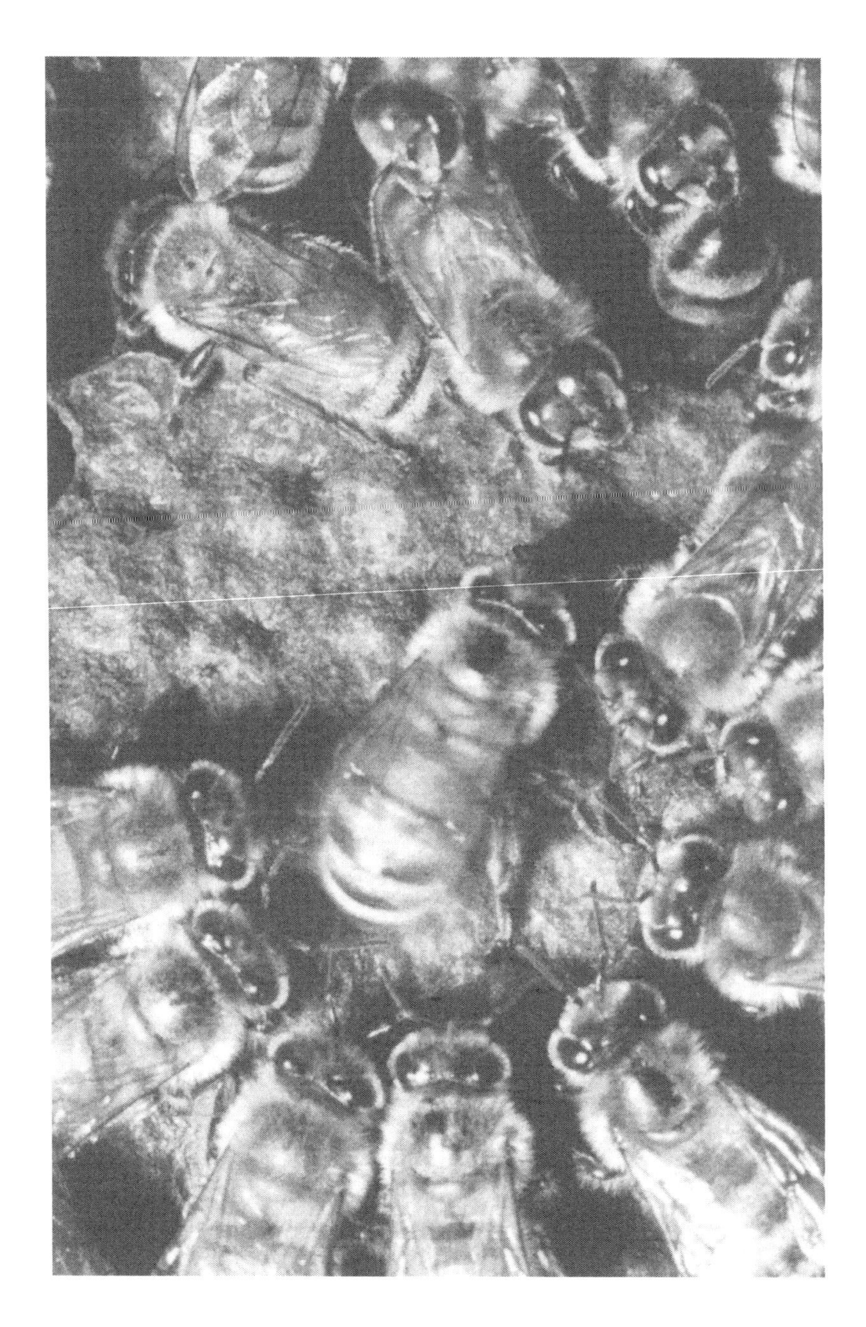

waggle run. The farther the target, the longer the waggle run, with a rate of increase of about 75 msec per 100 m. Workers can detect the buzzing sound produced during a waggle run, so it seems likely that dance followers measure the duration of a dancer's waggle run by noting the duration of the sound associated with each waggle run.

Besides information about direction and distance, a dancing bee also communicates the odor of the flowers at her forage site. This scent is partly carried back in the forager's waxy cuticle, but often a stronger source of the scent is the food she brings home—the loads of pollen on her hind legs or the nectar she regurgitates to the dance followers. Recruits appear to draw upon their knowledge of the food source's odor to help pinpoint its location after using the dance's vectorial information to arrive in the general vicinity. If a recruitment target lacks significant odor—for example, if it is a water source or a clump of weakly scented flowers—then bees will mark the site with scent from their Nasanov's glands.

2.6. Food Collection and Honey Production

2.6.1. THE SUBSTANCES COLLECTED

Many of the bees landing at the entrance to a beehive carry on each hind leg a little ball of brightly colored material. Many balls are orange, some are yellow, and still others are red, brown, or even blue. All are loads of pollen, gathered from flowers in the surrounding countryside. Pollen provides bees with the amino acids and vitamins that they must have to achieve maturity, including the full development of their hypopharyngeal (brood food) glands. Pollen also fills the bees' requirements for fats. Upon her return to the hive, a forager bearing pollen enters the hive directly and deposits her pollen loads in a cell (Figure 2.11). Often she inspects a number of cells before finding one that is satisfactory, and usually it is just above or beside the broodnest. Pollen stored here is readily available to the nurse bees. Although the foragers themselves deposit their pollen loads in the storage cells, the younger hive bees do the rest of the processing of this food. They tamp the pollen loads down tightly, which helps to exclude air, and they incorporate with the pollen a little regurgitated honey, which is microbicidal. In these ways they inhibit the germination and bacterial spoilage of the pollen.

A careful look at the other bees landing at the hive entrance, those

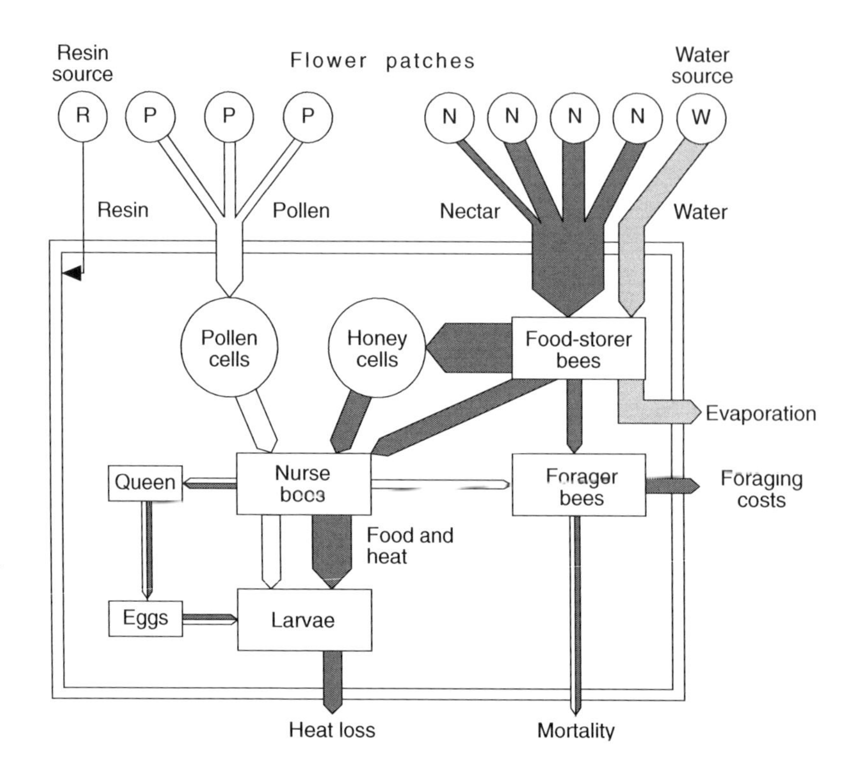

Figure 2.11 The flow of substances in a honey bee colony on a summer day. The width of each arrow is proportional to the amount of matter flowing along its route. Matter accumulates in the growing larvae, to increase the colony's population, and in the honey cells, to build up the energy store for winter.

without pollen loads, reveals many with noticeably swollen abdomens. If one captures one of these bees, holds her between gloved fingers and gently squeezes her abdomen, she will disgorge a droplet of clear or pale yellow liquid. Chemical analysis would reveal that most of the bees regurgitate a concentrated solution of sugar (Figure 2.12), mainly glucose, fructose, and sucrose. These are loads of nectar, the raw material from which honey is made, and the principal source of carbohydrates for the bees. But not all the swollen bees are nectar foragers returning with a load of energy; a few carry home liquid containing little or no sugar (Figure 2.12). These are the water collectors, returning from a puddle, stream, or whatever water source lies near the hive. Both nectar foragers and water collectors regurgitate the nectar or water in their honey stomachs to one or more hive bees (Figure 2.13). The returning bee opens her mandibles wide, with her proboscis retracted, and a drop of liquid appears on the upper surface of the base of her proboscis. The receiving bee stretches out her proboscis to full length and quickly takes the fluid. The recipients of nectar and water are usually middle-aged bees, the food-storer bees. They will either distribute the fresh nectar among other bees in the hive for immediate consumption or, more commonly, process the nectar into honey for storage and future consumption.

Food-storer bees usually "ripen" honey from nectar in the honey storage region above the broodnest. Here they manipulate the liquid in their mouthparts, repeatedly unfolding and refolding the proboscis, thereby exposing to the air an antennuated droplet of liquid in the angle between the two parts of the proboscis. A bee repeats the whole process for perhaps 20 min, reducing the water content of the liquid and adding to it more salivary gland secretions containing enzymes: invertase, to cleave sucrose into the more soluble sugars fructose and glucose, and glucose oxidase, to produce hydrogen peroxide to protect the honey from spoilage. Finally the food-storer bee deposits the liquid in a honey storage cell, which, when full, will be capped with a more or less airtight beeswax seal to reduce moisture absorption by the honey. Fully ripened honey contains only 16–20% water and so is hygroscopic.

Food-storer bees that receive water instead of nectar may spread it over the combs to cool the nest, especially the central broodnest, by evaporation. Alternatively, they will distribute it among the nurse bees. This happens most often in the early spring, when the bees are subsisting on their honey and stored pollen, and the nurse bees need water to dilute the thick honey to prepare the liquid food for the larval brood.

Pollen, nectar, and water are certainly the substances most commonly gathered by a colony's foragers, but from time to time during

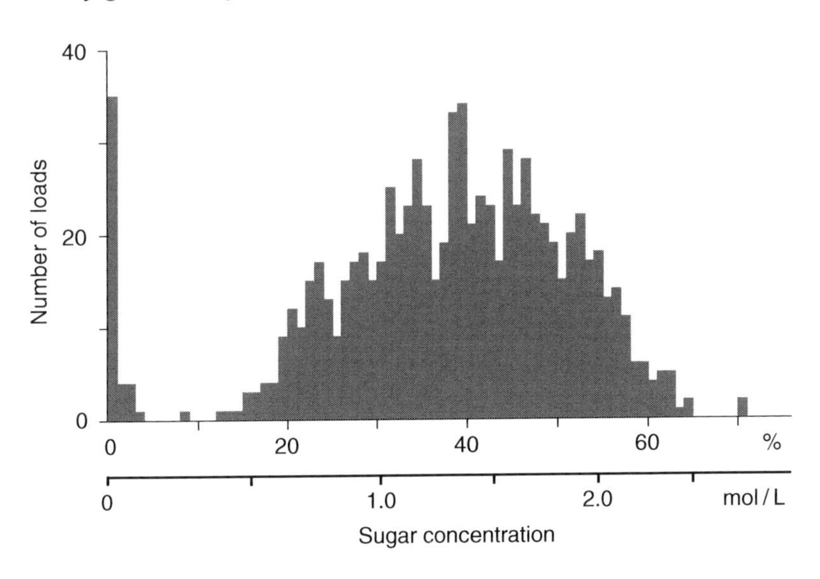

Figure 2.12 Sugar concentrations of the loads of liquid collected by bees. Although most bees returning with liquid are nectar foragers, carrying a sugar solution of concentration 0.5–2.5 mol/L, a small percentage—approximately 5% in this case—are water collectors, returning with water or an extremely dilute sugar solution. The data were collected on 12–16 May 1985 and 28–30 June 1989. The total number of loads assayed was 835. After Seeley 1986 and unpublished data of T. D. Seeley.

Figure 2.13 A nectar forager *(right)*, having returned to her hive, regurgitates her nectar load to a food-storer bee *(left)*. Photograph by K. Lorenzen.

the summer one also sees a few bees returning to the hive with shiny, brown loads of tree resin stuck in their pollen baskets. This resin is worked into cracks and holes in the walls of the colony's nest cavity, rendering it more weathertight and easier to defend. It is also applied as a smooth, clean coating over the nest cavity's walls. When fresh, the tree resins are so sticky that foragers returning with propolis must have their loads pulled off by other bees, but over time the resins dry and harden. They function not only mechanically but also chemically, for they too contain many microbicidal compounds.

2.6.2. THE QUANTITY OF FOOD NECESSARY FOR SURVIVAL

How much of the four substances just discussed must a colony gather each year? On an average foraging trip, a bee collects only about 15 mg of pollen or resin, or about 30 mg of nectar or water, but because a colony typically possesses several thousand foragers and each one can conduct multiple collecting trips each day, the total quantity of supplies assembled annually by a colony is impressive. Estimates vary, but on average a colony extracts from its environment each year some 20 kg of pollen, 120 kg of nectar, 25 L of water, and perhaps 100 g of resin. Most of the pollen and nectar are consumed during the spring and summer months, when brood rearing is most intense. To rear a single bee requires approximately 130 mg of pollen, and during a summer a colony will rear some 150,000 bees; hence a colony's annual pollen budget is approximately 20 kg. With respect to nectar, a colony consumes about 70 kg during the summer to provide food for the brood, keep the broodnest warm, and fuel the foraging operation. The other 50 or so kilograms of nectar is converted to some 20 kg of honey for eventual consumption during the cold, flowerless months. A colony needs such a large store of energy-rich food because of its means of winter survival, which is unique among insects. Instead of cooling down and becoming dormant, like most insects, a bee colony fights the cold by maintaining a warm microclimate inside the hive. To do so it contracts into a tight cluster and generates enough heat inside this cluster to keep the outermost bees above about 10°C, their lower lethal temperature. The bees generate this heat by isometrically contracting their powerful flight muscles. All told, a colony's heat production in midwinter is on the order of 40 watts, enought to keep the surface bees from perishing, even in the face of ambient temperatures of –30°C or less. Such intense heat production is energetically costly, however, requiring nearly a kilogram of honey each week for fuel all winter long.

2.6.3. COPING WITH LARGE AND RAPID CHANGES IN FORAGE ABUNDANCE

Nature does not provide a honey bee colony with a steady, dependable supply of food. Instead, a colony must deal with a boom-and-bust food supply, as summer turns to winter and, during the summer, as times of profuse forage ("nectar flows") alternate with times of dearth. To cope with the tremendous variability in the food supply, a colony maintains a large forager force that can be rapidly deployed to exploit fully the times of abundance. Also, it stockpiles much of the nectar and pollen gathered at rich times to carry it through the poor times.

A clear picture of this variability in food supply is obtained by placing a hive of bees on a set of scales and taking weight readings at regular intervals. The records for one colony weighed weekly for 3 years

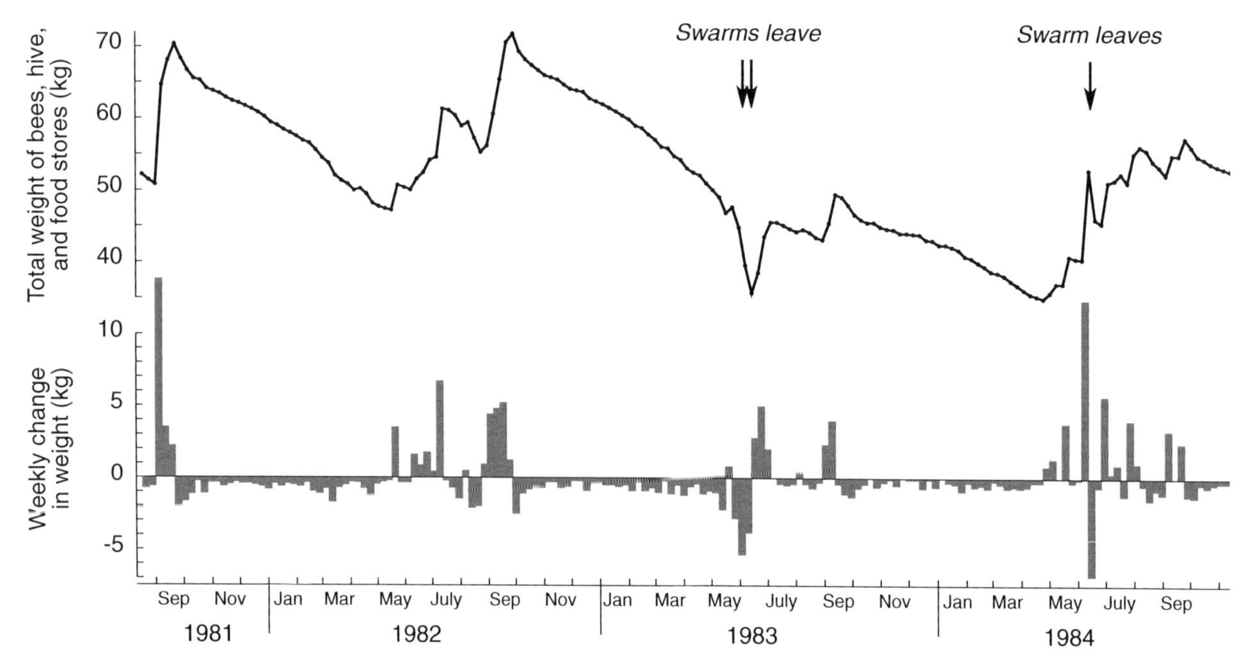

Figure 2.14 Three-year record of weekly weighings of a honey bee colony. Changes in weight reflect mainly the collection of fresh nectar (weight gain) and the consumption of stored honey (weight loss). Occasionally a departing swarm also caused a large weight loss. The colony gained weight rapidly, though sporadically, for just a few weeks each summer and then gradually lost approximately 25 kg of weight over the winter period, October to May. Based on unpublished data of T. D. Seeley.

are shown in Figure 2.14. They reveal that nectar was sufficiently abundant for the colony to gain weight for only about 10 weeks each year, and that for the remainder of the time the colony gradually lost weight as it drew on its stores of pollen and honey. What is perhaps most noteworthy is that even during the warm months, May through September, when it seems that some plants are always in flower, the colony experienced many weeks of meager food intake and net weight loss. Clearly, the availability of nectar can vary greatly from week to week. The pattern shown in Figure 2.14 is typical for colonies in Europe and North America; so evidently the pattern of long periods of sparse nectar punctuated by shorter intervals with profuse nectar is common for bee colonies.

To fully appreciate the extreme variability in the nectar supply of a colony, however, one needs to see the pattern of weight changes from a colony that is weighed nightly, after the colony has finished its foraging for the day. One such pattern is shown in Figure 2.15, which depicts the records for a colony in northeastern Connecticut in the late spring and early summer of 1986. At this time and place there was a succession of intense blooms by plants that provide copious nectar: first dandelions *(Taraxacum officinale)*, then black locust trees

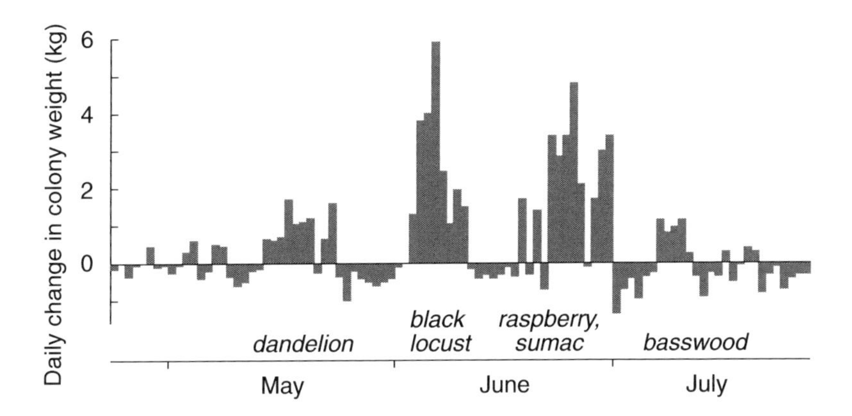

Figure 2.15 Day-to-day fluctuations in a colony's nectar collection. During a 3-month period the colony experienced four nectar flows, periods of intense nectar production by one or more species of flowering plants. Between these times of abundant forage and large weight gains, the colony lost weight, either because flowers were scarce or because poor weather prevented foraging. Based on unpublished data of T. D. Seeley.

(Robinia pseudoacacia), next raspberry (*Rubus* spp.) and sumac *(Rhus typhina)* plants, and finally basswood *(Tilia americana)* trees. Accordingly, the colony experienced tremendous swings in its nectar collection. For instance, at the end of May, after the dandelion bloom, the colony's weight was *falling* by about half a kilogram per day, which indicates that it was gathering essentially no nectar, but then in early June the black locust trees came into flower and the colony's weight was *rising* by up to 6 kg a day. This transition represents a 100-fold or more surge in the colony's rate of nectar collection. Foraging at this high intensity lasted only a few days, however, so that by the end of the second week in June the colony was again finding little or no nectar and was again losing weight. In later chapters, we will see that a honey bee colony possesses sophisticated mechanisms for coping with such immense fluctuations in the nectar supply.

3 The Foraging Abilities of a Colony

We ordinarily think of a colony of bees as a group of insects living inside a hive. A moment's reflection will disclose, however, the important fact that during the daytime many of the bees in a colony—the foragers—are dispersed far and wide over the surrounding countryside as they toil to gather their colony's food. To accomplish this, each forager flies as far as 10 km to a patch of flowers, gathers a load of nectar or pollen, and returns to the hive, where she promptly unloads her food and then sets out on her next collecting trip. On a typical day a colony will field several thousand bees, or about one-quarter of its members, as foragers. Thus in acquiring its food, a colony of honey bees functions as a large, diffuse, amoeboid entity which can extend itself over great distances and in multiple directions simultaneously to tap a vast array of food sources. If it is to succeed in gathering the 20 kg of pollen and 120 kg of nectar its needs each year, it must closely monitor the food sources within its foraging range and must wisely deploy its foragers among these sources so that food is gathered efficiently, in sufficient quantity, and with the correct nutritional mix. The colony also must properly apportion the food it gathers between present consumption and storage for future needs. Moreover, it must accomplish all these things in the face of constantly changing conditions, both outside the hive as different foraging opportunities come and go, and inside the hive as the colony's nutritional needs change from day to day. In this chapter we will see that a honey bee colony succeeds in meeting all these challenges.

3.1. Exploiting Food Sources over a Vast Region around the Hive

One of the most amazing attributes of a honey bee colony is its ability to project its foraging operation over an immense area around the hive: at least 100 km^2. This capacity for widespread foraging arises because each of the colony's foragers can find her way to and from flower patches located 6 or more km from home. Flying bees cruise along at about 25 km/hr, so a 6-km trip takes only about 15 minutes, but if one considers the small size of a bee, then one realizes that a foraging range of this magnitude is thoroughly impressive. A 6-*kilo*meter flight performed by a 15-*milli*meter bee is, after all, a voyage of 400,000 body lengths. A comparable performance by a 1.5-m tall human would be a flight of some 600 km, such as from Boston to Washington, Berlin to Zürich, or Bangkok to Rangoon.

Lying in the grass beside a beehive, gazing upward at the foragers soaring off against the blue sky, one has little indication that their activity extends so far from home. One begins to perceive the tremendous scope of a colony's foraging operation if one marks the foragers with paint, fluorescent dust, or a genetic marker, and then combs the countryside for these labeled bees. Alternatively, one can place magnets over the entrance of a hive, then go out into the fields, capture bees on flowers, and glue a small, metal identification disk to each captured bee's abdomen, keeping record of where in the countryside each disk was fastened to a bee. When the foragers from the study hive return home, the magnets automatically collect the identification disks (Gary 1971). Studies using one or the other of these two approaches show that most of a colony's foragers are found on flowers within 1 km of the hive, but that they will fly 14 km to reach flowers if none are closer (Eckert 1933; Levin 1961; Gary, Witherell, and Lorenzen 1978). Both approaches, however, can yield a distorted picture of the spatial distribution of a colony's foraging efforts because the picture they provide reflects not only where a colony's foragers go to find flowers, but also where human beings go to find bees. And unfortunately the people may not go everywhere the bees go. Also, the studies using these two approaches typically have been conducted where forage is unusually plentiful—such as alfalfa fields and almond orchards in full bloom—hence these studies are not likely to depict the full spatial scale of foraging by colonies living in nature.

Starting in the spring of 1979, I undertook with Kirk Visscher a study aimed at generating a sharper, more accurate picture of the spatial patterns of foraging by a colony living under natural conditions (Visscher and Seeley 1982). Our approach was to map out, day by day, the forage sites of one colony living in a forest setting. How could we acquire this overview of a colony's foraging operation? The technique of directly tracking a colony's thousands of foragers to their work sites would certainly not succeed. One cannot track even one bee as she flies away from the hive, let alone thousands. So we turned to an indirect, but powerful, technique pioneered by one of Karl von Frisch's students—Herta Knaffl (1953)—some 30 years earlier: let the bees inform us where they are going by means of their recruitment dances. (These dances are easily observed if the colony is living in a glass-walled observation hive.) The beauty of this technique is that one can determine where a colony's foragers are going from observations made entirely at the colony's hive, even if the foragers are commuting to sites several kilometers away. The one drawback of this technique is that it may not reveal all a colony's forage sites for any given day, because on each day only the foragers returning from the most profitable sites will advertise them with recruitment dances (see Section 5.2). It is likely that during each day of active foraging a colony exploits some flower patches that merit *continued* exploitation but not *greater* exploitation. If so, then some of a colony's forage sites will not be advertised by recruitment dances, and of course those sites that are not announced in the hive cannot be detected by someone watching the dancing bees. Nevertheless, this technique provides an accurate picture of the spatial scale of a colony's foraging operation, the primary goal of our study, since all a colony's forage sites will be represented by recruitment dances during the initial, build-up phase of their exploitation.

The first step in our investigation was to construct an observation hive suitable for this study (Figure 3.1). The hive needed to be large enough to house a full-size colony of bees, and it had to have a wedge in the entrance tunnel to force all foragers to enter the hive from one side of the comb. Because returning foragers generally perform their dances shortly after entering the hive, directing all the traffic to one side of the comb created a well-defined dance floor area near the entrance on one side of the hive. Over the dance floor we positioned a sampling grid so that we could select dancing bees at random for observation. Next we installed a colony of approximately 20,000 bees in the hive and then moved it to the Arnot Research For-

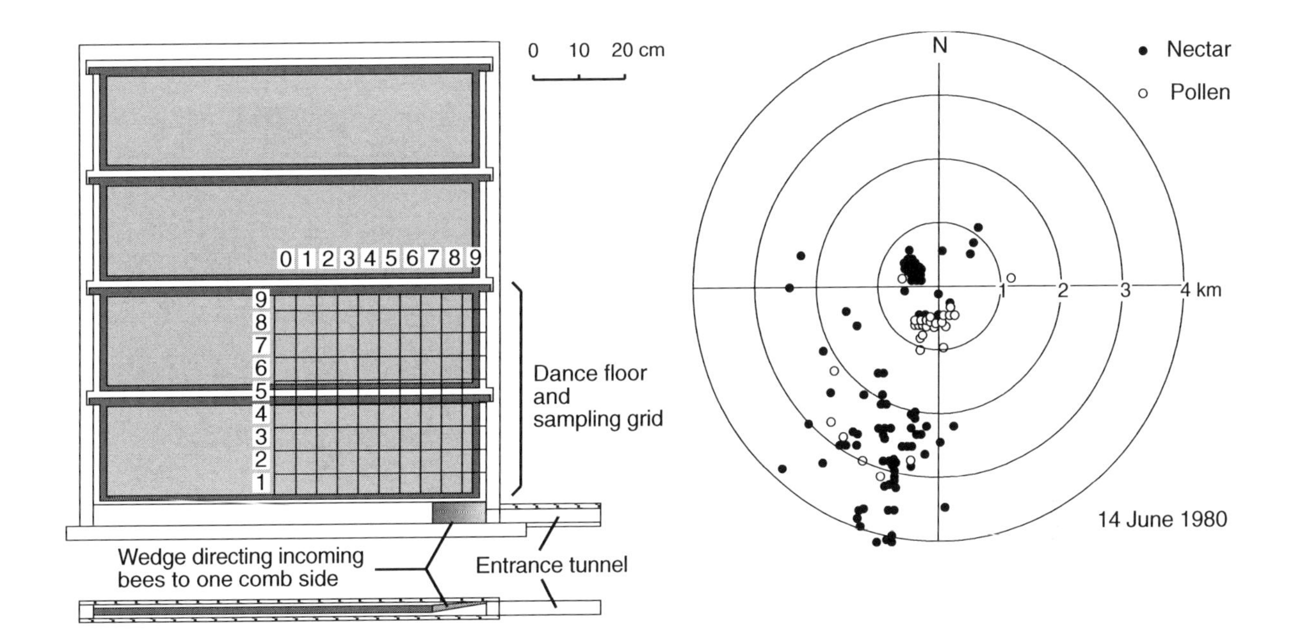

Figure 3.1 Determining where a colony's foragers are gathering food by reading their recruitment dances. The figure on the left depicts the large observation hive used for observing the dances of the foragers. The figure on the right depicts the results of one day's data collection, with each dot representing the estimated location of one forager's work site. On this day the colony was gathering nectar from two areas, one 2–4 km to the SSW and the other 0.5 km to the NW, and was gathering pollen from an area 0.5 km to the S. The total number of points plotted was 117. After Visscher 1982 and Visscher and Seeley 1982.

est of Cornell University, a region of abandoned agricultural fields and mature hardwood forests outside Ithaca, New York, where the bees could live in a reasonably natural habitat. A few days later, we began collecting data on the dancing bees. For each randomly selected dancer, we measured the angle and duration of her waggle runs, we noted what color pollen she carried (if any), and we recorded the time of day of her dance. With this information we could estimate the location of each dancer's forage site and the type of forage available there. Finally, we plotted each dancer's forage site on a map to give us a synoptic picture of the colony's richest forage sites for the day (Figure 3.1).

In Figure 3.2 is shown the distribution of distances to forage sites based on observing 1871 dancing bees during four nine-day periods spread over the summer of 1980. This shows clearly that a colony living under natural conditions conducts much of its foraging within several hundred meters of the hive, but also that it regularly forages at sources several kilometers from the hive. The modal distance from hive to forage site was 0.7 km, the median distance was 1.6 km, the mean distance was 2.2 km, and the maximum distance was 10.9 km. Perhaps the most important property of this distribution is the location of the 95th percentile, which falls at 6.0 km. This indicates that a

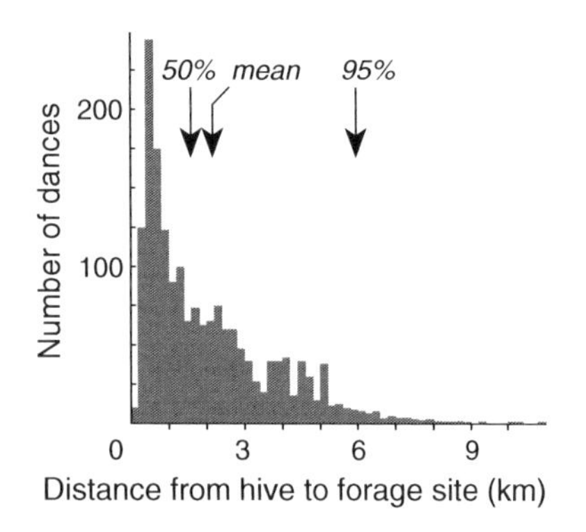

Figure 3.2 Distribution of the distances to a colony's forage sites, based on analysis of 1871 recruitment dances performed over four 9-day periods between 12 June and 27 August 1980. After Visscher 1982.

circle large enough to enclose 95% of the colony's forage sites would have a radius of 6 km, hence an area greater than 100 km^2. Such widespread foraging by a colony is evidently typical, for two other investigators have also plotted a colony's forage sites by reading the dances of its foragers and they too report that bees visit flowers mainly within 2 km from their hive, but frequently exploit blossoms up to 6 km away, and occasionally travel even 9 or 10 km to obtain food (Knaffl 1953; Vergara 1983, cited in Roubik 1989, p. 87).

Why does a honey bee colony collect its food over so vast an area? Hamilton and Watt (1970) point out that an animal group with a large biomass and energy budget, such as a bee colony, will often need to range widely to have an adequate resource base, especially if the food resources in the environment are highly patchy, which is evidently the case for honey bees (Figure 3.1). However, an alternative explanation also seems relevant to a bee colony. It is that the large foraging radius may not be energetically essential to a colony, but nevertheless may be advantageous to it, because the larger the foraging range, the larger the array of food sources from which the colony can choose to forage. This wider choice could raise the average richness of the food sources which a colony exploits and so raise the colony's foraging efficiency. As we shall see shortly, colonies are highly skilled at choosing among different food sources, selectively exploiting those that are the most profitable. No doubt there is a minimum foraging range which colonies require for an adequate resource base, but I suspect that colonies go well beyond this for enhanced efficiency in food collection.

3.2. Surveying the Countryside for Rich Food Sources

To profit fully from its immense foraging range, a honey bee colony must be able to find the richest flower patches that arise within this expanse. Moreover, a colony must be able to discover these flowers shortly after they come into bloom, lest the most rewarding blossoms be missed or lost to another colony. How effective is a colony's surveillance of the surrounding countryside for rich new patches of flowers? To address this question, I presented honey bee colonies with a treasure hunt in which the hidden treasures were lush patches of flowering buckwheat plants *(Fagopyrum esculentum)* dispersed over a forest, and I measured each colony's success in finding these prize food sources (Seeley 1987). The layout of this experiment consisted of

Figure 3.3 View of one of six patches of buckwheat flowers used to study the extent of a colony's reconnaissance for rich forage sites. This patch, like all the patches, covered a 10 m × 10 m area and was surrounded by a barbed wire fence for protection against damage by white-tailed deer *(Odocoileus virginianus)*. Photograph by T. D. Seeley.

four clustered colonies of bees and six widely spaced patches of buckwheat, each 100 m^2 in area (Figure 3.3) and planted 1000 to 3600 m from the hives (Figure 3.4). I carefully timed my planting of the buckwheat so that it would blossom when little other forage was available—in late June, after the raspberry (*Rubus* spp.) and sumac (*Rhus* spp.) blooms, or in mid-August, before goldenrod plants (*Solidago* spp.) bloom—hence at a time when the colonies would probably be searching vigorously for food and would certainly be eager to exploit my buckwheat flowers. Once the patches were in full blossom, I went to each patch, daubed paint of a patch-specific color on 150 of the approximately 200 bees foraging in each patch, and then dashed back to the hives to monitor their entrances for foragers bearing my paint marks. If one or more bees were seen entering or leaving a hive with paint representing a particular patch, I could conclude that the colony had discovered that patch.

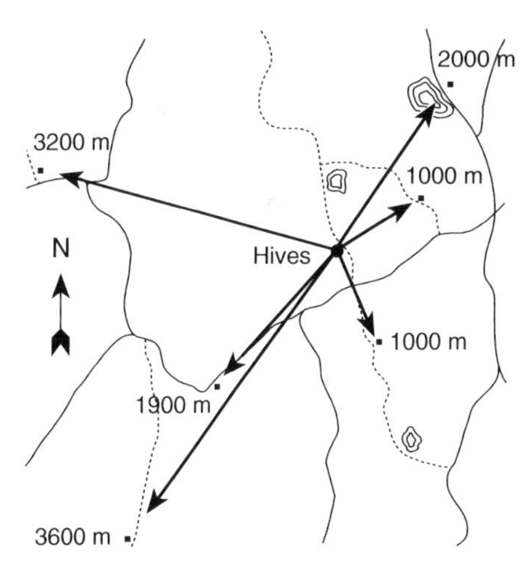

Figure 3.4 Map of the experimental area in the Yale Forest used to evaluate the effectiveness of a honey bee colony's search for food sources. The Yale Forest is a heavily wooded region in northeastern Connecticut where few alternative food sources were available at the time of the study.

The results of this experiment (Table 3.1) indicate that each colony had a high probability of discovering a given buckwheat patch within 2000 m of its hive (1000 m: $P = 0.70$; 2000 m: $P = 0.50$) but a zero probability of finding a particular patch at 3200 m or beyond. It should be noted, however, that these probabilities certainly underrepresent the actual surveillance ability of a colony because my method for determining which colonies had discovered each patch could not detect all

Table 3.1. Results of the experiment analyzing the ability of honey bee colonies to discover 100 m^2 patches of buckwheat flowers planted at various distances from their hives. "X/4" denotes that X out of the 4 test colonies discovered the patch in this trial of the experiment. The totals indicate the probability that a colony will discover a particular patch of flowers located at the distance shown. After Seeley 1987.

	Hive-to-patch distance (m)					
Trial date	1000	1000	1900	2000	3200	3600
August 1984	2/4	—	—	—	—	—
June 1985	3/4	3/4	1/4	2/4	0/4	0/4
August 1985	4/4	2/4	1/4	4/4	—	0/4
Totals	14/20 = 0.70		8/16 = 0.50		0/12 = 0.00	

the discoveries. For instance, if a colony found a patch but sent few foragers there because the patch was already heavily exploited by bees from other colonies, probably I would have failed to detect the colony's discovery of this patch. Despite this conservative bias, the results from this treasure-hunt experiment reveal an impressive ability by honey bee colonies to monitor their environment for rich food sources. A patch of flowers 100 m^2—about half the size of a tennis court—represents less than 1/125,000 of the area enclosed by a circle with a 2-km radius, yet remarkably a honey bee colony has a probability of 0.5 or higher of discovering any such flower patch located within 2 km of its hive. In Chapter 5 we will consider how a colony achieves such effective reconnaissance.

3.3. Responding Quickly to Valuable Discoveries

Having located a patch of blossoms laden with nectar or pollen, a colony must speedily dispatch foragers to the site to harvest its bounty before competitors arrive, darkness falls, the weather deteriorates, or the blossoms themselves fade. Time is of the essence in a colony's food-collection operation. Accordingly, honey bees have evolved their famous waggle dance behavior, which enables a bee that has discovered a rich patch of flowers to share information about its location and scent, and thereby recruit other colony members to the flowers (von Frisch 1967, Gould 1976). Just how speedy is this process of recruitment? Every beekeeper knows that once a single bee discovers an exposed honeycomb numerous other bees are apt to appear there a few minutes later. However, such rapid recruitment is probably atypical, since in this situation the object of attention is just a few meters from the hive of the recruiting bees. More relevant to understanding what occurs in nature are the reports of Charles Darwin (1878) and others (Butler 1945; Weaver 1979) that a flower patch several hundred meters from a beehive can be devoid of honey bees one day and then be heavily visited by bees the next, presumably as a result of strong recruitment to the flowers.[1] These reports, though, probably do not portray the full ability of a colony to rapidly deploy its foragers since the flowers observed may not have offered nectar or pollen rewards great enough to stimulate the bees to dance with maximum intensity.

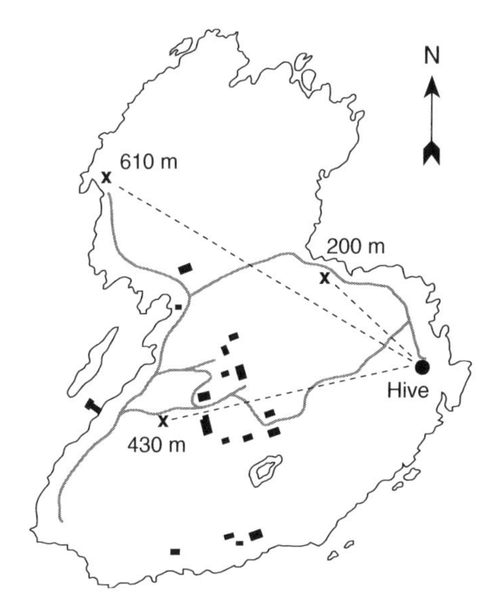

Figure 3.5 Map of the experimental layout on Appledore Island, Maine. A hive of bees was placed on the eastern side of the island, and a small patch of flowers was established at each of three remote points 200–610 m to the north and west. Black rectangles denote buildings.

A better picture of a colony's ability to respond quickly to the discovery of highly desirable flowers comes from an experiment designed specifically to measure this ability (Seeley and Visscher 1988). The setting was a small (39-ha), rocky island—Appledore Island—situated 10 km off the coast of Maine (Figure 3.5). This site was selected for the experiment because it has no resident honey bees and, in late summer, little natural forage for bees. The experimental plan Kirk Visscher and I devised called for locating a beehive on one side of the island, creating rich flower patches at various sites around the island, and, at each such site, recording how long it would take a scout bee to discover the flowers and how quickly thereafter her hivemates would appear. Because we would be operating on an island, thereby limiting the area over which the bees could search, we hoped that we would not have to wait a long time for the bees to discover the flowers at each site. So in August 1979, Visscher and I ferried out to this island a colony of bees and a portable patch of flowers, consisting of 14 mature borage plants *(Borago officinalis)* in flowerpots. The numerous blossoms on a borage plant normally offer rich nectar rewards to bees, but to make sure they provided a highly attractive food source, we injected a 10-μL droplet of concentrated sucrose solution

1. Darwin's own words paint a vivid picture: "I watched for a fortnight many times daily a wall covered with *Linaria cymbalaria* in full flower, and never saw a bee even looking at one. There was then a very hot day, and suddenly many bees were industriously at work on the flowers . . . As in the case of the Linaria, so with *Pedicularis sylvatica, Polygala vulgaris, Viola tricolor,* some species of Trifolium, I have watched the flowers day after day without seeing a bee at work, and then suddenly all the flowers were visited by many bees. Now how did so many bees discover at once that the flowers were secreting nectar?" (Darwin 1878, p. 424).

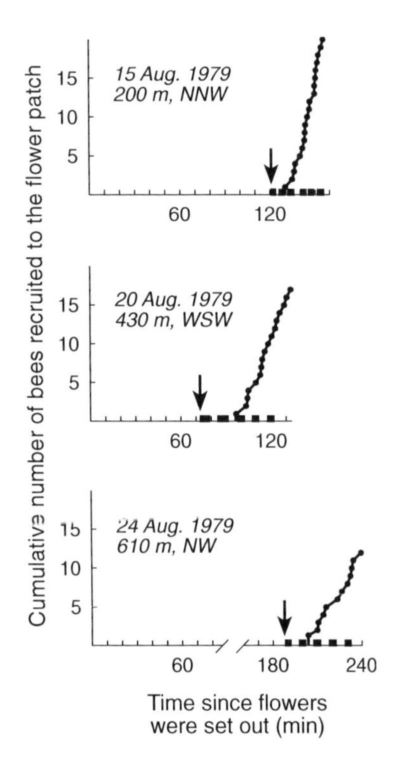

Figure 3.6 Rapid rise in the number of foragers at a rich flower patch following its discovery by a scout bee. Black bars on the abscissa in each plot indicate when the scout bee was at the patch of flowers. After Seeley and Visscher 1988.

into each borage flower at the start of each trial. We also brought along paints to daub on each bee upon arrival at the flowers, so that we could detect the appearance of new recruits. Three trials of this experiment were performed, spaced several days apart (Figure 3.6). Depending on the trial, it took 74 to 200 min for a scout bee to discover the flowers, but then only another 9 to 22 min for the first recruit to reach the flowers. Within an hour of the discovery of each borage patch, the total number of foragers working the borage flowers had risen to 10 to 20 bees, and was continuing to rise rapidly. Clearly, a colony of bees is capable of generating an extremely speedy buildup of foragers at a newly discovered patch of flowers.

To fully characterize the bee colony's ability to rapidly mobilize its foragers, however, this experiment needs to be extended to include trials with the flower patch at several thousand meters from the hive since the bees frequently forage at such distances. Under these conditions, forager deployment probably will be slower than that shown in Figure 3.6, but because the distances involved will be much greater, the overall picture of a colony's response speed will probably remain highly impressive.

3.4. Choosing among Food Sources

The ability to rapidly deploy foragers will lead to foraging success only if coupled with the ability to selectively direct foragers to rich forage sites, thereby enabling the colony to keep its foragers focused on highly profitable sites. This prerequisite is fulfilled, as is shown by the following simple experiment. On 19 June 1983, two groups of 30 labeled bees were fed simultaneously with sucrose solution at two feeders 500 m from the bees' hive. The feeders were positioned in opposite directions from the hive—north and south—so that the colony would have no difficulty distinguishing the two forage sites. One feeder (the "reference feeder") contained a 2.25-mol/L solution, while the other (the "test feeder") contained a 1.50-mol/L solution. A person stationed at each feeder captured recruits, recognized as unlabeled bees, upon arrival at the feeder. Between 11:00 and 3:00, 76 recruits were captured at the highly rewarding reference feeder, whereas only 9 recruits were captured at the less profitable test feeder (Figure 3.7). Hence the colony preferentially directed its foragers to the richer, reference feeder.

When this experiment was repeated 15 times over the next several weeks, with the test feeder loaded each day with a different solution in the range of 1.00 to 2.25 mol/L (nearly the full range of nectar; Figure 2.12), the pattern shown in the lower part of Figure 3.7 emerged (Seeley 1986). The colony steeply downgraded its recruitment to the test feeder as the sucrose concentration there was lowered from 2.25 to 1.00 mol/L, ultimately reaching a point where no recruits were dispatched to the test feeder. Even when the test feeder was loaded with a solution just one-eighth of a molar unit (2.125 mol/L) below that of the reference feeder's standard (2.25 mol/L) solution, it received 30% fewer recruits. Moreover, a difference between test and reference feeders of only 0.25 mol/L elicited a full 50% difference in recruitment rate. Clearly, this colony demonstrated high skill at selectively steering its recruits toward the richer of two forage sites. Why it did not direct all its recruits to the better site is something I will consider later on (Sections 5.10 and 5.14).

A fuller picture of a colony's ability to choose among forage sites comes from an experiment performed several years later, when instead of simply measuring differences in recruitment rates, I undertook the technically greater challenge of measuring differences in the number of foragers allocated to different forage sites (Seeley, Camazine, and Sneyd 1991). To accomplish this, I worked with a colony in which all 4000 of the workers had been painstakingly labeled for individual identification (see Chapter 4 for the details of the labeling technique). After labeling the bees over a 2-day period, 1–2 June 1989, I moved this carefully prepared colony to a special study site, the Cranberry Lake Biological Station. This lovely field station is located deep in a heavily forested region of the Adirondack Mountains in northern New York State. It is especially attractive for bee research because there are no feral bee colonies to disrupt experiments and,

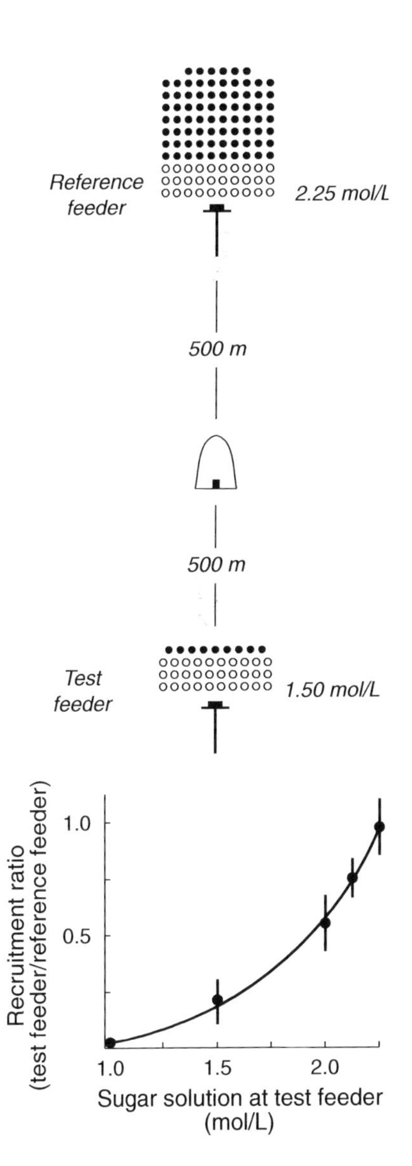

Figure 3.7 Differential recruitment to two feeders containing sugar solutions of different concentrations. *Top:* Experimental layout. The reference feeder always contained a 2.25-mol/L sucrose solution, while the concentration at the test feeder was adjusted between 1.0 and 2.25 mol/L (in the example, 1.50 mol/L). Thirty bees were trained to forage at each feeder and these bees were labeled to identify them as the recruiters (*open circles*) to each feeder. All recruits (*filled circles*)—recognized as unlabeled bees—were captured upon arrival at the feeders. *Bottom:* Summary of results. Whenever the sugar solution of the test feeder was less concentrated than that of the reference feeder, the colony showed strong discrimination between the two feeders, dispatching many fewer recruits to the test feeder. After Seeley 1986.

owing to the dense forest cover, there is little natural forage to entice bees away from the sugar water feeders. To start the experiment, my assistants and I trained two groups of approximately 10 bees each to two feeders positioned north and south of the hive, each one 400 m from the hive. During this initial training period, both feeders contained a rather dilute (1.00-mol/L) sucrose solution that motivated the bees visiting each feeder to continue their foraging but did not stimulate them to recruit any hivemates. The critical observations began at 7:30 on the morning of 19 June, following a 10-day period of cold, rainy weather. At this time, we loaded the north and south feeders with 1.0- and 2.5-mol/L sucrose solutions, respectively, and began recording the number of different individuals visiting each feeder. By noon the colony had generated a striking pattern of differential exploitation of the two feeders, with 91 bees engaged at the richer feeder and only 12 bees working at the poorer one (Figure 3.8). The positions of the richer and poorer feeders were then switched for the afternoon, and by 4:00 the colony had fully reversed the primary focus of its foraging, from the south to the north. This ability to choose between forage sites was again demonstrated the following day during a second trial of the experiment. Thus, when given a series of choices between two forage sites with different profitabilities, the colony consistently concentrated its collection efforts on the more profitable site. The net result was that the colony steadily tracked the richest food source in a changing array.

Perhaps the most remarkable feature of these experimental results is the high speed of the colony's tracking response. Within 4 hours of the noon reversal of the positions of richer and poorer forage sites, the colony had completely reversed the distribution of its foragers. That a colony can respond so swiftly suggests that in nature colonies need such speedy responses in order to track closely the best foraging opportunities in the surrounding countryside. Certainly the array of floral resources available to a colony changes from day to day as different flower patches bloom and fade, and probably the resource array changes even within a day, as sunlight and soil moisture conditions vary at each patch and the flowers accordingly alter their nectar and pollen production (reviewed in Shuel 1992). But just how dynamic is the spatial distribution of the best forage sites, and hence how severe a tracking problem do colonies face? The magnitude of the day-to-day change was revealed by the study described earlier in which Visscher and I monitored the recruitment dances within a

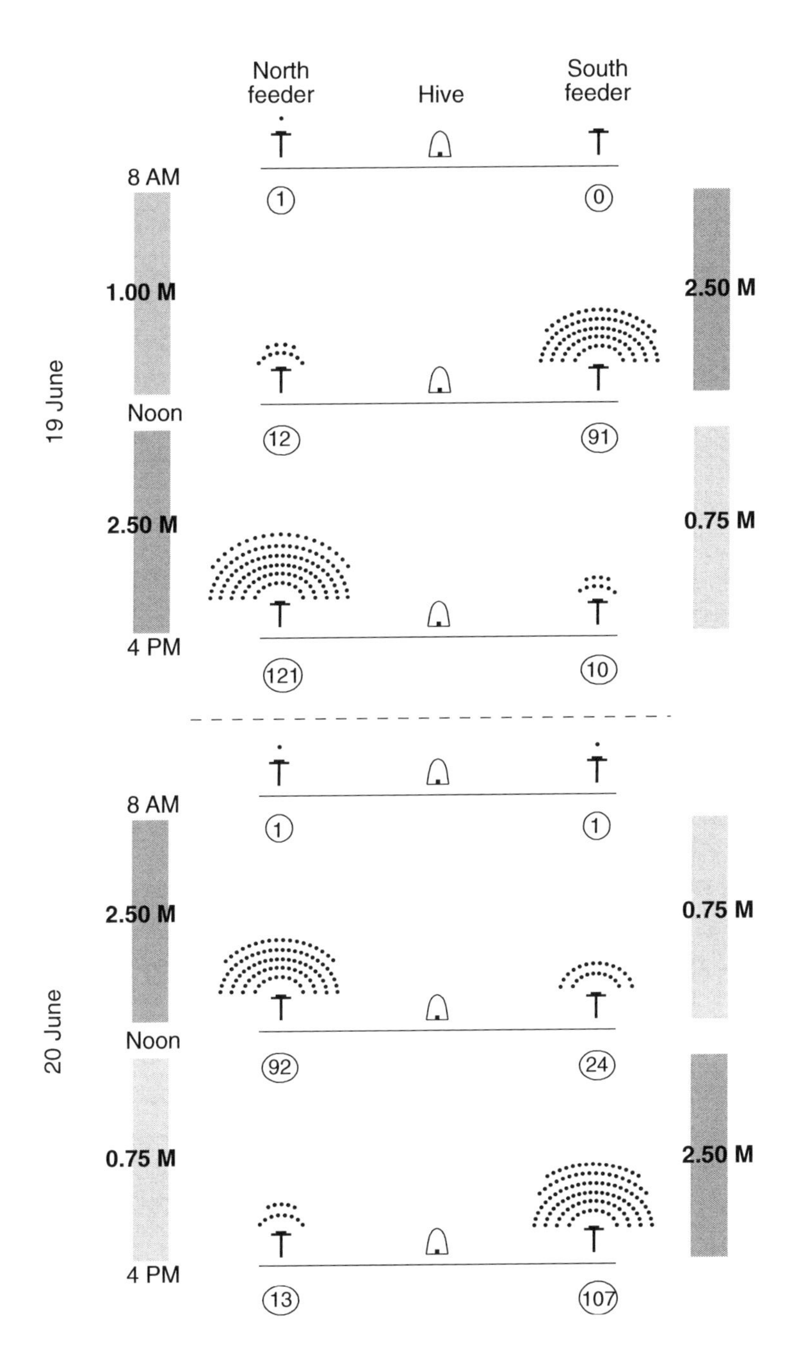

Figure 3.8 The ability of a colony to choose between forage sites. The number of dots above each feeder denotes the number of different bees that visited the feeder in the half hour preceding the time shown on the left. For several days prior to the start of observations, a small group of bees was trained to each feeder (12 and 15 bees for the north and south feeders, respectively); thus on the morning of 19 June the two feeders had essentially equivalent histories of low-level exploitation. The feeders were located 400 m from the hive and were identical except for the concentration of the sugar solution. After Seeley, Camazine, and Sneyd 1991.

colony and then plotted the colony's recruitment targets each day (1982). As we shall see in detail later (Sections 5.2 and 5.4), only bees visiting top-quality forage sites perform strong recruitment dances; hence each day's map of the colony's recruitment targets provides us with a daily picture of the spatial distribution of the colony's most profitable food sources.

These maps, several of which are shown in Plates I–VI, show clearly that the spatial pattern of a colony's top-ranked food sources can change dramatically from one day to the next. Indeed, *each* of our 36 days of dance recordings yielded a markedly different plot of recruitment targets. The following day-by-day commentary summarizes these dynamics for the 6-day period represented in Plates I–VI.

13 June (Plate I): Good weather. The targets of recruitment are clearly indicated. The richest, most desirable forage sites are those 0.5 km SSE and SSW of the hive, yielding mainly yellow and yellow-gray pollen, and a large site 2–4 km SSW of the hive, yielding mainly nectar. Also profitable enough to elicit dances are a site yielding orange pollen 1 km to the NE and a site with yellow-gray pollen 4 km to the NE.

14 June (Plate II): Good weather again. Nevertheless, the source of yellow-gray pollen 4 km to the NE now arouses few if any dances (none were "contacted" in our sampling of the dances). A nectar source 0.5 km to the NW becomes extremely profitable, stimulating many bees to dance. The pollen sources 0.5 km SSE and SSW of the hive and the nectar sources 2–4 km SSW remain highly rewarding.

15 June (Plate III): Good weather continues. The large nectar source 2–4 km to the SSW becomes less rewarding, as does the nectar source 0.5 km to the NW. The two sites yielding mainly yellow and yellow-gray pollen 0.5 km to the SSE and SSW remain highly attractive.

16 June (Plate IV): The weather this day is cool with intermittent rain, and the bees forage relatively little and only fairly close to the hive. The source of yellow-gray pollen 0.5 km to the SSW remains attractive but the adjacent source of yellow pollen in the SSE arouses little if any dancing. The source of orange pollen 0.5 km to the NW, which had elicited only one dance before, becomes highly attractive. The richest source of nectar is now a site 0.5 km S of the hive.

17 June (Plate V): The weather clears, providing good foraging conditions once again. The nectar source 2–4 km to the SSW, which had rated highly before the poor weather yesterday, does not regain its

Plates I-VI

Maps of a colony's forage sites, as inferred from reading the recruitment dances performed by the colony's foragers. Each symbol represents the location indicated by one bee's dance. Black dots indicate sites yielding nectar; all other symbols indicate sites yielding pollen of the type indicated in the key. For the time period shown, only a small fraction (2%) of the dances indicated sites beyond 4 km, and these sites are not shown. Bees advertise only highly profitable sites with their dances; hence each day's map indicates only the locations of the richest forage sites that the colony knows about on each day. Comparisons of the maps reveal large day-to-day changes in the locations of the richest foraging opportunities, which implies that colonies rely heavily upon their ability to rapidly redirect their foraging efforts. After Visscher and Seeley 1982.

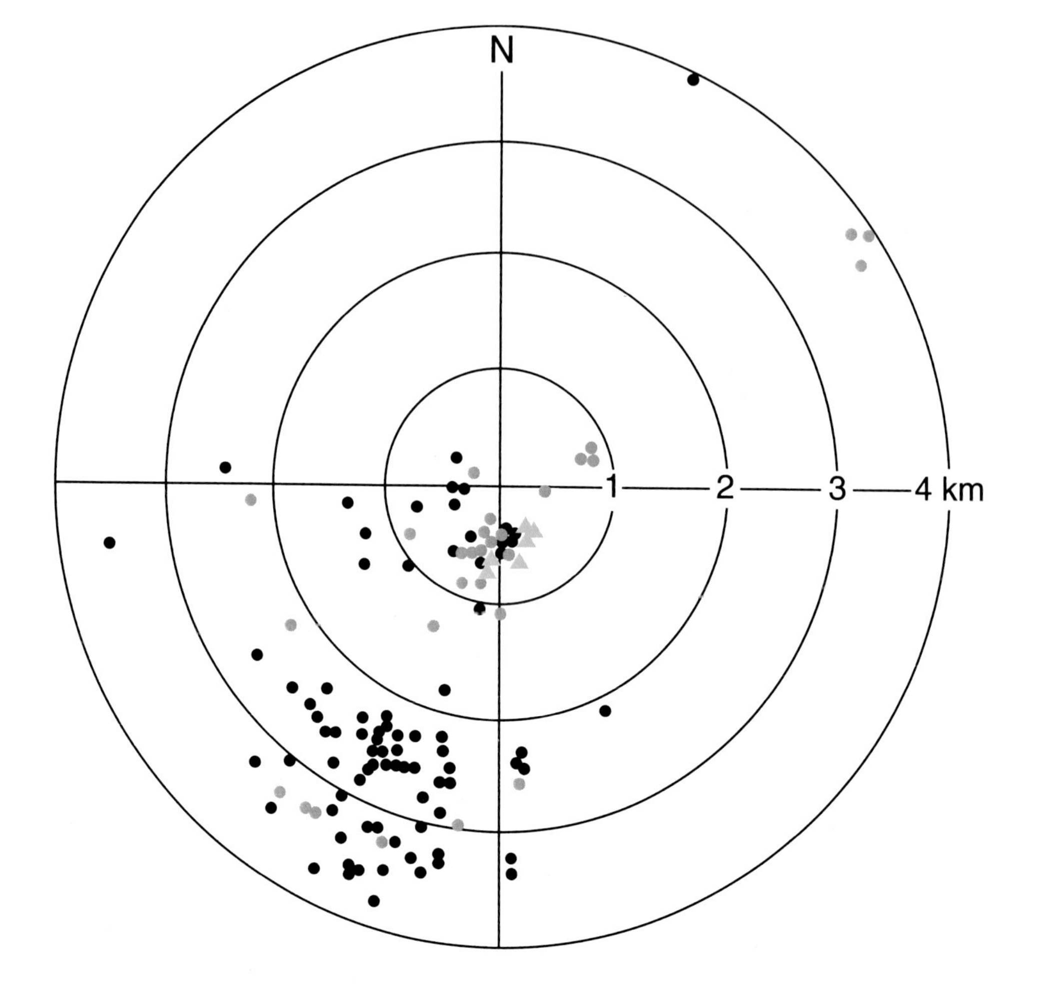

Plate I Forage sites on 13 June 1980.

nectar •
yellow-gray pollen •
yellow pollen ▲
orange pollen •
yellow-orange pollen ▲
brown pollen •

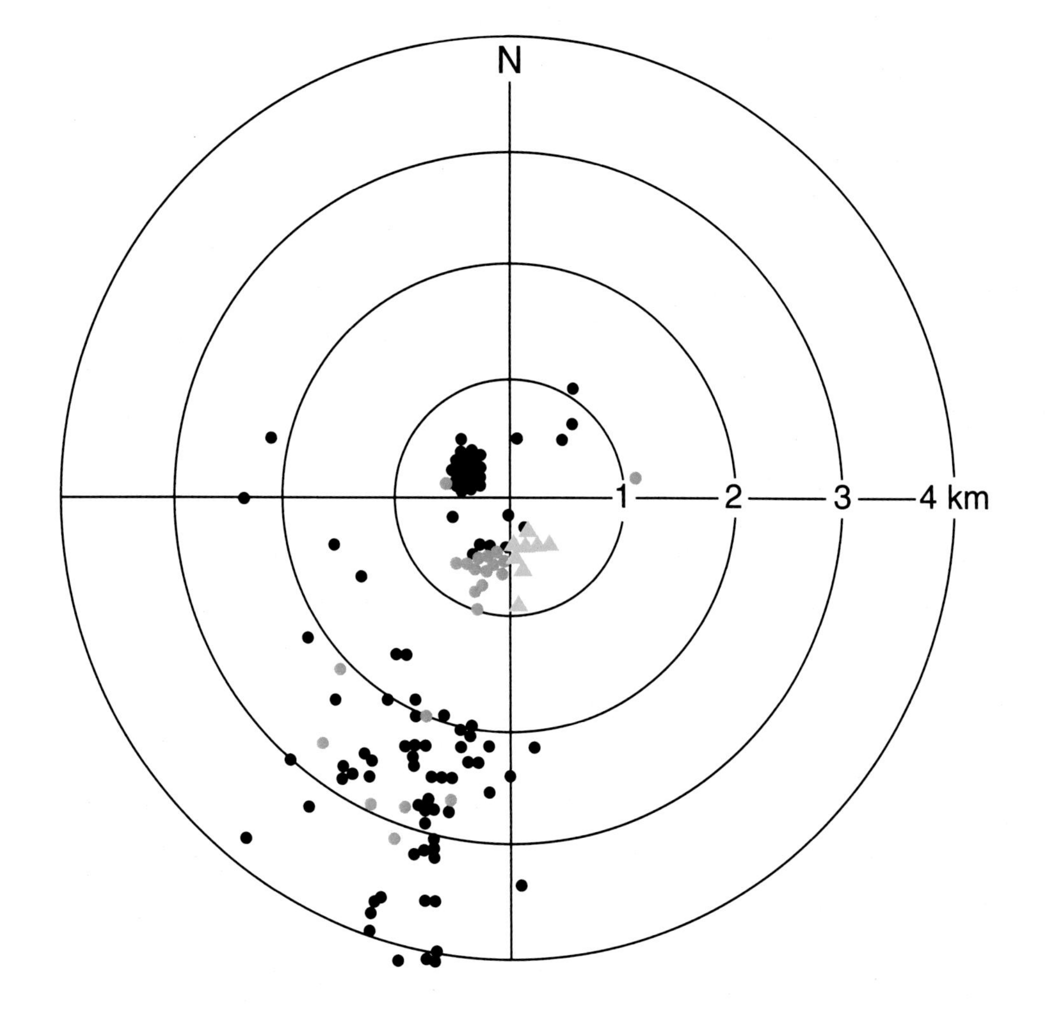

Plate II Forage sites on 14 June 1980.

nectar •
yellow-gray pollen •
yellow pollen ▲
orange pollen •
yellow-orange pollen ▲
brown pollen •

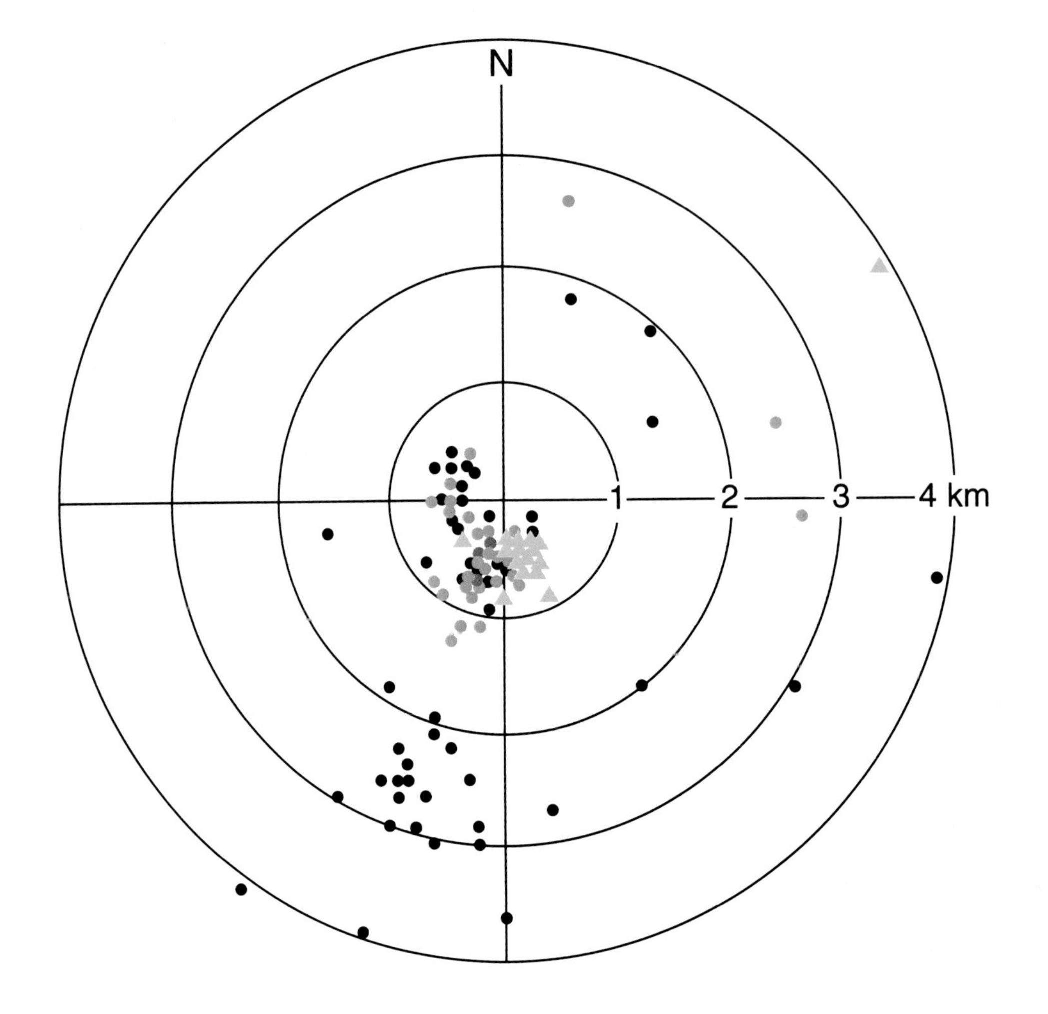

Plate III Forage sites on 15 June 1980.

nectar •
yellow-gray pollen •
yellow pollen ▲
orange pollen •
yellow-orange pollen ▲
brown pollen •

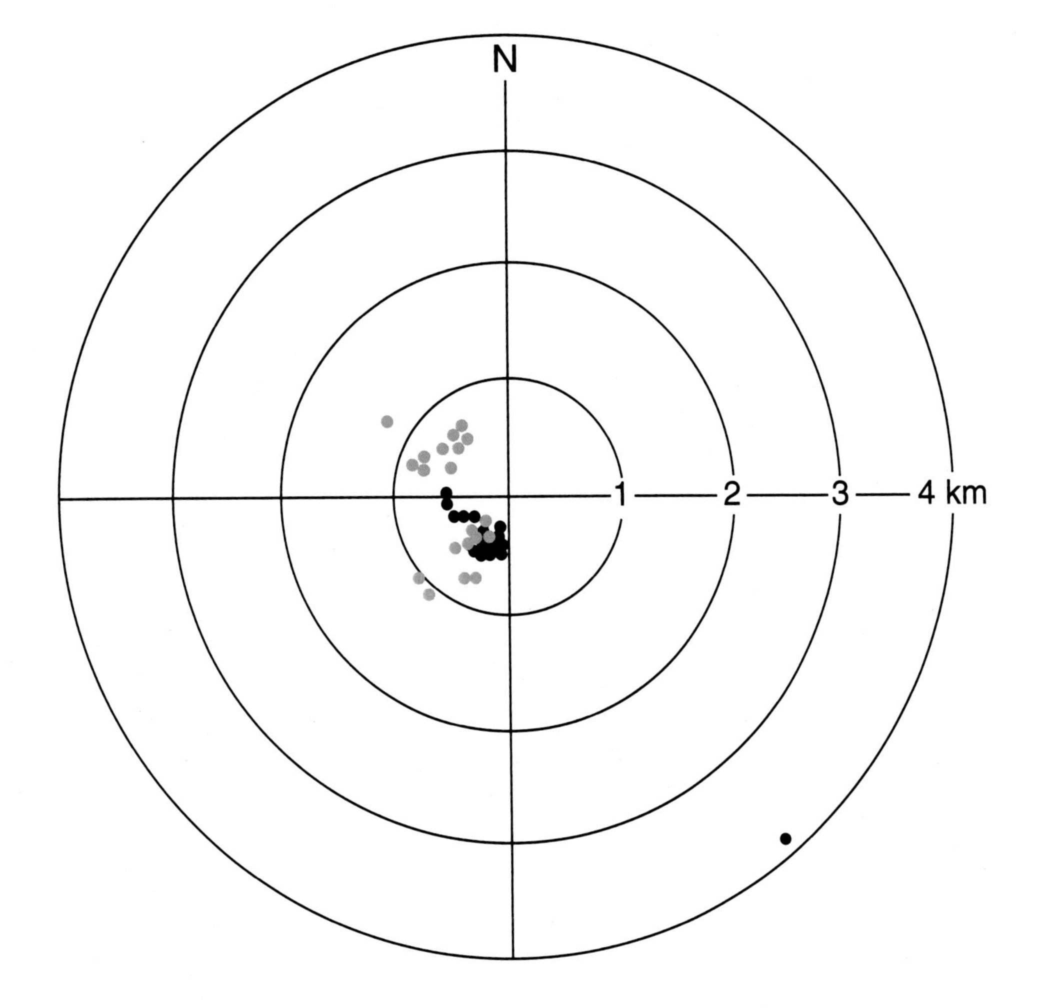

Plate IV Forage sites on 16 June 1980.

nectar •
yellow-gray pollen •
yellow pollen ▲
orange pollen •
yellow-orange pollen ▲
brown pollen •

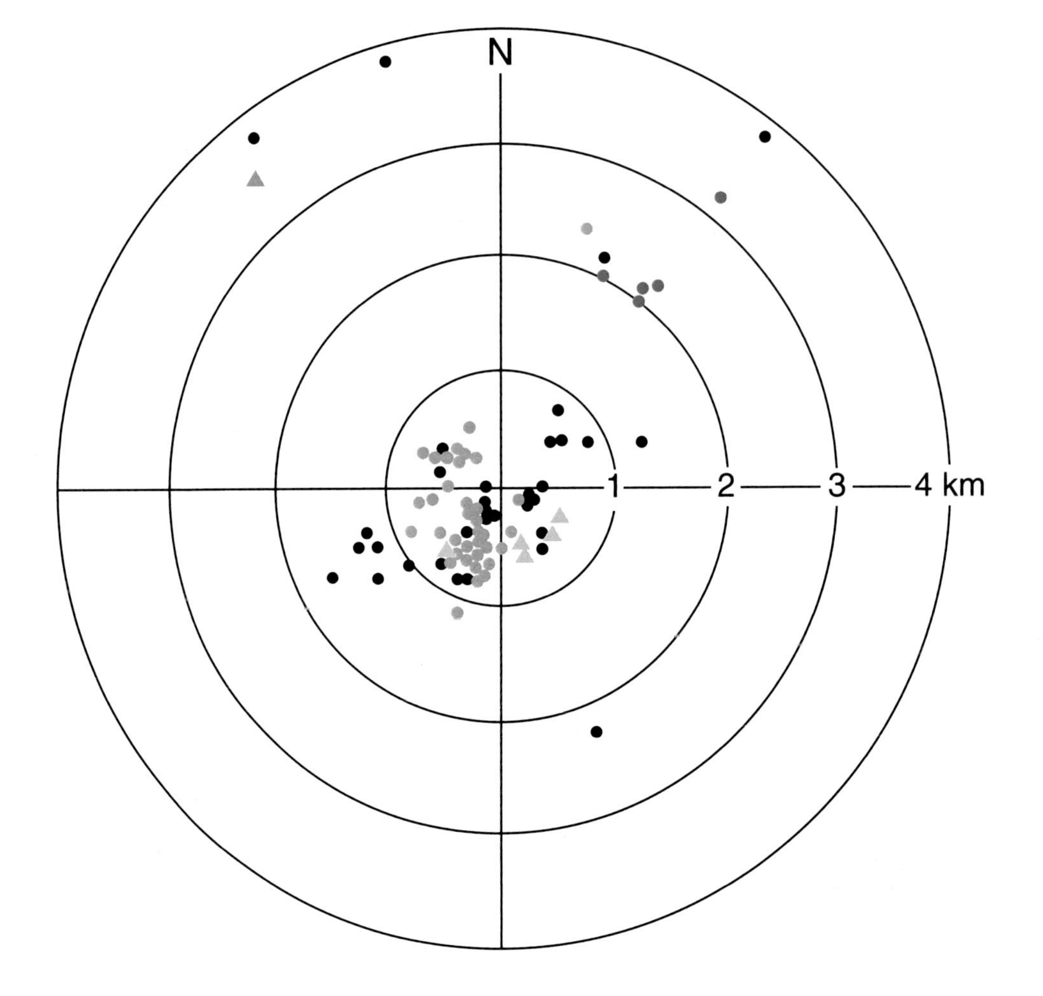

Plate V Forage sites on 17 June 1980.

nectar •
yellow-gray pollen •
yellow pollen ▲
orange pollen •
yellow-orange pollen ▲
brown pollen •

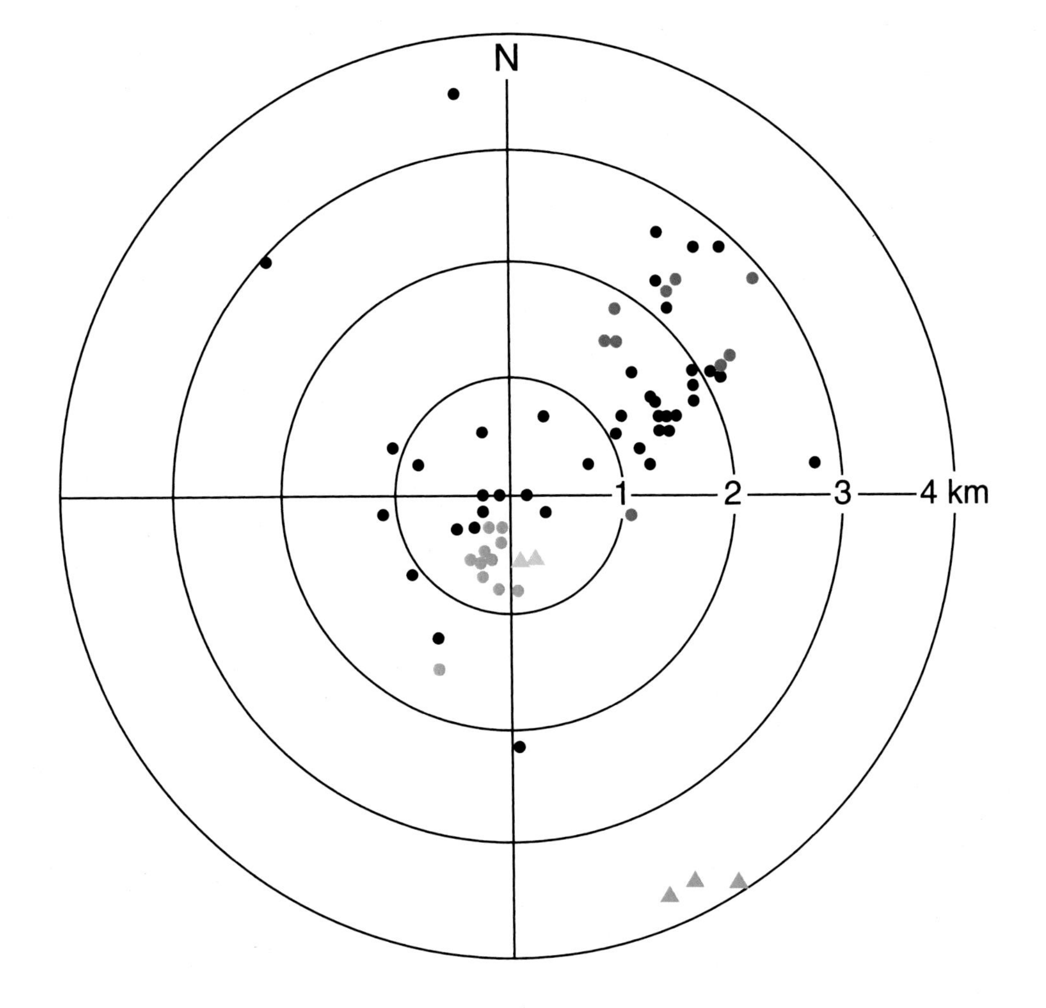

Plate VI Forage sites on 18 June 1980.

nectar •
yellow-gray pollen •
yellow pollen ▲
orange pollen •
yellow-orange pollen ▲
brown pollen •

high rating and is no longer advertised on the dance floor. The source of yellow pollen 0.5 km to the SSE again becomes highly rewarding, and the source of yellow-gray pollen 0.5 km to the SSW remains so, as does the site with orange pollen 0.5 km to the NW. Several sites to the NE, near and far, yielding nectar and brown pollen, begin to increase in attractiveness.

18 June (Plate VI): Good weather. The orange pollen source in the NW disappears from the dance floor, and the pollen sources 0.5 km SSW and SSE of the hive start to lose their high ratings. The nectar and brown pollen sources to the NE increase in quality, and a rich new source of yellow-orange pollen appears 3.5 km SE of the hive.

Daily sets of observations such as these surely do not disclose the full dynamics of a colony's food sources. Therefore, an important goal for a future study is a higher-resolution picture of a colony's recruitment foci. Achieving this will require videorecording all the dances in a hive and plotting the colony's recruitment targets hour by hour, not just day by day. Such a detailed analysis is likely to reveal major shifts in the foraging opportunities over the course of a day. This is to be expected since previous studies have shown that in many species of flowering plants nectar and pollen production is strong during only a portion of the day (Southwick 1983, reviewed in Shuel 1992). Moreover, during our days of dance watching, Kirk Visscher and I frequently noted that the colony's recruitment targets differed dramatically between morning and afternoon. Given these facts, and given the tremendous day-to-day changes in the foraging landscape just described, it now seems clear that the ability of a colony to swiftly redirect its foraging efforts is crucial to a colony's foraging success.

3.5. Adjusting Selectivity in Relation to Forage Abundance

We have seen that a colony's external supply of food is not constant, but varies greatly from week to week (Figure 2.14), or even from day to day (Figure 2.15), as the blooms of different nectar and pollen sources come and go, and as the weather varies between warm days and cool spells. One way that a colony copes with this variation in forage abundance is by adjusting its selectivity among food sources. Honey bee colonies—like people and other animals (Schoener 1971)—are less fastidious about their food during times of scarcity than during times of plenty. Stephens and Krebs (1986) explain the

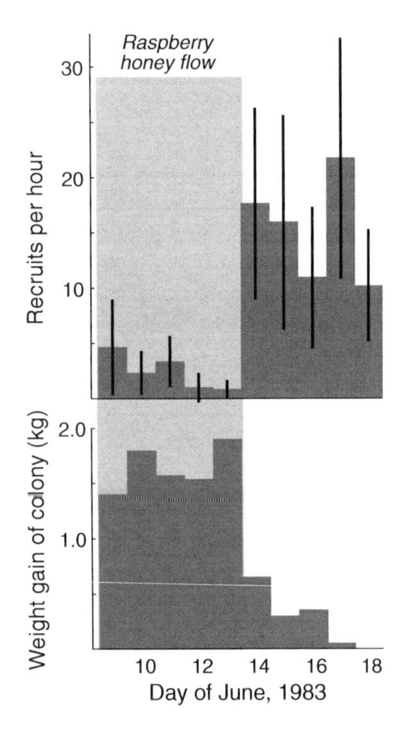

Figure 3.9 The ability of a colony to adjust its selectivity among food sources in relation to forage abundance. The top panel shows a colony's recruitment rate to a sugar water feeder with fixed properties (500 m from hive, 2.25-mol/L sucrose solution), while the bottom panel depicts the daily weight gain of a colony on scales, which indicates the abundance of forage from natural sources, mainly raspberry blossoms. For the first 5 days natural forage was abundant and the colony operated with a high acceptance threshold, one so high that the colony showed little interest in the sugar water feeder. Then during the second 5 days natural forage became sparse and the colony dropped its acceptance threshold, and began to direct many foragers to the feeder. Based on unpublished data of T. D. Seeley.

functional significance of this adjustment as follows: by lowering the acceptance threshold when food becomes sparse, an individual (be it an organism or a colony) continues to acquire food and so minimizes its probability of starvation, and by raising this threshold when food becomes abundant, an individual utilizes low-yield sources only when necessary and so maximizes its foraging efficiency.

One indication of a colony's ability to adjust its choosiness is the common experience of beekeepers that on a warm spring day when the dandelions are in full bloom, a honey-filled comb can be drawn from a hive and left exposed in the apiary without any danger of it being robbed by bees from the nearby hives. The bees prefer working in the fields with their blossoming flowers, rather than stealing food from a honeycomb, which normally requires dangerous fighting with guards at the entrance of the target colony's hive. But on an equally warm day in late fall, after a hard frost has killed the last wildflowers, a honeycomb removed from a hive will quickly be covered by a crowd of fighting bees, each one struggling fiercely for a load of honey. Clearly, a colony in the fall, when food is scarce, will exploit less attractive food sources than will a colony in the spring, when food is abundant.

A quantitative demonstration of a colony's ability to adjust its acceptance threshold comes from observations that I made in June 1983 while analyzing a colony's ability to selectively direct foragers to the better of two alternative forage sites (Figure 3.7). In the course of this work, I witnessed a sudden, surprising surge in the colony's attraction to the "reference feeder" (Figure 3.9). As indicated earlier (Section 3.4), this feeder was fixed at a site 500 m from the hive and provided food of constant quality: always a 2.25-mol/L sucrose solution. I allowed 30 labeled bees to forage at this feeder and captured unlabeled bees upon arrival, to measure the rate at which the labeled bees recruited hivemates. For 5 days starting on 9 June, recruitment to the feeder was disappointingly slow, only 1–5 bees per hour. Evidently, my feeder ranked barely above the colony's acceptance threshold, and the colony did not choose to devote much additional labor to it. Then, on the morning of 14 June, I received a surprise: recruits now began pouring in at the feeder at the rate of 20 per hour. For some reason, recruitment had multiplied nearly tenfold, even though nothing whatsoever had changed about the feeder. My first guess was that it had been discovered by bees from another hive, but a check of the 30 labeled bees visiting the feeder proved that all belonged to the study colony. The actual cause of this recruitment surge became obvious that evening when I took the daily weight reading of a nearby colony. This showed a gain markedly smaller than gains on previous days, indicating that the nectar flow from the wild raspberry (*Rubus* spp.) had markedly slowed. Evidently, with forage now much sparser, the colony had lowered its acceptance threshold and so directed more foragers to my feeder. In short, the colony's behavior followed the adage "Beggars can't be choosers."

The tuning of a colony's food-source selectivity in accordance with forage abundance is documented most precisely by determining the threshold concentration of sugar solution that elicits waggle dances under different conditions of forage abundance. (It should be noted that a colony's dance threshold also represents its acceptance threshold because only if a forage site is advertised by recruitment dances does it receive foragers, and only if a forage site receives foragers is it exploited (accepted) by a colony.) This approach was pioneered by Martin Lindauer (1948), who estimated for each day throughout the summers of 1945 and 1946 the threshold sugar concentration that would elicit dancing by the bees visiting a particular experimental feeder. He found, for example, that in June, when copious nectar was available from such sources as red clover *(Trifolium pratense)*, rape *(Brassica napus)*, and linden trees (*Tilia* spp.), the threshold was 2.0 mol/L, whereas in August, a time of meager forage, the dance threshold was far lower, just 0.125 mol/L. Forty-five years later I repeated Lindauer's observations, using more sophisticated techniques for assaying both the dance threshold and the forage abundance, and observed the same phenomenon (Seeley 1994). For example, on 11 July 1990, when nectar was moderately abundant from raspberry blossoms (*Rubus* spp.) and a colony on scales *gained* 1.20 kg a day, the dance threshold for bees visiting a sucrose solution feeder 400 m from their hive was 1.7 mol/L; but 11 days later, when virtually all natural sources of nectar had faded and the colony on scales *lost* 0.70 kg a day, the dance threshold for bees visiting the same 400 m feeder was only about 0.5 mol/L (see also Figure 5.12).

3.6. Regulating Comb Construction

To avoid starvation, not only must a honey bee colony collect some 120 kg of nectar during the summer, it must also stockpile a third of this nectar in concentrated form, as some 20 kg of honey, for consumption during the winter. This requirement of large-scale honey storage means that an important part of the food-acquisition process is the production of sufficient beeswax comb to hold the colony's honey stores. Beeswax production is energetically expensive, however, requiring at least 6 g of sugar for every gram of wax synthesized (reviewed in Hepburn 1986), so colonies do not simply produce all the comb they will ever need immediately upon occupying a new nest site. Rather, a colony first builds a small set of combs that meets its initial need for comb in which to rear brood and store food, and then later builds additional comb as required for expanded brood rearing and greater honey storage. Thus within a few days of moving into a new nest site a colony will have constructed an initial set of combs consisting, on average, of some 20,000 cells, but by the end of the summer a successful colony will have enlarged its combs to some 100,000 cells, most of which will be filled with honey, and in subsequent years it may expand its combs still further (Seeley and Morse 1976; also my unpublished observations). One striking feature of comb construction is the way that it is tightly regulated so that a colony builds new comb only when it is truly needed—when a colony's foragers are collecting copious nectar and most of the colony's storage cells are already brimming with honey. Beekeepers know, for example, that they can induce a strong colony to build new, honey-filled combs by removing most of the colony's combs at the start of a nectar flow (Killion 1992). The bees will quickly fill the few remaining combs with honey and will then start building additional comb for honey storage.

This tight control of comb production was recently demonstrated experimentally by Kelley (1991). He established a colony of some 4000 bees in a three-frame observation hive in which the bottom frame served as the colony's broodnest region (the queen was restricted to the bottom frame), the middle frame its honey storage region, and the top frame provided a space for comb building. He then made controlled, daily feedings of sugar solution to the colony (to simulate a nectar flow) and monitored the fullness of the colony's honey storage comb and its construction of new comb. Figure 3.10 depicts the results of one trial of this experiment. For the first 6 days he did not feed the colony sugar solution so its honeycomb remained largely empty, with fewer than 20% of the storage cells containing honey, and he witnessed no comb construction. Then for the next 7 days he fed the colony 65 to 350 ml of sugar solution daily, and for the first 4 days he again observed no comb construction. On the fifth day, however, by which time nearly 80% of the storage cells contained honey, the colony suddenly began building comb in which to store honey. Next he ceased feeding the colony for three days, whereupon it quickly ceased building comb, even though its storage cells remained as full as before. Only when he resumed the feeding did the colony again produce additional comb.

It seems clear, therefore, that two conditions that are necessary for comb construction are high nectar intake and nearly full honeycombs. This makes good sense. High nectar intake alone would be an inappropriate stimulus for comb building because a colony can experience a high nectar influx and still have plenty of empty comb in the hive for honey storage. Likewise, comb fullness alone would be a poor stimulus because a colony could experience nearly full honeycombs and yet have no more nectar coming into its hive. Building new comb under such circumstances would be a less prudent use of the colony's energy resources than waiting until nectar again comes into the hive, at which time the need for additional comb becomes certain.

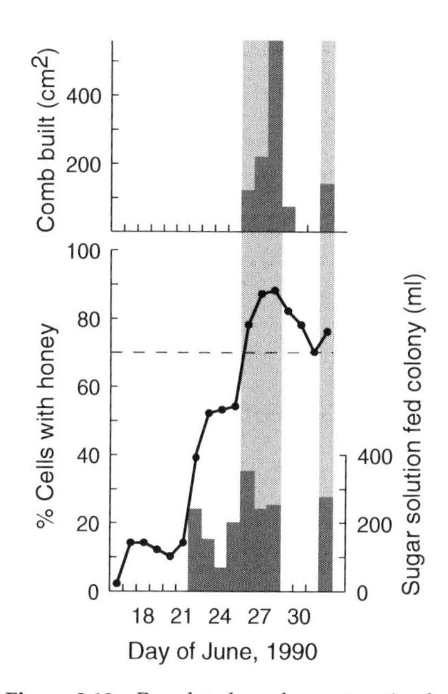

Figure 3.10 Regulated comb construction by a honey bee colony. A colony was installed in an observation hive with three frames: *bottom*, brood comb; *middle*, honey storage comb; *top*, empty space for new comb. The colony was fed sugar solution in the amounts shown, to simulate a nectar flow, and the fullness of its middle comb and the construction of its top comb were monitored. Note that comb building was essentially limited to those days (*lightly shaded areas*) characterized by a specific combination of conditions: (1) strong influx of sugar solution and (2) more than 70% of the storage cells partially filled. After Kelley (1991).

3.7. Regulating Pollen Collection

Honey bees are called "honey bees" rather than "pollen bees" because of a major difference between the controls on nectar collection and those on pollen collection. Whereas a colony stops gathering nectar only after it has completely filled its hive with honey-filled combs, it ceases collecting pollen as soon as it has accumulated a modest reserve of pollen. Thus under favorable foraging conditions a colony will amass 50 or more kg of honey in its hive, but at the same time it will store up less than 1 kg of pollen (Jeffree and Allen 1957; Fewell and Winston 1992). Why should there be such a striking difference between nectar and pollen storage? The answer is easily understood in terms of optimal inventory policies for energy and protein (Winston 1991). A colony consumes energy rapidly throughout the year, which of course includes a winter period lasting several months when bees cannot extract energy from the environment (Figure 2.14). To buffer itself against this long-term break in energy acquisition, a colony needs a large reserve of honey. In contrast, a colony consumes protein (for brood production) mainly over the summer, and during

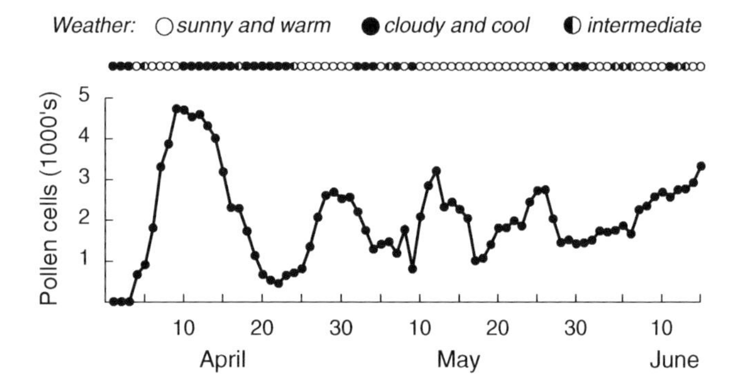

Figure 3.11 Dynamics in the pollen reserves of a colony. On 1 April 1991 a colony of 10,000 worker bees and a queen was installed in an eight-frame observation hive with empty combs. The hive's combs consisted of some 55,000 cells; those containing pollen were counted at the end of each day. After an initial 3-day period when no foraging was possible, the colony intensively collected pollen from pussy willow *(Salix discolor)* and built up a substantial reserve of nearly 5000 cells (approximately 750 g) of pollen. This was drawn down over the next 2 weeks of cool weather and little foraging. Following this, the colony worked to rebuild the reserve whenever possible, and so always had stored pollen to cover its protein needs during spells either of bad weather (such as 1–4 May and 27–31 May) or of good weather but sparse forage (such as 13–17 May). Although the colony's pollen reserve remained small, never exceeding 10% of the cells in the hive, it was always sufficient to buffer the colony against the brief periods when no pollen could be gathered. Based on unpublished data of S. Camazine.

this time it experiences periods lasting only several days when pollen cannot be gathered, such as times of rainy or cool weather (Figure 3.11). Since these breaks in pollen acquisition are short-term, a colony will be adequately buffered against them by only a small reserve of pollen. Presumably the reason colonies do not build larger pollen stores is that the increased benefits (greater independence from swings in pollen intake) would not offset the increased costs (greater investment in comb, or fewer cells available for brood production and honey storage, or both).

To maintain its small but effective pollen reserve, a bee colony must modulate its pollen collection, increasing it when the reserve has been drawn down and decreasing it when the reserve has been rebuilt to its proper size. That colonies have this ability is demonstrated by an experiment performed by Scott Camazine (unpublished). He established two colonies side by side, each one consisting of some 4000 bees occupying a three-frame observation hive. In each hive, the upper two frames were filled with brood and honey, while the bottom frame was reserved for manipulations of the colony's pollen stores. The two colonies received opposite experimental treatments. When one was prevented from building up a pollen reserve, by removing each day its bottom frame and replacing it with a frame of empty cells, the other was kept well stocked with pollen, by giving it each day a new bottom frame filled with pollen. Each colony's rate of pollen collection was measured by counting the pollen foragers entering the hive. As shown in Figure 3.12, each colony adjusted its pollen foraging in a manner appropriate for regulating the quantity of pollen stored in its hive. The response of colony 1 to the addition of a frame of pollen on the evening of 20 August is especially striking. Within 1 day the colony's rate of pollen collection plummeted from 26 to 8 pollen foragers/min, and within 2 more days it had fallen to below 1 pollen forager/min, even though flowers yielding pollen remained plentiful outside the hive, as demonstrated by the rising rate of pollen collection by colony 2. Lindauer (1952) likewise reports that preventing bees from building a pollen reserve in their hive increases the amount of pollen collected by a colony, while Barker (1971), Free and Williams (1971), and Fewell and Winston (1992) report that providing pollen has the opposite effect. Thus it is clear that a honey bee colony can adaptively modulate its pollen collection so that it maintains an appropriate reserve of pollen inside its hive.

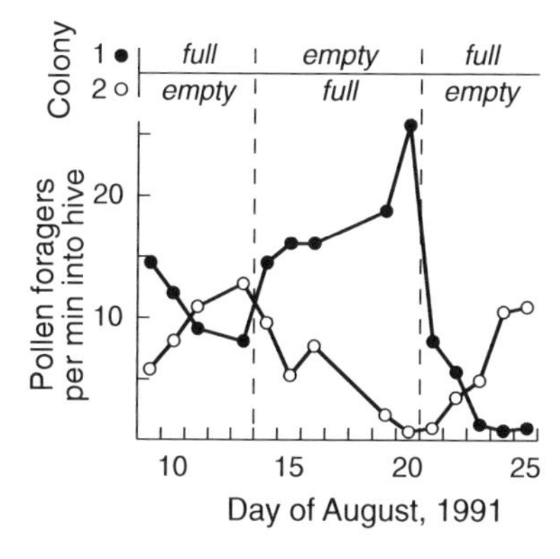

Figure 3.12 Modulation of pollen foraging in relation to the amount of pollen stored in the hive. Two colonies in observation hives received opposite treatments—provision or elimination of stored pollen—and their rates of pollen foraging were monitored. The colony that was provided with a large pollen reserve steadily lowered its pollen intake, whereas the colony that was prevented from building a pollen reserve steadily increased its pollen intake. Based on unpublished data of S. Camazine.

3.8. Regulating Water Collection

A colony utilizes water for two quite different purposes. When there are many larvae to be fed, the nurse bees must produce large volumes of liquid food, which requires a copious supply of water. In addition, when there is danger of the brood overheating, the bees must spread water over the brood combs for evaporative cooling. Often a colony's water need is covered passively by the water gathered in nectar, but there are times when a colony's water demand far exceeds the supply from nectar and it must actively gather water. This need to gather water can come from two opposite weather conditions: hot days, when a colony faces a threat of lethally high temperatures inside the hive, and chilly days, when a colony faces reduced nectar intake due to diminished foraging. The following two examples illustrate the ability of a bee colony to regulate its water collection in accordance with its circumstances.

In the winter of 1951, Martin Lindauer (1954) moved a colony of bees living in an observation hive to a greenhouse, where he also provided a drinking place for the bees. This setup enabled him to count the number of water-collection trips made by the colony's foragers, and so monitor precisely the colony's rate of water collection. He then imposed a heat stress on the colony by directing an infrared lamp toward one of the observation hive's glass walls. The colony's water collection response is graphed in Figure 3.13. During the hour before the heat stress, the colony's hive was quite cool inside and only 2 bees visited the drinking fountain, presumably to get water for thirsty nurse bees. But once the heater was turned on and the hive temperature began to rise rapidly, the colony boosted its water intake, ultimately reaching a level some 30 times the baseline rate. Eventually

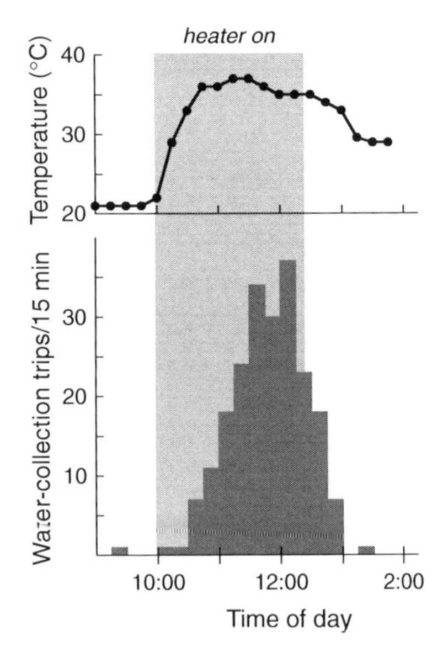

Figure 3.13 Regulated water collection by a honey bee colony. A colony in an observation hive was installed in a greenhouse, where its rate of water collection from an artificial water source was easily monitored. When an infrared lamp was turned on beside the hive, threatening the colony with lethal overheating, the colony boosted its water intake to begin evaporative cooling inside the hive. This stabilized the hive's interior temperature. When the heat stress was removed, the colony promptly lowered its water intake. After Lindauer 1954.

the heater was turned off and within an hour the colony had ceased collecting water. Clearly, this colony was able to quickly and accurately adjust its water collection to cope with a severe danger of overheating.

A second set of Lindauer's observations (1954) documents the ability of a colony to modulate its water collection to adjust for changes in its nectar influx. This time Lindauer placed his study colony in the open countryside, but in a location where the nearest natural water source was 200 m away. He also established an artificial drinking place—a trickle of water running over a gently sloping board—4 m from the hive, which the colony's water collectors quickly adopted. Each bee visiting the drinking spot was labeled for individual identification with paint marks. This allowed Lindauer to record the number of different bees visiting the water source each day, which he did from 6 April to 18 September 1951. The records from the end of April are especially informative (Figure 3.14). The 3 days of 28–30 April were characterized by unusually cold weather, which prevented the colony's nectar foragers and pollen foragers from flying from the hive. On 28 April the temperature rose briefly to 10°C in the afternoon, at which time a few dedicated water collectors made the short trip to the drinking fountain. On 29 April, the air warmed to only 6°C, and not even one bee came for water. This cold spell came at a time of massive brood rearing by the colony, and consequently the colony's water shortage must have become acute during these two days with virtually no nectar or water collection. Then on the afternoon of 30 April, when the temperature rose to 9°C (barely warm enough for bee flight), bees from the colony stormed the water source. These individuals must have been maximally motivated to gather water, for they were attempting to do so under frightfully poor conditions. Many sank into a chill coma when they drank the cold water and would have died from hypothermia had Lindauer not warmed them in his hands and gently carried them back to their hive. Such observations graphically illustrate both that colonies can experience severe water shortages in cool weather and that they can adjust their water intake to solve this problem.

Summary

1. A colony is able to project its foraging operation over an immense area—at least 100 km^2—for its foragers will exploit food

sources 6 km or more from the hive. Such a large foraging area ensures an adequate resource base. Colonies may also range this widely to provide themselves with a large choice of potential food sources.

2. A colony is able to conduct effective surveillance for rich new patches of flowers over its vast foraging range. In one experiment, for example, a rich food source arising within 2 km of a colony's hive had a probability of at least 0.5 of being detected by the colony.

3. Having discovered a patch of flowers laden with nectar or pollen, a colony is able to dispatch foragers to the site quickly, before its treasure is lost to competitors, bad weather, or darkness. A colony accomplishes this by having foragers recruit nestmates to rich finds. In one experiment involving highly rewarding flowers located 430 m from the hive, for instance, a colony was able to field 18 bees to the flowers within an hour of their discovery.

4. When presented with multiple food sources of different profitabilities, a colony is able to choose among them, concentrating its collection efforts on the more profitable sites. Moreover, a colony is able to track effectively the richest food sources in a rapidly changing array. This is demonstrated by an experiment in which the positions of two sucrose solution feeders, one more profitable than the other, were reversed at noon and by 4:00 the colony had fully reversed the principal focus of its foraging. Such a speedy response is needed to track closely the best foraging opportunities in nature. The array of floral food sources in the countryside around a colony's hive can change dramatically from day to day, and may even vary markedly from hour to hour.

5. A colony adjusts its selectivity among food sources in relation to forage abundance, becoming less fastidious during times of scarcity. For example, when nectar is available in abundance from natural sources, bees will show little interest in an artificial sugar water feeder, but once the nectar flow is over, a colony will massively exploit the feeder. By lowering the acceptance theshold when forage becomes sparse, a colony minimizes its probability of starvation, and by raising this threshold when food becomes abundant, a colony utilizes low-yield sources only when necessary and so maximizes its foraging efficiency.

6. A colony controls tightly its production of the energetically expensive beeswax comb that it uses for brood rearing and food storage. After initially building a small set of combs, a colony builds additional comb only when absolutely necessary, when its foragers

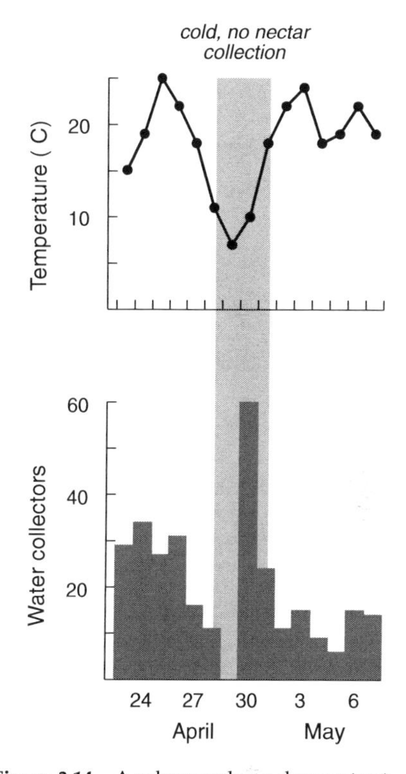

Figure 3.14 A colony makes a desperate attempt to gather water after cold weather prevents intake of fluids. Starting in early spring, Lindauer counted daily the number of a colony's bees collecting water from an artificial drinking place between 2:00 and 2:30 in the afternoon. When cold weather prevented the colony from collecting nectar and water for 2 days (28 and 29 April), the colony developed an acute shortage of water. Then on the third day of the cold spell (30 April), despite still cool temperatures, water collectors stormed the drinking place. As they attempted to suck up the cold water in the cool air, however, many of the bees fell into a chill coma, unable to return to the hive. Based on data in Lindauer 1954.

are gathering nectar at a high rate and its combs for honey storage are already nearly full.

7. A colony modulates its pollen collection to maintain the small pollen reserve needed to buffer the colony against the brief breaks in pollen collection that arise during the summer. When the reserve has been drawn down during a period of poor foraging, the colony raises its rate of pollen intake, and then when the reserve has been rebuilt, it lowers its rate of pollen collection.

8. A colony adjusts its water collection according to its internal needs. A strong need for water can arise both on hot days, when water is needed for evaporative cooling of the broodnest, and on cool days, when water is needed by nurse bees to produce food of high water content for larvae. At such times, a colony will greatly increase the number of foragers devoted to water collection, but as soon as the need has been met it will quickly reduce its water intake.

EXPERIMENTAL ANALYSIS

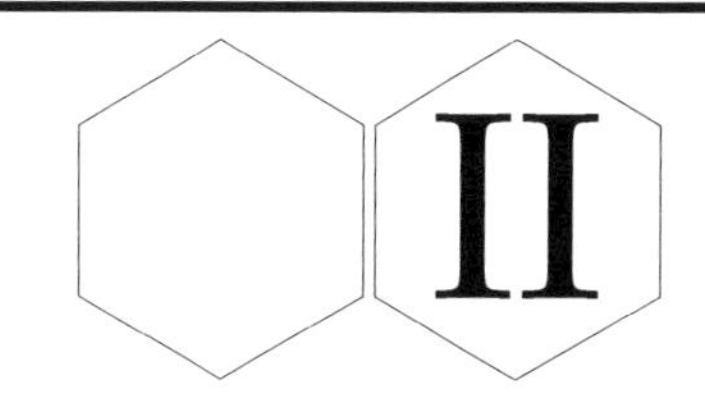

Methods and Equipment

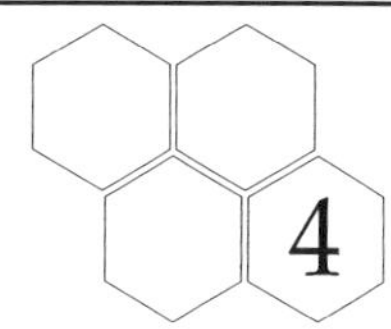

In this chapter, I describe the most important equipment and techniques used in analyzing the food collection process of honey bee colonies. They are presented here so that readers can develop a sense of the general methods used in studying the physiology of honey bee colonies, before examining specific experiments in the following chapters.

4.1. The Observation Hive

For studies requiring a normal strength colony, I have used a large observation hive, with four long frames (each 23 × 89 cm), that holds up to 20,000 bees (Figure 3.1). This hive is impressive but it is not well suited to experimental work because it is neither easily moved nor readily opened (the large sheets of glass covering each side are extremely heavy and difficult to handle). Also, in many of my experiments I have needed to work with a small colony in which each bee was labeled for individual identification. Thus a small observation hive, one that holds about 4000 bees, has proven better suited to my needs. There is no indication that abnormal behavior arises in a colony of this size, which after all is not so unusual since the populations of many colonies fall to this size by the end of winter (Jeffree 1955; Avitabile 1978).

Based on 20 years of experience, my small observation hive today appears as shown photographically in Figure 4.1 and diagramatically in Figure 4.2. The fundamental difference between this and a standard beehive is that the combs are arranged in a single layer behind

Figure 4.1 The small observation hive mounted in the laboratory. The protective covering has been removed and the two comb's, one above the other, are visible. The hive's entrance is connected to metal tubes through the building's wall by a transparent entrance tunnel through which bees walk while arriving or departing. This facilitates observation of labeled individuals. The glass wall shown has an opening covered with black nylon net over the dance floor region. This makes it possible to apply paint marks to bees inside the hive. Photograph by T. D. Seeley.

glass walls, allowing one to observe the entire surface of each comb and all of the bees' activities except those performed deep in the cells. It should be noted that the spacing between the opposite inner surfaces of the glass walls is critical. I find that an interval of 4.3–4.5 cm works best. If it is smaller, the bees cannot move freely between comb and glass, and wet paint on freshly labeled bees will smear on the glass. If the spacing is larger, the bees are apt to form two layers, one on the comb and one on the glass, or apply wax to the glass, both of which interfere with observations. The thin, flat shape of this hive does, however, make it harder for the bees to maintain a proper temperature inside, so in cool weather I place an outer pane over the glass walls (with a dead-air space between the two sheets of glass) to help prevent excessive cooling while I am making observations. Whenever observations are not being made, I cover the glass walls with insulation boards made of wood and a 2.5-cm layer of insulating styrofoam.

The hive entrance is located on one of the narrow sides, level with the hive floor. Bees arriving at the hive are guided to one side of the comb by means of a wedge in the entrance, which makes it easy to monitor all incoming bees while watching just one side of the hive, as is customary. To ensure that all waggle dances by incoming foragers are kept in view, it is necessary to plug all passageways between the two sides of the comb within 20–30 cm of the entrance, lest some bees crawl to the other side of the hive before starting their dances. I fill most such passageways with strips of wood and bits of wax, and leave open only a few, located far from the entrance, to permit traffic between the two sides of the hive's comb. For certain experiments it is necessary to label particular bees inside the hive, usually ones in the dance floor area near the entrance. To accomplish this, I replace the solid glass wall on the main observation side with a glass sheet that has a 25 cm wide × 16 cm high opening over the dance floor area (Figure 4.1). This opening is covered with black nylon net ("tulle," openings 3 mm in diameter) glued with silicon cement to the inner surface of the glass. The tip of a fine paint brush is easily inserted through this net to label the bees of interest.

Outside the hive entrance I mount a long (at least 20 cm), low (2.5-cm) tunnel whose sides and top are constructed of glass or clear plastic (Figure 4.1). Arriving and departing bees will walk down this tunnel, which facilitates the observation of labeled bees. I encourage the bees to walk on the tunnel's floor, so that their labels are most easily seen, by applying a thin coating of grease to the sides and ceiling of the tunnel for about 4 cm at each end.

For most experiments, I set up the hive with a brood-filled comb in the bottom position, an empty comb in the top one, and enough adult bees to cover both combs. I also insert a sheet of queen-excluder material between the upper and lower combs and thereby restrict the queen to the lower comb. This arrangement has several advantages. First, it limits the colony's brood production to the bottom comb, so that the colony does not quickly outgrow the small hive. Second, it reserves half the cells in the hive for honey storage. Only if a colony has plenty of empty cells for honey will its nectar foragers work vigorously and be strongly motivated to dance. When nectar is in good supply, however, the bees can quickly fill with honey all the cells in the upper comb. Because the colony's food storage and brood rearing are segregated between the upper and lower combs, I can remove the full honeycomb and replace it with an empty one without disrupting the colony's brood production. When it is necessary to feed the colony, I uncork one of the openings atop the hive and invert over it a jar of sucrose solution which has had small holes drilled in its lid. During transport of the hive, I provide the bees with the necessary ventilation by taping screening over the entrance opening at the bottom and both feeder openings on top.

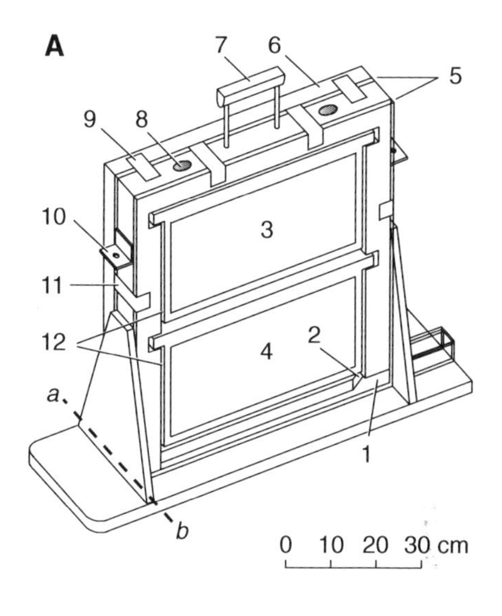

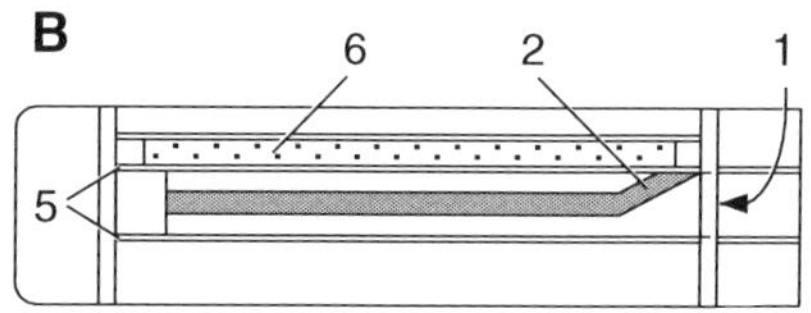

Figure 4.2 Observation hive (*A*) with section corresponding to the line *ab* (*B*): (*1*) entranceway; (*2*) wooden wedge inside the hive entrance to guide incoming bees to one side of the comb; (*3*) and (*4*) upper and lower combs (only the wooden frames are shown); (*5*) glass walls of 6 mm plate glass; (*6*) insulated cover of wood and Styrofoam (only one of two covers is shown); (*7*) handle for easy transport; (*8*) ventilation and feeder opening, covered with screening or closed with a cork as appropriate; (*9*) Velcro tape for fastening the insulated cover; (*10*) metal bracket for attaching the hive to a threaded suspension rod; (*11*) masking tape for securing the glass walls; (*12*) gap to allow movement of bees between the two sides of the comb.

The bees become accustomed to light inside their nest, even quite bright light from artificial illumination, after a few days of living on lighted combs. However, I leave the observation hive closed for a day or two after installing a colony to maintain the normal association between bright light and the entrance opening, and this seems to help the bees learn to orient to their new home.

4.2. The Hut for the Observation Hive

Direct sunlight on the glass walls of the observation hive will cause overheating of the colony. Indirect sunlight from the blue sky also causes problems, because foragers trying to get outside the hive will move toward this bright light and so will end up struggling against the glass walls rather than passing freely out the entrance. Therefore

Figure 4.3 The small observation hive in the portable hut. Bees enter the hive through a tunnel leading from the window in the far wall of the hut. The hive is suspended by threaded rods from a metal bar passing overhead. The hut's roof is constructed of translucent fiberglass to admit diffuse light. Shuttered windows provide additional light when necessary. The roof is hinged on one side so that it can be tilted open to provide ventilation on hot days; fresh air enters though louvers near the base. The hut is bolted together, and hence can easily be dismantled for transport. Photograph by T. D. Seeley.

it is best for the hive to receive diffuse illumination during observations. Whenever possible, I mount the hive inside a building and connect the entrance tunnel either to a board with a suitable opening mounted in a window or to special tubes through the building's wall. Often, however, there is no building available where the hive needs to be located for an experiment, and in these situations I use a small (1.2 × 1.2 × 2.0 m) hut which can be disassembled for transport.

The design of this observation hut has undergone many refinements over the years. The current version (Figures 4.3 and 4.4) protects the hive from the weather, provides diffuse illumination for observing the bees, affords the observer a comfortable work site, and is light enough for easy transport. It is constructed of four sheets of plywood 6 mm thick, braced where necessary with solid lumber and fastened together with bolts. The translucent roof of white, corrugated fiberglass mounted on a wooden frame transmits suitably diffused skylight. Additional light is obtained by raising the shutters on screened windows. Inside, the walls are painted white to reflect light, except those opposite the hive's glass walls, which are painted black to minimize background reflections and so provide a clear view of the bees through the glass. The hive is suspended by threaded rods from a metal bar overhead, making it easy for the observer to enter and leave the hut without jarring the hive. The hive's transparent entrance tunnel attaches to a window to facilitate monitoring the return of important bees. Finally, a comfortable chair, together with the good ventilation provided by the tilting roof and adjustable ventilation louvers, makes it possible to watch the bees for hours on end, which is required for many experiments.

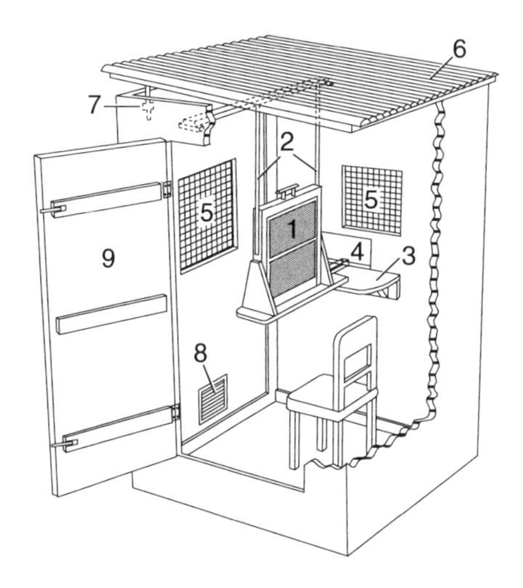

Figure 4.4 Diagrammatic view of the portable observation hive and hut (with walls partially cut away): (*1*) hive with transparent entrance tunnel; (*2*) threaded rods by which the hive is suspended, attached to a metal bar overhead; (*3*) wooden platform to which the hive is anchored; (*4*) window at the end of the entrance tunnel; (*5*) shuttered windows for light in addition to that provided by the translucent roof; (*6*) roof of white fiberglass, in tilted position for ventilation; (*7*) bracket for adjusting the roof opening; (*8*) adjustable ventilation louvers; (*9*) hinged door. Based on a drawing by B. Klein.

4.3. The Bees

In the vast majority of my experiments, I have used colonies of bees of the Italian race *(Apis mellifera ligustica).* Hence, unless I expressly state otherwise, the experiments described here involved Italian bees. These bees, originally purchased from one of the commercial bee breeders in the southern United States, are not, however, purely of the Italian race. The honey bees in North America are a complex mixture of the various races imported from Europe by American beekeepers beginning in the 1600s. Besides the Italian race, these include the English-German race *(A. m. mellifera),* the Carniolan race *(A. m. carnica),* and the Caucasian race *(A. m. caucasica)* (Ruttner 1988).

Fortunately, there is no evidence of strong differences between the European races in the major features of colony organization for food collection. This fact, combined with the fact that the bees studied probably contained traces of all these races, suggests that the results of my experiments apply to all the European races of honey bees.

In some cases, the objectives of an experiment required that I use two colonies whose members were readily distinguished. For these experiments, I used one colony of Italian bees, whose workers are naturally marked by their yellow-brown abdomen color, and one colony of Carniolan bees, whose workers are distinguished easily from the Italian bees by their black abdomen color. To be sure that my Carniolan colonies were pure Carniolan, and so consisted entirely of black bees, I used colonies that were headed by a Carniolan queen which had been instrumentally inseminated with the semen of Carniolan drones.

A colony of bees may be calm or excitable, reluctant or eager to

Figure 4.5 Bees sucking sucrose solution from grooves in the feeder. Photograph by T. D. Seeley.

sting. For the purposes of experimental work, especially those projects that involve opening the observation hive for manipulations, calm and mild-tempered bees are much preferable. Therefore, I selected peaceful bees as the inhabitants of my observation hive. The hive is stocked with bees simply by removing from a full-sized hive two frames of comb that are partly filled with brood and food and that are fully covered with adult bees (including the queen), installing them in the observation hive, and then transporting this hive to a new location so that the bees inside cannot return to their original home.

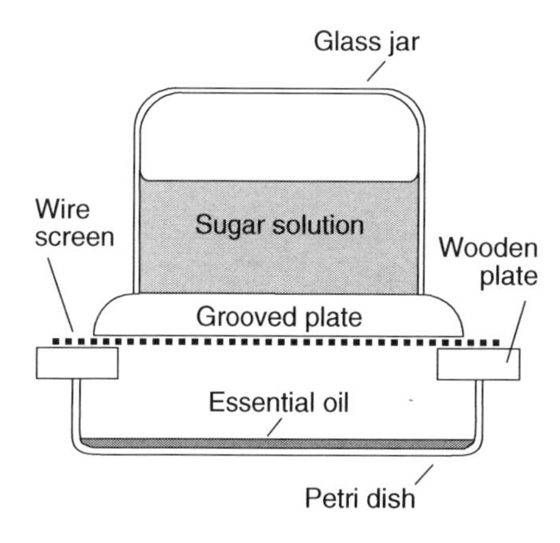

Figure 4.6 Sectional view of the sugar water feeder. The sucrose solution is enclosed so that its concentration remains constant. The feeder is marked with scent that evaporates from the bottom reservoir of essential oil and diffuses out through spaces beneath the grooved plate.

4.4. Sugar Water Feeders

For most experiments, one needs to control the quality and spatial distribution of the food sources available to the bees. Sometimes sufficient control can be achieved with specially planted patches of flowers (Sections 3.2 and 3.3), but generally this is best accomplished by providing the bees with an artificial food source filled with sucrose solution. The feeders that I use are modeled after one described by von Frisch (1967, figure 18). Each consists of a glass jar 4 cm high and 6 cm wide, filled with sugar water, which stands on a circular Plexiglas plate 5 mm thick, 7 cm in diameter, into which 24 radial grooves—each 10 mm long, 1 mm deep, and 1 mm wide—have been cut (Figures 4.5 and 4.6). This design works well for several reasons. It provides food ad libitum for up to about 50 bees simultaneously, but does not require frequent refilling. Also, because it is a closed, pneumatic device, there is no evaporation that would change the sugar concentration of the food and there is little danger of the bees soiling their wings with the sticky sugar solution. This feeding jar is mounted atop a brightly colored wooden plate into which has been cut an opening that is slightly smaller than the circular Plexiglas plate that is placed over it. Between the wooden and Plexiglas plates is a wire screen. The wooden plate is placed atop a petri dish, which is partially filled with essential oil. This oil evaporates slowly through the openings in the wire screen and so marks the feeder with a scent of constant intensity. As a rule, I mix the same essential oil into the sucrose solution; 50 μL of oil per liter of sugar solution is sufficient to mimic the natural scent of nectar. If the experiment calls for just one odor, I use anise, because my early experiments (unpublished) revealed that a feeder marked with this scent elicits stronger recruitment than does an otherwise identical

Figure 4.7 An assistant tending a sugar water feeder placed atop a small table. Photograph by T. D. Seeley.

feeder marked with peppermint, orange, lemon, clove, or vanilla. These are all satisfactory second scents. The feeding dish is placed atop a small, brightly colored wooden table that supports the feeder at a convenient working height for someone seated in a lawn chair (Figure 4.7).

Bees are trained to forage from the feeder with the techniques described by von Frisch (1967). The feeder's table is placed just outside the hive entrance, a wooden bridge is laid from entrance to table, and the feeder is placed on the bridge a few centimeters outside the entrance opening. Then drops of concentrated sugar solution (at least 2.0 mol/L) are placed in a line leading from the hive entrance to the feeder. Foragers quickly discover this line of sweet drops and are led by it to the feeder. Once 10 or so bees are simultaneously drinking food from the feeder, I begin to slide it along the bridge, initially moving it only a few centimeters at a time and always checking that several bees have returned to the feeder before advancing it to the next location. Eventually the feeder reaches the table, the bridge is taken away, and moves of a meter or so become possible. These moves can be stretched to 10 or 50 m once the feeder is about 50 m from the hive (see von Frisch 1967, pp. 17–18, for details). With these techniques, it is not difficult to establish multiple food sources several hundred meters from the hive, which approximates the normal spatial pattern of a colony's food sources (Figure 3.2). Because the feeders and the hive are generally separated by large distances, it is essential to the smooth performance of an experiment to have walkie-talkies for coordinating the activities of the various individuals stationed at the feeders and the hive.

Once a feeder has reached its destination, I begin labeling the bees visiting it with paint marks so that I can identify individuals and determine the number of bees visiting the feeder. Once this number has reached the desired size, I reload the feeder with a more dilute sugar solution to reduce recruitment and my assistants and I begin capturing the excess newcomers. For this, I prefer the technique devised by Gould, Henerey, and MacLeod (1970) of using Ziploc-type plastic freezer bags to minimize disturbance at the feeder. One simply places a plastic bag over a bee while she is feeding, waits for her to crawl up into it when she is fully loaded, and then seals up the bag when she is inside. With a little dexterity, one can capture some 20 or 30 bees in a single bag. Alternatively one can seize bees with forceps and place them in a stoppered vial of alcohol, but bees so caught will release alarm pheromone and the remaining bees visiting the feeder will become skittish.

The concentration of sugar solution needed to interest bees in the feeder will depend on the abundance of the natural food, as will be described in Chapter 5, and will vary from about 0.5 to 2.5 mol/L. Since the more dilute solutions can quickly spoil through fermentation, on any given day I use only solutions mixed the previous evening and refrigerated overnight. I store and carry the solutions into the field in 250- or 500-mL canning jars since these are easily washed and rarely leak. If nectar is available in great abundance from natural sources (during a nectar flow), then generally it is impossible to entice bees to forage from an artificial feeder, even one loaded with 2.5 mol/L sucrose solution. They prefer real blossoms. All one can do in this situation is postpone one's experiment until the nectar flow is over, or travel to a different study site with fewer flowers.

4.5. Labeling Bees

For most investigations, having bees labeled for individual identification is as important as having them live in an observation hive and forage from a controlled feeder. Often one needs only to label the small number of bees (10–30) visiting each feeder, and for this I apply paint marks with fine (000 or 0000) camel's hair or red sable paintbrushes. I apply a spot of paint to each bee's thorax or abdomen, or both, hence with five colors (yellow, white, red, blue, and green) I have 35 different labels of one or two colors. If it is necessary to distinguish bees *from different feeders,* I apply a second paint mark on each bee's abdomen, using a distinctive sixth color (usually orange, purple, or light gray) to denote each feeder. The paint I prefer is that described by von Frisch (1967): dry artist pigments mixed in clear ("white") shellac. The resulting paint dries rapidly, but not so quickly that it hardens on the brush before it can be applied to the bee. Also, because one mixes the pigments and base oneself, one can precisely control the consistency and color of each batch of paint. By trial and error one learns the proper consistency, neither too thin, lest it spread excessively on the bee, nor too thick, for then it does not penetrate the hairy cover of the bee's body and does not stick properly. Light colors, made by mixing white (titanium oxide) with colored pigments, are the ones most easily seen, especially when the bees are inside the observation hive. I keep the paints in rubber-stoppered vials inserted

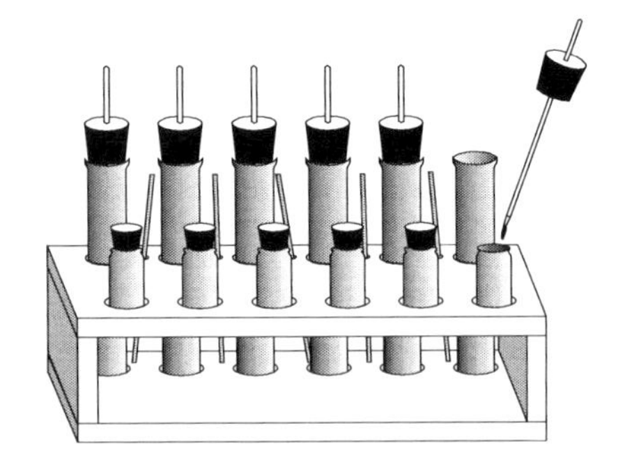

Figure 4.8 A paint set for labeling bees.

in a wooden holder. For each color there is a separate paintbrush, which is stuck through the hole of a rubber stopper and stored in a test tube with its bristles dipped in absolute undenatured alcohol (Figure 4.8). Thus both paint and brush are kept ready for use. This holder is customarily held between the knees, to leave both hands free, one to uncork a paint vial and the other to manipulate the paint brush. It is easy to daub paint on a bee while she is standing still and engrossed in an activity, either at the feeder sucking up the sugar solution or inside the hive giving up or receiving a load of nectar or water.

Sometimes an experiment requires an observation colony in which *each bee* is labeled for individual identification. I have created such colonies containing 4000 bees, and have found that the following procedure makes this a straightforward, though tedious, undertaking. First, using a large funnel, I shake several hundred worker bees off the frames of a standard beehive and into a wire cage (10 × 10 × 25 cm). From this cage I then shake small groups of approximately 50 bees into plastic bags and place them in a refrigerator to immobilize them. After at least 15 min of cooling, a bag of bees is removed from the refrigerator and poured onto a container of "reusable ice," where the bees stay chilled during the labeling operation. A plastic bee tag (Opalithplättchen; with 500 number and color combinations, manufactured by Chr. Graze, Endersbach, Germany; see Figure 1.8) is glued on the thorax of each bee, and a dot of one of eight different colors of paint is applied to its abdomen. The labeled bees, still chilled, are then gently poured into a cage containing a sugar water feeder and their own queen, who is confined to a smaller cage. Here they warm up, cluster around their queen, and feed. This procedure is repeated until 4000 bees have been labeled. With practice, one person can label 100 bees per hour, so a team of four people can complete the job in a day. The following day, I transfer the bees to an observation hive by first placing the caged bees in a standard beehive with just two combs, then opening the cage and releasing the queen so that the bees may crawl onto the combs, and finally, after several hours, transferring the bee-covered combs to the observation hive. The stage is now set for examinations of colony organization with extremely high resolution. One can prolong the period during which all of the colony's adult bees are labeled, and so derive the most from one's setup efforts, by providing the colony with combs that initially contain either no brood or brood limited to eggs and young larvae.

4.6. Measuring the Total Number of Bees Visiting a Feeder

Often it is important to know how many different bees are visiting each feeder in an experimental array. Counting them is easily accomplished if one is working with a colony in which all the bees are labeled for individual identification, as decribed above. One equips an assistant stationed at each feeder with roll call sheets listing the identification codes of all 4000 bees in the study colony, and he or she then simply crosses off the identification code corresponding to each bee seen at the feeder. As a rule, I have the assistants fill out a different sheet every half hour. Each sheet therefore shows which bees have visited a particular feeder at least once every 30 min, and how many different bees have done so.

4.7. Observing Bees of Known Age

To investigate the role of bee age in colony organization, it is necessary to know the ages of bees within the observation hive. This is most easily accomplished by introducing bees of known age. They are obtained by removing combs containing mature brood from other colonies, brushing off all the adult bees, and placing the combs overnight in an incubator set at 34°C. The next day one gathers 0-day-old bees off these combs, labels them with bee tags or paint to indicate their age, and places them in a small, wire-mesh cage mounted over one of the ventilation holes atop the observation hive, with only a screen separating them from the bees below. After about 4 hours the glue or paint on these young bees will be thoroughly dry. Also by this time the bees will have absorbed sufficient odors from the colony below so that they can be released into the observation hive with little risk of being attacked as intruders. It helps too to feed the young bees some sugar solution, by painting it on the walls of their wire-mesh cage, before letting them climb down into the observation hive.

4.8. Recording the Behavior of Bees in the Hive

Most data are collected by directly recording observations in a notebook or computer, but sometimes these recording techniques are augmented by videorecording behaviors for subsequent analysis with a slow-motion videoplayer, audiorecording oral descriptions of rapid events, and graphically recording spatial information on a sheet of glass or plastic taped over the glass wall of the observation hive. Close observations of bees are facilitated by the use of 3.5X magnifying lenses mounted in a headset.

Several variables of the within-hive behavior of foragers have been repeatedly recorded in different experiments over the years. One is the *search time* (also called the "time to start of unloading" in Seeley 1986, 1989a, and Seeley, Camazine, and Sneyd 1991). This is the amount of time that a nectar forager, upon return to the hive, spends searching for a food-storer bee who will accept her nectar load. Using a stopwatch, I measure search time as the interval between when the bee enters the hive and when she begins unambiguously to transfer nectar to a food-storer bee, which is indicated by a food-storer bee's placing her tongue between the forager's mouthparts for a prolonged period (at least 3 sec). The *delivery time* (also called the "time to end of unloading") is the interval between the time a forager enters the hive and the time she finishes unloading to food storers, which usually is indicated clearly by the forager's extensively grooming her mouthparts. Another variable is the *maximum number of food storers contacted simultaneously:* the maximum number of food-storer bees that simultaneously insert their tongues between the mouthparts of a nectar forager, that is, the greatest number of bees simultaneously unloading the forager.

Another important aspect of the behavior of foragers is their dancing: whether they perform a waggle dance, a tremble dance (Section 6.3), or no dance at all upon return to the hive. The proportion of foragers performing dances is determined by following foragers one at a time inside the observation hive from time of arrival to time of departure. If a bee performs a waggle dance, it is often important to count the total number of waggle runs produced during her time inside the hive between foraging trips. I refer to this count as the *dance duration.* The locations of dances are recorded with wax pencil tracings on a glass sheet taped against the glass wall of the observation hive.

4.9. The Scale Hive

Often I wish to know how much nectar is available to my observation hive colony from the natural blossoms in the surrounding countryside. This amount can change markedly from one day to the next, and when it does the behavior of foragers is strongly affected; hence it is an important environmental variable to monitor. I do this by positioning one or more full-size hives of bees within 100 m of the observation hive, mounting each one on platform scales (Detecto model 4510), and weighing the hives each evening, after all their foragers have come home. A colony's daily weight change represents mainly the nectar gathered by the colony minus the honey it consumed (basically a constant) during the previous 24 hr; hence it provides a reliable indicator of the availability of nectar from natural sources.

4.10. Censusing a Colony

The population size of a colony living in an observation hive is easily measured. Over each side of the hive I place a glass sheet on which I have drawn a grid of 5-cm-sided squares; in the case of the two-frame observation hive, the grid is 10 squares wide and 9 squares high. Each of the grid squares on each side of the hive is assigned a number. Then, using a random number table, I randomly select one grid square at a time and count each bee whose thorax lies inside the grid square. This procedure is repeated until 20% of the grid squares have been sampled; then the number of bees counted is summed, and this sum is multiplied by 5 to yield an estimate of the total size of the colony's population. An unpublished analysis of this censusing procedure, by Kirk Visscher, showed that one must sample at least 20% of the grid squares to obtain an accurate estimate of a colony's population.

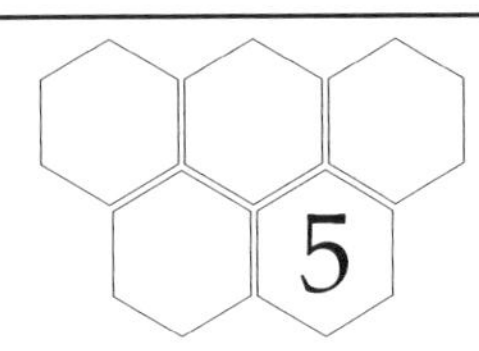

Allocation of Labor among Forage Sites

Whenever flowers are in bloom, a honey bee colony must solve the difficult problem of wisely deploying its foragers among the kaleidoscope of flower patches in the surrounding countryside. This entails acquiring information about the foraging opportunities in the environment, combining this with information about the colony's needs for nectar and pollen, and generating an appropriate distribution of foragers among the patches of flowers. Success in solving this problem means that the colony's foragers will gather nectar and pollen quickly, efficiently, and in the correct proportions. In general, flower patches that are large and offer easy collection of a strongly needed food should receive many bees, whereas patches that are small, contain sparse forage, or provide a little-needed food should receive few foragers, perhaps none at all.

In this chapter, I will skirt the complexities in this labor allocation problem that arise when one considers the joint collection of pollen and nectar. As noted earlier (Section 3.7), a colony's need to gather pollen can change strongly from day to day, depending on the size of its pollen reserve, whereas a colony's need to gather nectar remains high, hence is stable, until the hive becomes filled with honey. Thus if we were to consider the complete puzzle of labor allocation among both pollen and nectar sources, we would have to consider the highly dynamic variable of colony demand for pollen. I will deal with this complication in Chapter 8, but for now I will sidestep it by focusing on the smaller, but still sizable, puzzle of how a colony achieves a proper allocation of foragers just among nectar sources (Figure 5.1). Previously I have shown that a colony is able to exploit selectively different sources of nectar, concentrating its efforts on the most rewarding ones, and that it is able to rapidly redistribute its foragers among these sources to track changes in the resource array (Section 3.4). Now I will investigate how this ability to exploit nectar sources wisely arises from the actions and interactions of the individual bees composing a colony, considering first how a colony acquires information about the current foraging opportunities, and second how it acts upon this information to generate an effective allocation of labor.

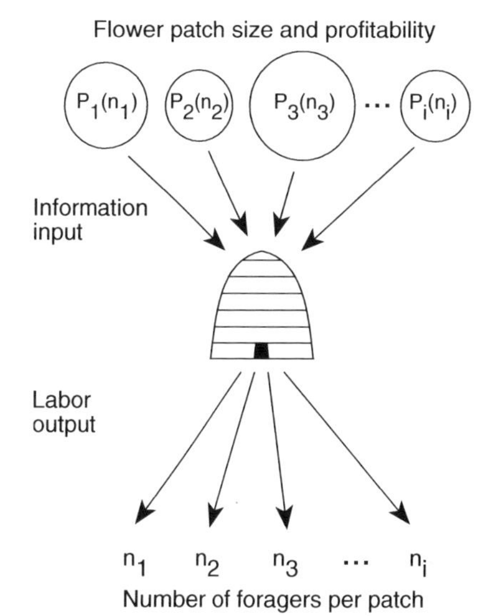

Figure 5.1 The labor allocation problem that a honey bee colony must solve whenever it is collecting nectar. The colony gathers nectar by deploying several thousand foragers across an array of flower patches in the surrounding environment. These patches differ in size and in profitability, with the profitability of each patch a decreasing function of the total number of bees foraging there. If the colony is to exploit effectively this array of nectar sources, it must constantly gather information about the flower patches and must accordingly adjust its distribution of foragers among the patches.

How a Colony Acquires Information about Food Sources

5.1. Which Bees Gather the Information?

Every time a forager bee returns to her hive from a patch of flowers, she brings home not only food stored in her pollen baskets and honey stomach, but also information about her food source stored in her brain. She can share this knowledge with her nestmates by means of the waggle dance communication behavior. Hence a colony acquires information about food sources from all its members that are actively gathering food (hereafter called "employed foragers"). Under favorable conditions, a typical colony of some 25,000 bees will have approximately one-quarter of the members engaged in food collection; thus a colony can easily have several thousand bees bringing in information about food sources. Such a large group will monitor dozens of forage sites scattered far and wide around the hive, and will maintain a strong flow of information about the foraging opportunities into the hive (Figure 5.2).

For a colony to achieve long-term foraging success in a rapidly changing flower market, it must receive both updates on old food sources and reports on promising new ones. How is the collection of these two kinds of information organized? At any given moment, nearly all the foragers returning to a hive are employed foragers that have been *exploiting* food sources discovered some hours or days before; hence most foragers bring in information about older sources (Figure 5.3). Only a small fraction of the returning foragers have been *exploring* for new food sources, and so may bear news of fresh discoveries. These explorers, or scouts, come from the ranks of the unemployed foragers in a colony. Unemployed foragers are bees that need to locate a forage site, either because they are just beginning their forage careers (novice foragers) or because they have recently aban-

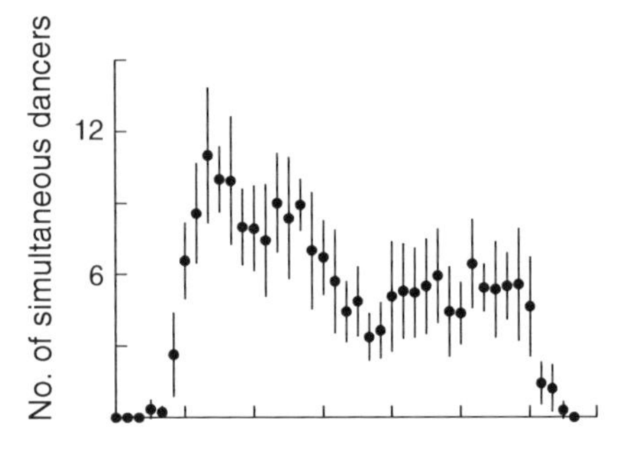

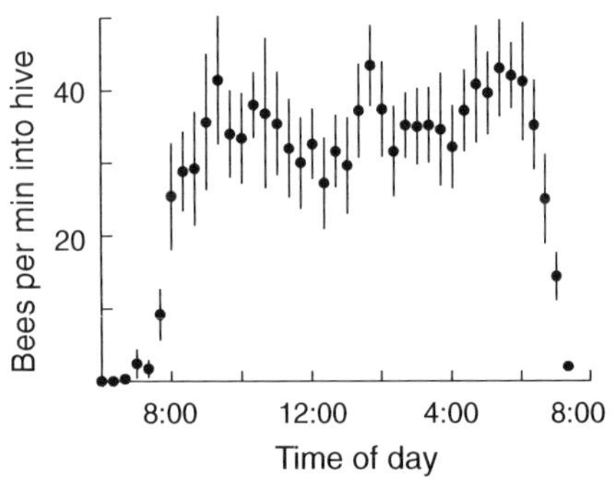

Figure 5.2 Levels of dancing and forager influx for one small colony on 12 July 1989 ($x \pm$ SD). The colony consisted of approximately 4,000 worker bees plus a queen, and occupied the two-frame observation hive depicted in Figure 4.1. Although the details of the dancing and foraging patterns will differ between days and between colonies, the important features of this figure—multiple dances performed simultaneously and a steady influx of information-bearing bees—are typical. After Seeley and Towne 1992.

doned a depleted patch of flowers (experienced foragers). Most such unemployed foragers follow the recruitment dances of their nestmates to find a forage site, hence are recruits; but a few do so by searching on their own, hence are scouts.

Lindauer (1952) first assessed what fraction of unemployed foragers locate their next food source by scouting. He focused on novice foragers. Between 31 July and 3 August 1949, he numbered 390 newly emerged bees and introduced them into a colony living in an observation hive; and on 22 August he began recording all episodes of dance following by these bees as well as all instances of returning to the hive with food. By 11 September, 159 of the bees had flown from the hive and returned with a load of nectar or pollen, hence had found a first forage site, and only 9 (6%) had done so without following any recruitment dances, that is, by scouting.

I repeated Lindauer's study in the summer of 1980 (Seeley 1983), but with attention on the experienced rather than the novice foragers. My procedure involved setting up a colony in a two-frame observation hive on Appledore Island, Maine (to avoid interference from other colonies), training a group of 15 labeled foragers to a feeder, letting them forage there for 2 days, and then shutting off the feeder and watching the bees closely in the hive to determine how many located their next food source without following dances. In the first two trials of this experiment, performed between 29 June and 11 July, fully 10 out of 28 bees (36%) scouted for a new food source, while in the second two trials, performed between 21 July and 31 July, only 1 out of 22 bees (5%) did so. Correlated with this striking decline in the percentage of scouts was an equally dramatic rise in the intensity of dancing in the hive. It rose from less than 1 feeble dance to nearly 5 strong dances, on average, at any given time, as the wild catnip *(Nepeta cataria)* began to bloom profusely on Appledore Island in late July. This suggests that the proportion of unemployed foragers that locate their next forage site by scouting is strongly influenced by the availability of dance information. If so, then this implies that a colony can adaptively modulate its exploration effort, increasing it whenever information about rich food sources, expressed in recruitment dances, becomes scarce in the hive. This hypothesis is consistent with the results of a methodologically similar study (Seeley and Visscher 1988) in which Visscher and I measured the percentage of scouts among experienced foragers during a period of meager forage in the Yale Forest (16 June to 24 July 1986) and found that this percentage was again quite high. Five sets of observations yielded data on 58 foragers, of which 14 (24%) searched independently and 44 followed recruitment dances to locate a new food source. Averaging across times of high and low forage abundance, I estimate that approximately 10% of a colony's foragers go scouting when they need to find a new forage site.

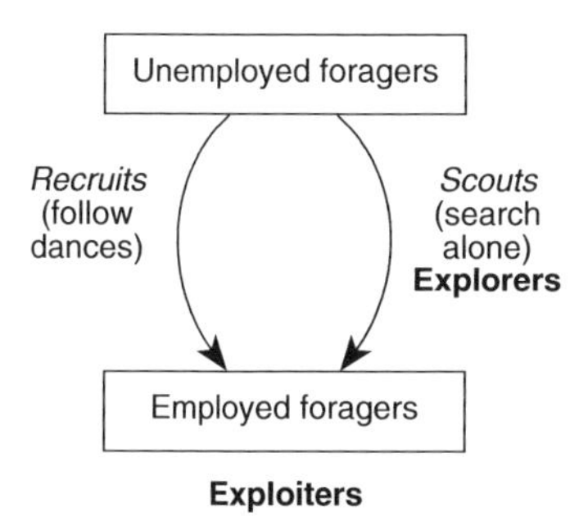

Figure 5.3 Explorers and exploiters among a colony's foragers. The explorers are unemployed foragers that are searching independently (not following recruitment dances) to locate a new food source; they provide news of freshly discovered food sources. The exploiters are all the employed foragers; they provide updates on previously discovered food sources.

The figure of approximately 10% scouts among unemployed foragers enables one to estimate how many bees explore for new food sources each day from a typical colony. Assume that the colony contains 25,000 bees, that 25% of its members are foragers, and that each day some 20% of the foragers become unemployed (need to locate a forage site) as novice foragers join the foraging operation and as experienced foragers abandon depleted flower patches. These figures lead to an estimate of 1250 foragers per day that need to locate a work site (25,000 bees × 25% × 20% = 1250 bees). If 10% do so by scouting, then the colony fields each day some 125 explorers—a respectable search force.

Unfortunately, little is known about how an individual scout bee conducts her search for a new source of food, because the search process is not easily observed. For example, when I would sit beside a rich patch of borage flowers on Appledore Island and watch it closely for the arrival of a scout bee (Section 3.3), all I could see were the final moments of a scout's search. These always consisted of the scout bee's flying low over the ground, stopping briefly at each bright flower along her way to inspect it for food, and finally chancing upon the nectar-rich borage flowers which Kirk Visscher and I had set out. Clearly, scouts sometimes conduct flower-by-flower searchs of a particular area. And it is clear that scouts often conduct such searchs far from the hive, for they often discover important food sources several kilometers from home. But the precise spatial pattern of an individual scout's search remains a deep mystery. At best one can make a rough first estimate of how large an area (S) a single scout bee explores in a day. If all n scouts in a colony search independently over a total area A, then the probability that at least one scout will discover a particular flower patch within A is $P = 1 - (1 - S/A)^n$. Solving this for S, using the estimate of $n = 125$ bees calculated above, and the finding (Section 3.2) that a colony has a probability of 0.5 of locating a particular flower patch within 2 km of the hive (hence A = 12.6 km^2), yields a value for S of approximately 70,000 m^2, or about one-sixteenth of a square kilometer. Perhaps a scout bee accomplishes

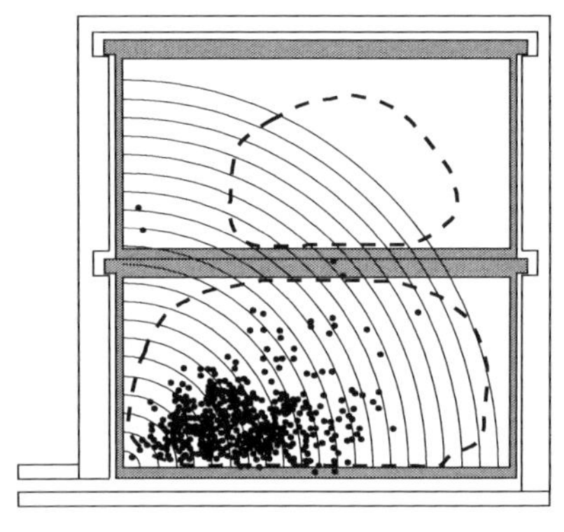

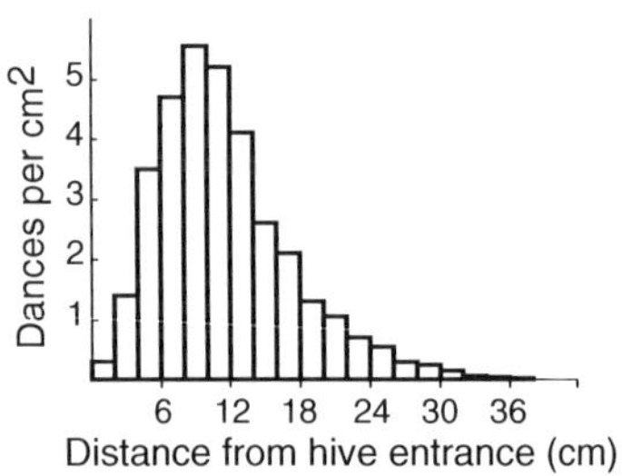

Figure 5.4 The dance floor in an observation hive. *Top:* Locations of 437 dances observed with scan sampling at 2-min intervals between 9:00 and 10:30, 12 July 1989. Inside the hive, regions of beeswax comb are shown as *white,* while wooden surfaces (comb frames) are denoted by *shading.* The *dashed lines* mark the boundaries of the brood-filled cells on each comb. The *quarter-circle lines* centered on the entrance mark the 2-cm-wide bands used to measure dance density. *Bottom:* The density of dances as a function of distance from the hive entrance. Calculations of dance density were based on the data shown above plus comparable data gathered on two other days; $n = 1224$ dance locations. After Seeley and Towne 1992.

such a broad reconnaissance by flying steadily along at moderate height over the vegetation, stopping to perform a detailed examination only when she spies bright flowers below.

5.2. Which Information Is Shared?

Although every returning forager brings home information about her food source's location and profitability, only bees returning from highly profitable sources perform dances and so *share* their information with their nestmates. This selective reporting is easily demonstrated experimentally. For example, on 14 July 1990, I established two sugar water feeders 400 m from an observation hive, with 30 bees from the hive visiting each feeder (details in Seeley and Towne 1992). The two feeders contained different concentrations of sucrose solution, 1.0 and 1.5 mol/L; bees from the richer feeder showed a high probability of dancing ($P = 0.73$), whereas those from the poorer feeder showed only a low probability of dancing ($P = 0.08$).

In nature, the fraction of the returning foragers that perform a dance is low, generally less than 10%. Once, for instance, I wanted to determine how often bees working natural food sources perform dances, so I monitored 58 bees gathering nectar and pollen from flower patches, following each bee every time she returned to my observation hive. Over a 2-day period I recorded 153 returns to the hive, and in only 11 instances (7%) was a dance performed (data from bees followed in the study by Seeley and Visscher 1988). Clearly, an important feature of the bee colony's information-acquisition process is a strong filtering out, early on, of information about low-yield nectar sources. The pool of shared information within the hive consists, therefore, almost exclusively of information about highly profitable food sources.

5.3. Where Information Is Shared inside the Hive

Foragers do not report on their food sources throughout the hive, but instead concentrate their announcements in the area just inside the hive entrance, the dance floor (von Frisch 1967). For example, plots of dance locations inside one observation hive revealed a band of high dance density approximately 4–18 cm in from the entrance opening, with 94% of the dances performed within 24 cm of the entryway (Figure 5.4). Similar spatial patterns have been reported by other investigators (Körner 1939; von Frisch 1940; Boch 1956), so the general form of this pattern appears to be typical. Such conspicuous clustering of dances may be merely a by-product of foragers minimizing their time and travel inside the hive between foraging trips, but it may also benefit the overall process of a colony's food collection by facilitating the flow of information. In particular, an unemployed forager's task of finding a dancer inside the dark hive, and so securing information about foraging opportunities, is surely simplified by the existence of a special region of the hive with a high density of dancing bees.

At any given time, the dances in a hive will represent several distinct forage sites separated by hundreds, if not thousands, of meters (Plates I–VI). How is this information about spatially segregated sites mapped onto the dance floor? Is there a clear spatial separation of dances for different sites, or are dances for different sites mixed together completely at random? Neither of these two extreme possibilities matches reality. Consider the pattern shown in Figure 5.5, which shows the locations of dances for two sites in approximately the same direction from the hive but spaced more than 6 km apart. There is no clear-cut spatial segregation of the two sets of dances, but there is a small, statistically significant ($P < 0.001$) difference between the mean horizontal distance from the hive entrance for the two sets of dances. This kind of difference, also reported previously by Boch (1956), evidently arises because a dancing bee, upon finishing one waggle run and circling back to start another, rarely travels all the way back to the starting point of the previous waggle run. Hence in performing a dance, a bee slowly drifts across the dance floor, moving in the same general direction as she steers her waggle runs (Figure 5.6). This drift is probably especially pronounced for bees who advertise distant forage sites and so perform long waggle runs. This explains, for example, why the raspberry-patch dances shown in Figure 5.5 were performed deeper in the hive than the feeder dances. The tendency to drift while dancing may be adaptive, for it helps a bee broadcast her dance information over much of the dance floor. At the same time, however, the spatial distributions of dances for different forage sites can overlap broadly. This means that an information-seeking bee standing in the middle of the dance floor has ready access to information about any of a number of food sources. Exactly how she samples this information will be considered below (Section 5.10).

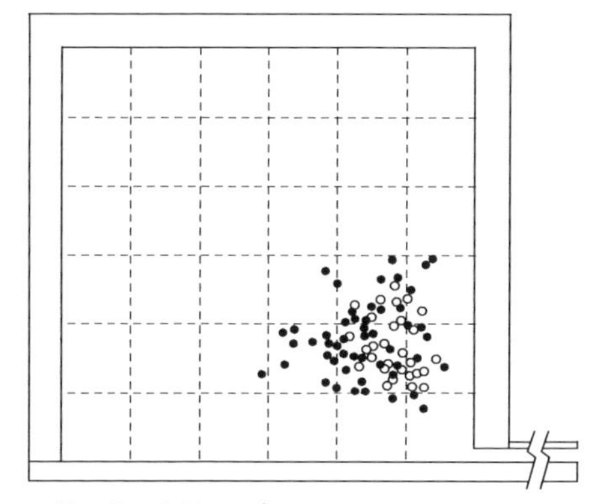

Figure 5.5 Spatial distributions of dances for two widely separated forage sites. Dances were plotted in an observation hive during scan samples made at 2-min intervals over a 20-min period. Arrows at the bottom right specify the waggle run direction for the dances of each site. After Seeley 1994.

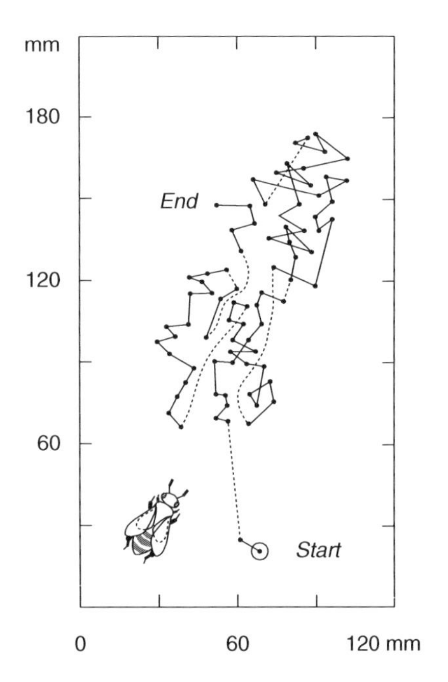

Figure 5.6 The spatial pattern on the dance floor of 81 waggle runs produced during a single dance. Each dot marks the position of the bee's thorax at the start of a waggle run. *Solid lines* connect the waggle runs within a continuous bout of dancing; *dashed lines* connect the last and first waggle runs of two consecutive bouts of dancing. The bee (drawn to scale) shows the orientation of the waggle runs in this dance. After Seeley 1994.

5.4. The Coding of Information about Profitability

One fundamentally important feature of the reporting process is the steep grading of each dance's strength in accordance with the profitability of the nectar source it represents. In principle, the strength of a communication signal can be adjusted by changing either its duration or its intensity, or both. In the case of the bee's recruitment signal, it now seems clear that the principal way in which signal strength is varied is by adjusting signal duration, that is, by controlling the number of waggle runs per dance.

5.4.1. MODULATION OF SIGNAL DURATION

Karl von Frisch (1967, p. 45) first documented this manner of adjusting dance signal strength when he recorded the average duration of dancing by four bees visiting a sucrose solution feeder whose concentration he raised in steps over the course of a day. A more detailed picture of the process is presented in Figure 5.7, which comes from a recent experiment (Seeley and Towne 1992). Thirty bees were trained to a feeder from each of two colonies, one experimental and one control. The two colonies' hives were positioned side by side and their feeders were established 420 m away, but separated from each another by 300 m (see Figure 5.17). When, on 9 July 1987, the sugar solution in the experimental colony's feeder was raised in stages from 0.5 to 2.5 mol/L, the number of waggle runs per dance for this feeder and the rate of recruitment to this feeder both increased markedly. The recruitment rate to the control colony's feeder showed no significant change during the course of the experiment, indicating that the ambient conditions were stable. It is important to note that the variance in dance duration also increased as the profitability of the experimental colony's feeder was raised, so that there is tremendous overlap among the dance distributions for the different profitabilities. This implies that even though the mean duration of a group of dances for a nectar source is a highly accurate indicator of its profitability, the duration of any one dance does not provide precise information about the profitability of the nectar source it represents.

The dances represented in Figure 5.7 varied in strength by nearly two orders of magnitude, from 1 to 69 waggle runs, and in nature the range of signal strength can be wider still. I have seen extremely powerful dances lasting more than 200 waggle runs (257 maximum) performed by pollen foragers in a colony starved for pollen following an 8-day period of cool, rainy weather when no bees could leave the hive.

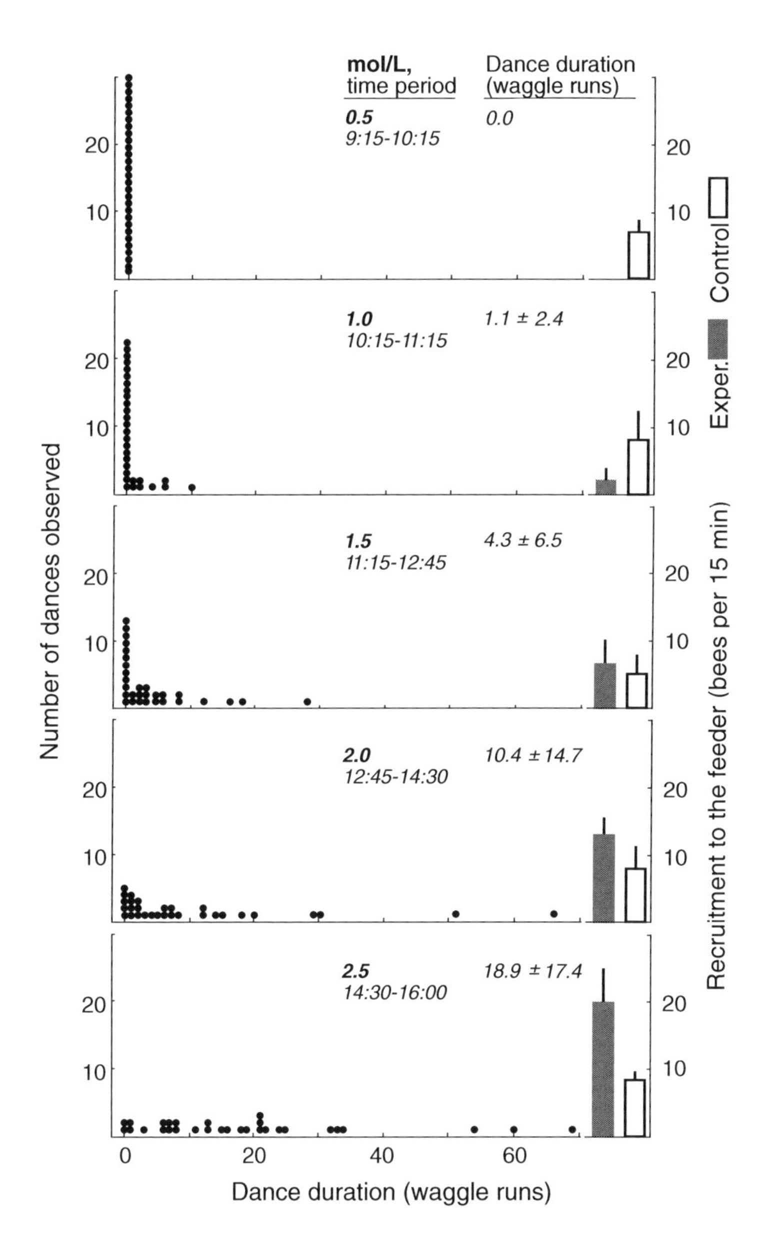

Figure 5.7 Dance duration in relation to nectar-source profitability. Thirty bees were trained to a feeder from each of two colonies—one experimental and one control—then over the course of one day the concentration of the sugar solution in the experimental colony's feeder was raised in stages from 0.5 to 2.5 mol/L. At each concentration, the duration of 30 dances and the rate of recruitment *(filled bars)* to the feeder were measured. Bees that did not dance were given a dance duration of zero. The sugar solution in the control colony's feeder was held constant at 1.5 mol/L and the rate of recruitment *(open bars)* to it was also measured. After Seeley and Towne 1992.

Figure 5.8 provides a detailed picture of the variation in dance signal strength that can occur in nature, based on a videorecord of all the dances performed one day in a small observation hive. Here again the range spanned more than two orders of magnitude, from 1 to 117 waggle runs. This figure also depicts the relative contributions of dances of different durations to a colony's pool of shared informa-

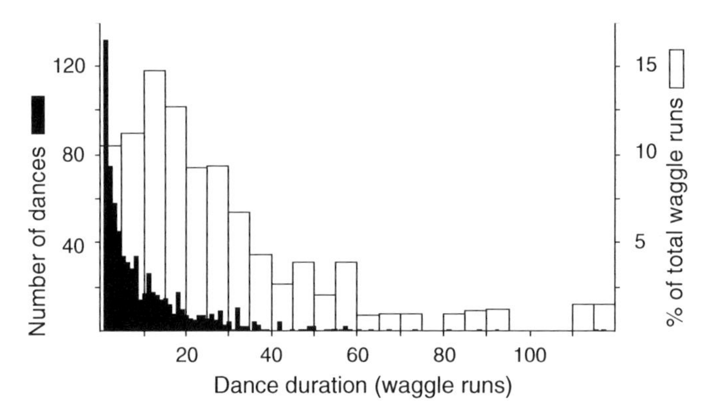

Figure 5.8 Natural variation of dance duration and the informational importance of dances of different durations. *Filled bars:* frequency distribution of dances of different durations for all 725 dances performed in a two-frame observation hive on 10 July 1989. *Open bars:* percentage distribution of waggle runs among dances of different durations for all 7756 waggle runs contained in the 725 dances. After Seeley and Towne 1992.

tion. At first glance, it appears that short dances are the most important, because the median dance duration was just 6 waggle runs and 90% of the dances contained fewer than 27 waggle runs. However, the information contribution of the dances of any given duration depends not only on how frequently they occur, but also on how long they last, since the longer a dance the greater its communication effects. Thus a better picture of the importance of the dances of a given duration is produced by calculating the fraction of the total waggle runs that derive from such dances. As is shown in Figure 5.8, these calculations reveal that extremely short dances are not very important, because the vast majority (90%) of the 7756 waggle runs came from dances with more than 5 waggle runs. In summary, it is clear that in a beehive the information flow about nectar sources occurs through dances spanning the range of approximately 1 to 100 waggle runs, and that this variation in dance duration is used to express information about nectar-source profitability.

5.4.2. MODULATION OF SIGNAL INTENSITY

Do bees code nectar-source profitability in dance intensity as well as dance duration? Several authors (Lindauer 1948; Boch 1956; von Frisch 1967) have stated that richer sources seem to elicit livelier—not just longer—dances than do poorer sources. I too have this impression, in the sense that the dances announcing desirable sources are usually highly energetic whereas those for marginal sources are often rather feeble. These subjective assessments are corroborated by Esch's (1963) examination of sound (carrier frequency 250–300 Hz) production during waggle runs in relation to nectar-source profitability. He found that waggle runs performed for a feeder filled with a sugar solution well above the dance threshold concentration are nearly always accompanied by strong sound, while those for a feeder filled with a threshold-level solution are frequently silent.

This finding does not, however, tell us whether or not bees visiting nectar sources with profitabilities *above the dance threshold* modulate the liveliness of their waggle runs in accordance with profitability. To address this question, Esch (1963) trained bees to a sugar water feeder and recorded the 250–300 Hz sounds of their dances at several levels of sugar concentration. He found that the frequency of sound pulses during a waggle run rises with increasing sugar concentration. Waddington (1982) and Waddington and Kirchner (1992) investigated the matter further and reported that not only the pulse frequency, but also the carrier frequency, amplitude, and duration of the sounds produced during a dance, plus the rate of circling, are all positively correlated with the sugar concentration at the feeder. They point out, though, that the changes they report are unlikely to convey information about nectar-source profitability, because there is no evidence that the range of values observed for each variable is large enough to have effects on the dance-following bees. For example, Waddington and Kirchner (1992) report that the dance sounds for their lowest and highest sugar concentrations had carrier frequencies of approximately 250 and 270 Hz, but in an earlier study Kirchner, Dreller and Towne (1991) found that bees neither discriminate nor show differential sensitivity to sounds of different frequencies across this narrow range. I suggest that the small changes reported in the dance-liveliness variables are all incidental by-products of the bees changing their foraging tempo (and body temperature) in relation to nectar-source profitability, as described by Stabentheiner and Hagmüller (1991). In support of this view, Esch (1963) and Wenner, Wells, and Rohlf (1967) report that the pulse frequency in a dance's sound is strongly affected by the dancing bee's temperature.

There is also direct evidence that bees do *not* express information about profitability by varying the intensity of the dance signal, and do so instead simply by varying dance duration. This evidence comes from an experiment in which I established two sugar water feeders 400 m away and in opposite directions from an observation hive, with one feeder highly profitable and eliciting strong dancing, and the other less profitable and so eliciting weaker dancing (Seeley and Towne 1992). I allowed 30 labeled bees to visit each feeder and captured all additional bees recruited to the feeders. I also measured the

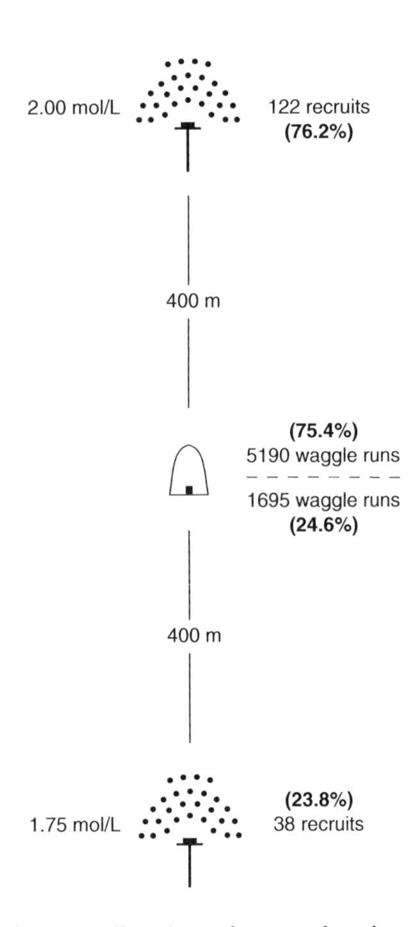

Figure 5.9 Experimental array and results from one trial of the test of whether waggle runs for richer and poorer food sources differ in effectiveness in arousing recruits. Based on data in table 2 of Seeley and Towne 1992.

amount of dancing (total number of waggle runs) for each feeder (Figure 5.9). If the waggle runs for the richer and poorer feeders are performed with equal vigor, and so are equally effective in arousing recruits, then the proportion of recruits to each feeder should be predicted by the proportion of the waggle runs for each feeder. But if the waggle runs for the richer feeder are performed more vigorously, and so are more effective in arousing recruits, then the proportion of recruits to the richer feeder should exceed its proportion of the waggle runs. The results were clear-cut: recruitment was always closely proportional to the number of waggle runs. For example, in the trial conducted on 11 July 1990, the two feeders were loaded with 2.00- and 1.75-mol/L sucrose solutions. Over a 5-hr period they were advertised by a total of 6885 waggle runs, with 75.4% of them for the richer feeder, and during this time period the feeders received a total of 160 recruits, with 76.2% of them arriving at the richer feeder (Figure 5.9). Ten other trials of this experiment were performed, and in each case the proportion of waggle runs for each feeder accurately predicted its proportion of the recruits (the results will be presented in greater detail in Section 5.10). It is important to note that in several trials the difference in profitability between the two feeders was quite large, as on 16 July, when fully 93% of the waggle runs were for the richer feeder, and yet even here there was no indication of a difference in strength of the individual waggle runs for the two feeders. Recruitment remained strictly proportional to the number of waggle runs. Thus the evidence indicates that the bees rely principally, if not completely, on modulation of dance signal duration to express information about nectar-source profitability.

5.5. The Bees' Criterion of Profitability

By what yardstick does a worker bee measure the profitability of a nectar source? Since nectar is basically a sugar solution, it provides the bees mainly with energy; thus it is likely that foragers assess nectar sources by some criterion of energetic profitability. Most standard models in foraging theory assume that animals should maximize their net rate of energy intake while foraging (Stephens and Krebs 1986), because time is often in short supply and this situation should favor rapid acquisition of energy. Other models suggest instead that animals should maximize their net energetic efficiency while foraging (Schmid-Hempel, Kacelnik, and Houston 1985), which would be appropriate for animals whose foraging success is limited by energy expenditure rather than time availability. If during an average foraging trip an animal collects G units of energy, expends C units of energy, and spends time T, then to maximize the net rate of energy collection it should maximize $(G - C)/T$, and to maximize net energetic efficiency it should maximize $(G - C)/C$.

To determine whether bees use either of these two criteria for evaluating nectar sources, I performed an experiment with the following protocol: (1) train a group of 20 bees to each of two feeders at different distances (250 and 550 m) from the hive; (2) determine a concentration of sucrose solution for each feeder such that the bees judged the two feeders equally profitable (indicated by the two groups performing the same mean number of waggle runs per dance); and finally, for the conditions that elicit equal dancing, (3) calculate the profitability of each feeder according to different hypothetical measures, including the two already described (Seeley 1994). Should any one of these measures yield equal values for the two feeders, this result would suggest that this *possible* measure is a good approximation of the bees' *actual* measure of nectar-source profitability.

Two trials of this experiment have been performed. I started both by filling, in the morning, the far (550 m) feeder with a 2.50-mol/L sucrose solution and the near (250 m) feeder with a 1.75-mol/L sucrose solution. As is indicated in Table 5.1, there was a higher mean

Table 5.1. Results of the experiment analyzing the bee's criterion of nectar-source profitability. After Seeley 1994.

	Trial 1: 15 July 1992			Trial 2: 17 July 1992		
	Near feeder	Far feeder	Ratio	Near feeder	Far feeder	Ratio
Waggle runs/trip						
A.M. (N: 1.75; F: 2.50)	11.0 ± 2.0	7.0 ± 1.8	1.57	8.8 ± 1.1	5.6 ± 1.2	1.57
P.M. (N: 1.25; F: 2.50)	10.1 ± 1.7	11.4 ± 2.0	0.89	4.7 ± 1.1	8.1 ± 1.3	0.58
Sucrose solutions for equal dancing	1.32	2.50		1.46	2.50	
Mean gain per trip (J)	360	725		415	725	
Mean cost per trip (J)	7.8	15.7		7.2	12.3	
Mean time per trip (sec)	246	440		223	359	
Possible criteria						
gain – cost (J)	352	709	0.50	408	713	0.57
(gain – cost)/time (J/sec)	1.43	1.61	0.88	1.83	1.99	0.92
(gain – cost)/cost (J/J)	45.1	45.2	1.00	56.7	57.9	0.98

Note: N = near feeder, F = far feeder. The associated numbers specify the concentration of the sucrose solution, in mol/L, at each feeder. Each value of "waggle runs/trip" represents the mean of 60 measurements of dance duration (x ± SE).

number of waggle runs per dance for the near feeder in both trials. Then during the afternoon of both trials the far feeder was kept at 2.50 mol/L and the near feeder was filled with a less concentrated solution, 1.25 mol/L. Dance measurements now indicated a lower mean number of waggle runs per dance for the near feeder in both trials. By interpolating between the results from the morning and the afternoon, I estimated a concentration of sucrose solution for the near feeder that would elicit dances with the same mean length as dances for the far feeder: 1.32 and 1.46 mol/L for the two trials. The mean amount of nectar loaded and the mean time budget of a foraging trip were also determined for both feeders under conditions eliciting equal dancing. This information, combined with the prior measurements of the metabolic rates of bees by Wolf and his colleagues (1989), enabled me to calculate the mean values of G, C, and T for a foraging trip to each feeder. These values are summarized in Table 5.1.

This table also shows that the net energy gain per foraging trip, $G - C$, was extremely different for the two feeders under conditions that elicited equal dancing. Likewise, the net rate of energy delivery to the hive, $(G - C)/T$, was significantly different ($P < 0.05$) for the two feeders, although here the difference was not so pronounced as for net energy gain. What is most remarkable, however, is that the net energetic efficiency, $(G - C)/C$, did not differ significantly ($P > 0.40$) for the two feeders. These results, although preliminary since they come from only two trials, suggest that a worker bee judges the profitability of nectar sources according to the criterion of energetic efficiency. This provisional result is consistent with the work of Waddington (1985), who found that bees adjust their rate of circling when performing "round dances" in a way that suggests that they assess a feeder's profitability in terms of a ratio of gain to cost.[1]

Why might natural selection have favored the energy efficiency criterion over other criteria, such as rate of energy delivery to the hive?

1. Traditionally, bee researchers have followed Karl von Frish (1967) in recognizing two types of recruitment dances in honey bees: *round dances,* which indicate simply that a nearby food source is available, and *waggle dances,* which specify the direction and distance of distant food sources. A recent study by Kirchner, Lindauer, and Michelsen (1988) has revealed, however, that the distinction between round and waggle dances is artificial. They showed that information about distance and direction is coded in all recruitment dances regardless of the distance to the food. Thus it now seems most appropriate to refer to all recruitment dances as waggle dances, noting that in dances for food sources at short distances (less than about 100 m) the waggle runs are exceedingly short and do not provide precise information about the location of the food.

There is growing evidence that a bee's lifetime foraging gains are limited by lifetime energy expenditure rather than the life span available for foraging (Neukirch 1982; Schmid-Hempel and Wolf 1988; Wolf and Schmid-Hempel 1989; but see also Dukas and Visscher 1994). If so, then a bee will maximize her total energy delivery to the hive by maximizing her energy delivery per unit of expenditure, that is, by maximizing her energetic efficiency while foraging. Now consider what a nectar forager must do to maximize energetic efficiency. First she must locate a forage site where the potential for efficient foraging is as high as possible; then she must behave in a way that maximizes the efficiency of her foraging at this site. A bee clears the first of these two hurdles by following a recruitment dance to locate a forage site (Section 5.1). This works because, as we shall see, recruitment dances are evidently graded in duration according to the criterion of foraging efficiency, and hence a forager is most likely to be recruited to a site offering highly efficient foraging. Once a forager has been recruited to such a site, she then clears the second hurdle by appropriately adjusting her behavior at the site. For example, nectar foragers often return to their hive before they have gathered a full load of nectar (Schmid-Hempel, Kacelnik, and Houston 1985). This partial loading strategy helps a bee maximize the energetic efficiency of her foraging because flying from flower to flower with a heavy, nearly full load of nectar would impose a large cost in energy expenditure (Wolf et al. 1989).

A cautionary note must be added. There is a strong possiblity that nectar foraging bees do not have a fixed criterion of nectar-source profitability, but instead use different criteria at different times of the year and under different colony conditions (Schmid-Hempel, Winston, and Ydenberg 1993). For example, bees in autumn may seek to maximize the rate of energy delivery to the hive rather than the efficiency of this energy delivery, since their foraging gains are severely limited by the time available before winter sets in. One indication of seasonal effects comes from a pair of studies by Wolf and Schmid-Hempel (1990) and Fewell, Ydenberg, and Winston (1991), which report that nectar foragers in small colonies gathered sugar water from feeders at nearly the maximum rate of energy gain in *autumn,* but not in *summer.* Perhaps nectar foraging honey bees usually operate with a goal of maximizing energetic efficiency but will switch to one of maximizing the rate of energy gain whenever their colony faces a severe threat of starvation (see also Section 7.3). Clearly, the question of what criterion a bee uses to assess the profitability of a nectar source, under various colony conditions, demands further study.

5.6. The Relationship between Nectar-Source Profitability and Waggle Dance Duration

To report on the profitability of a nectar source, a bee first registers the *stimulus* of "nectar-source profitability" (by integrating information about numerous variables of a flower patch) and then converts this into the *response* of "number of waggle runs." I will now consider several features of this stimulus-response relation that are important for understanding how a honey bee colony acquires information about the nectar sources outside the hive. All the findings discussed here are based on one set of experimental procedures: bees from an observation hive were trained to forage from a feeder whose profitability could be adjusted precisely, the dances of these bees were videorecorded at different settings of feeder profitability, and the videorecords were analyzed to determine the mean number of waggle runs per dance at each setting of profitability. In some experiments, it was essential to measure also the energy gained, energy expended, and time spent per foraging trip; so the data needed to calculate these variables were also gathered (details in Seeley 1994).

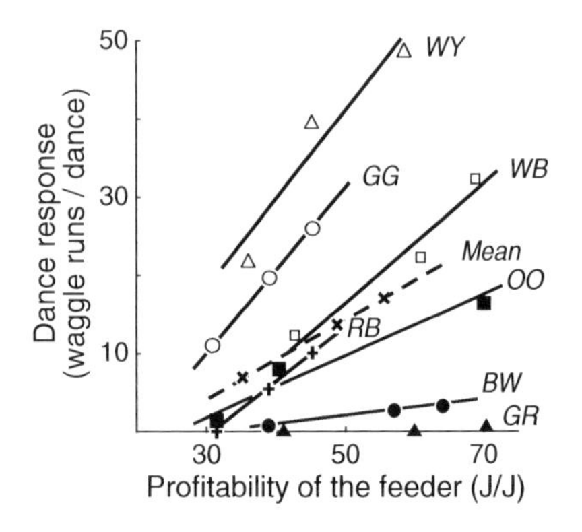

Figure 5.10 Stimulus-response functions for 7 bees, each identified by two initials, reporting on a sucrose solution feeder of variable profitability. Each bee made approximately 12 trips to the feeder at each of three different settings of profitability and at the end of each trip produced a dance response, which was videorecorded. After Seeley 1994.

5.6.1. The Linearity of the Stimulus-Response Function

Individual bees show a linear grading of their dance response as a function of nectar-source profitability. This became evident when a feeder was established 350 m from the observation hive, and was successively loaded with a sucrose solution of 1.50, 2.00, or 2.50 mol/L for 90 min each. Data gathered at each setting for 7 bees revealed the pattern shown in Figure 5.10. Different bees experienced different profitabilities for the same sucrose solution (because they differed in the amount of solution loaded, in body weight, and in foraging tempo) and danced to different extents, even for the same general level of profitability. Statistical analysis indicates significant variation in the slope of regression lines among the 7 bees; however, all the bees showed a clearly linear relation between profitability and dance response.

To understand the rationale behind this linearity, it is useful to consider the magnitude of variation for both suprathreshold stimuli and dance responses. As noted earlier, the dance responses for natural food sources range from 1 to about 100 waggle runs (Figure 5.8). One can estimate the range of stimuli to which bees give a dance response using the regression lines for the Figure 5.10 data to calculate for each bee her threshold and maximum stimulus values, that is, the levels of nectar-source profitability that will elicit 1 and 100 waggle runs. These calculations reveal a stimulus range (maximum/threshold) of 5–20, hence on the order of 10. Given that the range of stimuli that elicit responses (approximately 10) is far smaller than the range of responses (approximately 100), it is clear that a simple, linear stimulus-response function will allow a bee to report on the full range of suprathreshold stimuli and at the same time have good resolution of low-level stimuli. In contrast, if the stimulus range were far greater than the response range, a bee would need to make her dance response a logarithmic function of stimulus level to achieve the same ends.

This line of reasoning explains why a linear function is sufficient for the bees, but it leaves unanswered the puzzle of why a linear function is best for them. For example, why do they not grade their dancing nonlinearly such that only extremely profitable sources are advertised strongly? We will see in Section 5.7 that bees sometimes do restrict their reports to highly profitable food sources, but only when nectar sources are plentiful and hence the bees can be highly selective and can remain highly busy. We will see, too, that when they limit their advertising to extremely profitable sources they do so by raising the response (dance) threshold, not by adopting a nonlinear response function. Thus the ability to adjust the response threshold may have eliminated most of the benefits of a nonlinear response function.

5.6.2. NO OR SLOW ADAPTATION IN THE BEES' DANCE RESPONSE

An important property of any sensory unit is its rate of adaptation, or decrease in response to a constant stimulus, because this rate determines the kind of information the unit reports. A rapidly adapting ("phasic") unit provides mainly information about the *changes* in a stimulus, while a slowly adapting ("tonic") unit transmits information about the *level* of a stimulus (Young 1989). Following this reasoning, one would expect a dancing bee to show exceedingly slow (or no) adaptation, because a forager should report to her nestmates the level of profitability of her nectar source rather than just changes in its profitability. Presumably it is the current level of profitability, not how much the profitability has changed, that primarily determines the attractiveness of a nectar source. For example, if a flower patch provides constant rich foraging throughout a day, then all else being equal, a forager from this patch should perform a long recruitment dance every time she returns to the hive.

This is basically what one sees. On 16 June 1992, 10 bees were allowed to forage from a feeder located 350 m from their hive. They were initially exposed to a weak stimulus for 1 hr, then to a strong stimulus for 2 hr, and finally to the weak stimulus for 1 hr. Figure 5.11 illustrates the results for the 6 strongest dancers. There is no sign of adaptation to the strong stimulus. This is seen most clearly in the summed response of the 10 bees, where one sees actually a slight *increase* in the collective dance response over the 2-hr period of strong stimulus. This rise is probably not characteristic of responses to prolonged stimuli, but is instead simply a reflection of the air becoming warmer, hence the foraging conditions improving, over the course of the experiment. Alternatively, the bees might have been gradually raising their assessment of the feeder because they had experienced a sustained, high yield from this food source. A repetition of this experiment yielded essentially identical results (Seeley 1994).

Although bees show no detectable adaption in their dance response, and so do not need to experience a steadily improving nectar source to maintain a strong dance response, it should not be concluded that bees—in deciding how long to dance—are unresponsive to steady improvements (or deteriorations) in a nectar source. A recent study by Raveret Richter and Waddington (1993) reports that bees performing round dances for a feeder varied certain parameters of their dances—including rate of direction reversal, circuit rate, and speed—in relation to their past experiences at the feeder, not just their immediate experience there. For example, foragers performed live-

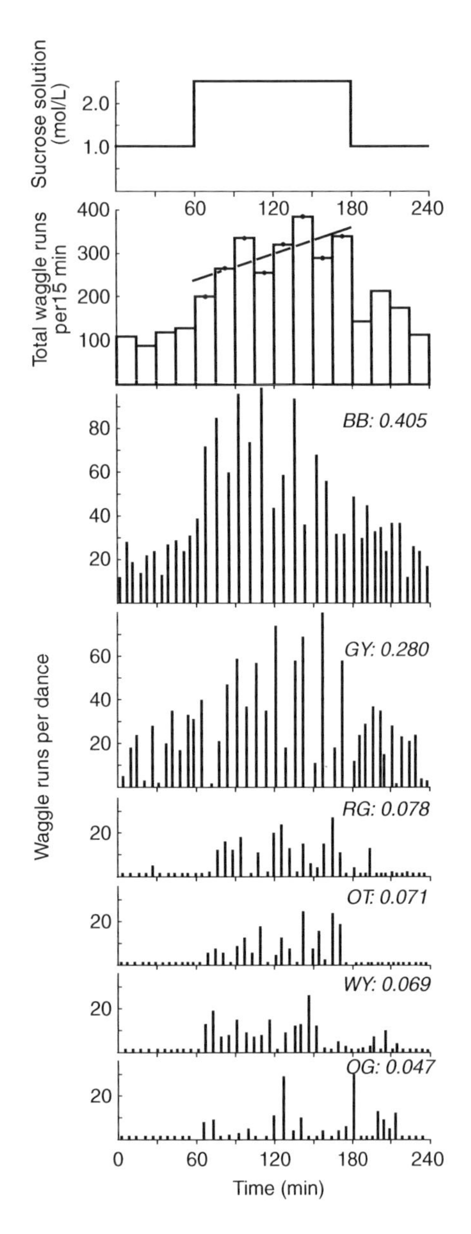

Figure 5.11 Durations of dances produced by 6 bees responding first to a weak and then to a strong stimulus, followed by another weak stimulus. The stimulus intensity was varied by changing the concentration of the sucrose solution in a feeder according to the schedule shown in the top panel, and each bee's dance responses were measured from videorecords. The summed response of the 6 bees (plus 4 more, their responses not shown) shows no sign of adaptation, even after 2 hr of strong stimulation. Numbers on the right denote the proportional contribution of each bee to the total waggle runs produced by all 10 bees during the 2-hr period of strong stimulation. After Seeley 1994.

lier dances for a given concentrationn of sucrose solution if the feeder had previously offered a lower concentration than if it had offered a higher concentration. It will be important to test whether bees performing waggle dances likewise vary the number the waggle runs not simply as a function of the profitability experienced on the current foraging trip, but also in relation to the profitability experienced on recent foraging trips. The most exciting food source, to a bee, may well be one that not only is highly profitable at present but also promises to be even better in the future.

5.6.3. STRONG VARIATION AMONG BEES IN THE DANCE RESPONSE

Figures 5.10 and 5.11 both illustrate a third striking feature of bees as sensory units, namely tremendous variation among bees in the dance response to a given stimulus of food-source profitability. For example, an analysis of variance performed on the dance data represented in Figure 5.11 for the 2 hr of strong stimulation reveals highly significant ($P < 0.001$) heterogeneity among the mean dance durations of the bees, with 74% of the total variation in dance duration due to variation among bees and only 26% due to variation within bees. Much of this variation in response reflects differences among bees in their dance thresholds. Consider the 6 bees represented in Figure 5.11. For 2 of them (BB and GY), even the 1.00-mol/L feeder provided a stimulus above threshold, while for 4 others only the 2.50-mol/L feeder gave a suprathreshold stimulus. Likewise, extrapolations of the stimulus-response lines in Figure 5.10 indicate marked differences among bees in the level of feeder profitability that was the threshold for dancing.

What could be the functional significance, if any, of this large variation in the dance response threshold? Assume that the variation among individuals is partly a result of their genetic variation, which largely reflects the queen's mating with multiple males (Section 1.2). One possibility is that the variation in dance threshold actually does not enhance a colony's foraging performance, but exists simply as a by-product of the queen's mating multiply to secure lots of genetic variation in her colony to cope with diseases (Sherman, Seeley, and Reeve 1988). But another possibility is that the variation among individuals is adaptive because it gives a colony a broad dynamic range in responding to nectar sources, that is, an ability to respond in a graded fashion to a broad range of stimuli. If bees lacked variation in their response thresholds, the dynamic range of the colony would be only as wide as the dynamic range of each bee.[2] This hypothesis can be tested by seeing whether the variation among individuals in dance thresholds does indeed have a genetic basis, and whether colonies headed by singly mated and multiply mated queens differ substantially in the range of nectar sources over which they can produce a graded recruitment response. If both prove true, then the variability in the dance threshold supports the idea that a honey bee queen mates multiply in order to improve her colony's ability to cope with a wide range of environmental conditions (Crozier and Page 1985).

5.7. The Adaptive Tuning of Dance Thresholds

The level of nectar-source profitability that is a bee's threshold for dancing is not a fixed trait, but is instead flexibly adjusted in relation to the colony's foraging status. Specifically, nectar foragers operate with lower dance thresholds when their colony's nectar intake is low than when it is high. This pattern of shifting the dance threshold is highly beneficial for the colony, for it helps a colony continue to acquire energy when nectar sources become sparse, since in this situation both low- and high-yield sources are advertised on the dance floor. It also helps a colony achieve high efficiency in its energy acquisition when nectar sources become abundant, since at these times only high-yield sources are advertised on the dance floor. As we shall see later in this chapter, only those nectar sources that have been well advertised by a colony's foragers are heavily exploited by the colony.

A clear picture of the mechanisms controlling the dance thresholds of nectar foragers turns out to be essential to understanding how a colony deploys its foragers among nectar sources. Unfortunately, these mechanisms have been the subject of considerable misunderstanding in the past, hence the need to examine them in some detail.

5.7.1. THE BASIC THRESHOLD-SHIFT PHENOMENON

Karl von Frisch realized during his earliest studies of the bees that they can change their dance thresholds, for he noticed that the will-

2. This assumes that the dynamic range of a single bee is relatively narrow, which will be the case if there is an upper limit on the dance response of each bee. Although in principle there is no upper limit to the duration of a bee's dance signal, in practice there does seem to be an upper limit of about 100 waggle runs (Section 5.4). Possibly dances longer than this are generally not performed because they require an excessively long diversion from food collection.

ingness of bees to exploit an artificial food source is strongly influenced by the ambient foraging conditions. "Success [in establishing an artificial feeding place] is threatened in two directions: in spring the natural honeyflow is so good that, even with concentrated sugar solution to which honey has been added, it is hard to get bees to come to the food dish. They prefer the field with their blossoming flowers. Often all we could do was to put off the beginning of experimentation for several weeks. But in late summer, after the natural honeyflow has ceased, the colonies are so eager for anything sweet that strangers from other hives may become a pest" (von Frisch 1967, p. 18; based on observations that he made as early as 1920). This pattern of seasonal change in the bees' choosiness about nectar sources was documented more precisely by Lindauer (1948), who determined for each day during two summers the lowest concentration of a sugar solution in a standard feeder that would elicit dances. He found that from April to early July, a period with many productive nectar sources, he had to offer a 1–2-mol/L solution, but in late July and August, when nectar sources were scarce, he generally needed to offer only a 1/8-mol/L solution to trigger dancing.

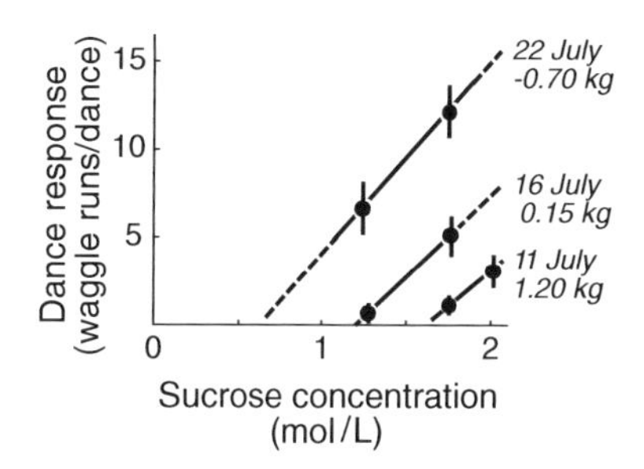

Figure 5.12 Shifting of the dance threshold in relation to foraging conditions. The bees' dance response as a function of sucrose concentration was determined for 3 days in July with very different foraging conditions. Numbers on the right denote the daily weight change of a beehive mounted on scales, a measure of the availability of nectar from natural sources. The weight *gain* of 1.2 kg on 11 July indicates moderately abundant nectar, while the weight *loss* of 0.7 kg on 22 July indicates a severe dearth of nectar. After Seeley 1994.

A graphic illustration of this tuning of the dance threshold in relation to the foraging conditions is provided in Figure 5.12. This figure is based on an experiment performed during July 1991 at the Cranberry Lake Biological Station (described in Section 5.10). Two feeding places were established north and south of an observation hive, each 400 m from the hive. The two feeders contained sucrose solutions of different concentrations and so elicited dances with different mean durations. My assistants and I measured the bees' dances for each feeder and thereby determined a sucrose concentration-dance duration function for each day of data collection (details in Seeley and Towne 1992). This was done on 11 days over a 24-day period, during which time the ambient foraging conditions changed greatly, as indicated by the daily weight changes of a nearby colony on scales. When the experiment began in early July, there was a surprisingly strong nectar flow from raspberry plants (*Rubus* spp.) and the scale colony *gained* about 1.5 kg each day; but by the end of the experiment in late July, the raspberry bloom had passed so that there was virtually no nectar available and the scale colony *lost* about 0.5 kg each day. Figure 5.12 shows the dance functions for 3 days with markedly different levels of nectar availability, ranging from high (11 July) to low (22 July). Clearly, as the ambient foraging opportunities dwindled, the dance threshold concentration of sucrose solution at the feeders dropped dramatically, from 1.7 mol/L to approximately 0.5 mol/L.

5.7.2. THE CAUSE OF SHIFTS IN THE DANCE THRESHOLD

In the summer of 1985, I performed an experiment designed to determine whether the shifts in dance threshold just described are triggered by changes in a colony's nectar influx or changes in some other variable closely correlated with nectar influx. For example, one could argue that bees are more fastidious dancers during a nectar flow because they sense that the nectar being gathered then is higher in *quality,* not just greater in *quantity.* Indeed, until recently, most bee researchers (including me) have explained the dance threshold shift as a response to a change in the quality, not the quantity, of nectar brought into the hive (Lindauer 1971; Rinderer 1983; Seeley 1985; Gould and Gould 1988). It was generally believed that nectar of especially high sugar concentration is brought into the hive during a nectar flow. Coupled with this was the belief that food storers preferentially unload foragers bearing nectar of high sugar concentration and that each nectar forager raises her dance threshold—hence lowers her tendency to dance—when she must search a long time before locating a food-storer bee willing to unload her. Assuming all this is correct, one can easily account for the observed pattern of reduced dancing for medium-level sucrose solutions during a nectar flow. For example, one would say that when a nectar flow starts, foragers bringing home medium-concentration nectar will have to search longer to find food-storer bees willing to accept their nectar (because the food storers now mainly unload foragers with high-concentration nectar), and thus these foragers will raise their dance thresholds and so will reduce their dancing. This set of beliefs was based on certain experimental results from the 1950s (Lindauer 1954; Boch 1956). But as we shall see shortly, although these earlier experiments did demonstrate that nectar foragers raise their dance thresholds when they must search longer to locate a food-storer bee, these experiments did not reveal what causes nectar foragers to search longer and hence what ultimately causes them to change their dance thresholds.

To distinguish between the two hypotheses for what ultimately causes nectar foragers to change their dance thresholds—change in nectar quantity versus change in nectar quality—I needed to moni-

Figure 5.13 View of the Cranberry Lake region in northern New York state. It lies within the vast (20,000 km^2) Adirondack Park, and so by law most of the land around Cranberry Lake must be kept "forever wild." The habitat—dense hardwood forests interspersed with lakes and bogs—provides extremely meager forage for bees. The Cranberry Lake Biological Station is situated on a point on the far shore. Photograph by T. D. Seeley.

tor the dances of bees foraging at a standard feeder while I varied the quantity, but not the quality, of the nectar flowing into their colony. If the bees danced more strongly whenever the colony's nectar influx was lowered this would demonstrate that change in nectar influx alone is sufficient to cause nectar foragers to shift their dance thresholds. I found an ideal place for this experiment at the Cranberry Lake Biological Station. The station, accessible only by a 10-km boat ride across the lake, is surrounded for nearly 20 km in all directions by pristine forests, bogs, and the open waters of Cranberry Lake (Figure 5.13). This environment provides exceedingly few natural sources of nectar. Indeed, colonies introduced here and placed on scales steadily lose weight at a rate of approximately 0.5 kg a day, except during the period of raspberry (*Rubus* spp.) blooms in late June or early July (Seeley 1989a). This dearth of natural nectar sources means that any experimental colony that I bring here must gather virtually all its "nectar" from my sugar water feeders, a situation which enables me

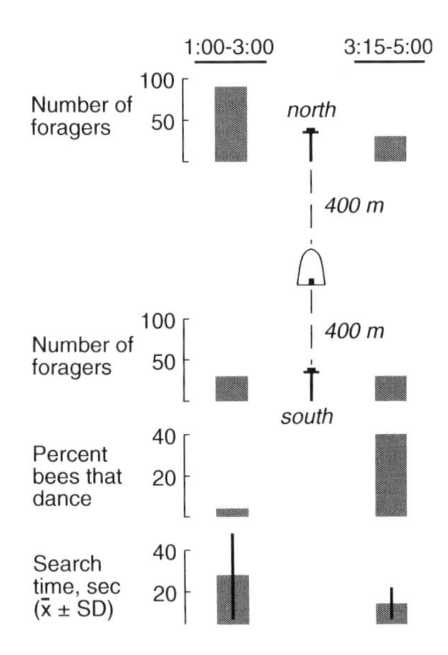

Figure 5.14 Experimental array and results of the test for what causes nectar foragers to shift their dance thresholds. Based on data in table 1 of Seeley 1986.

to tightly control the quantity and quality of the nectar collected by the colony under study.

Figure 5.14 depicts the layout and results of my experiment (details in Seeley 1986). On the morning of 18 June 1985, my assistants and I established two feeders 400 m north and south of the observation hive and filled both with a 1.50 mol/L sucrose solution. In the afternoon from 1:00 to 3:00, we allowed 90 bees and 30 bees to forage at the north and south feeders, respectively, and recorded several variables of the in-hive behavior of the bees visiting the south feeder, especially the intensity of their dancing. (All recruits to either feeder were captured to create stable conditions at the feeders throughout each period of data collection.) At 3:00, we quickly captured 60 of the bees visiting the north feeder, thereby halving the colony's rate of nectar intake, and repeated the in-hive observations on the south-feeder bees. Immediately the bees from the south intensified their dancing, with the proportion of bees performing a dance skyrocketing from 3% to 40%. Two repetitions of this experiment on 20 and 21 June yielded the same striking pattern: greatly strengthened dancing by bees from the south whenever the nectar flow from the north declined. It is important to note that in none of these three trials was there any change in the quality of the food brought into the hive; only the quantity changed. Thus it is unequivocally clear that nectar foragers do shift their dance thresholds in response to changes in the colony's nectar influx.

This finding leaves open the question of whether or not nectar foragers also shift their dance thresholds in response to changes in their colony's nectar quality (mean sugar concentration). It is now virtually certain that they do not, for there is no evidence that nectar foragers can even acquire information about changes in the mean sugar concentration of the nectar gathered by their colony. It had been supposed that a forager obtains this information by noting changes in how long she must search to find a food storer willing to accept her nectar (Lindauer 1954, Boch 1956). But, as is illustrated in Figure 5.14 (and will be explained in greater detail below, Figure 5.18), this search time reflects the quantity, not the quality, of the nectar flowing into a hive. Moreover, there is no evidence that a colony's nectar quality changes in a consistent way between times of rich and poor forage, hence between times of high and low dance thresholds. In Figure 5.15 we see, for example, that at the end of the dandelion *(Taraxacum officinale)* bloom in the Yale Forest in May 1985, the quantity of nectar collected daily by a colony dropped swiftly, and the dance thresholds of its foragers dropped dramatically (as indicated by a surge in recruitment to a standard feeder), but the quality of this colony's nectar did not decline at all.

5.7.3. HOW FORAGERS ACQUIRE INFORMATION ABOUT THEIR COLONY'S NECTAR INFLUX

Martin Lindauer (1948) first suggested that foragers might be informed of their colony's rate of nectar collection by how speedily they can unload their nectar upon return to the hive. On 13 April 1946, he set out six dishes containing 2.0-mol/L sucrose solution around an observation hive and let the number of bees visiting them increase through recruitment. Half an hour later, when about 40 bees were visiting the dishes, he observed that each forager would finish unloading her nectar within 20–50 sec of entering the hive. But 2 hr after the start of the experiment, when more than 100 bees were visiting the

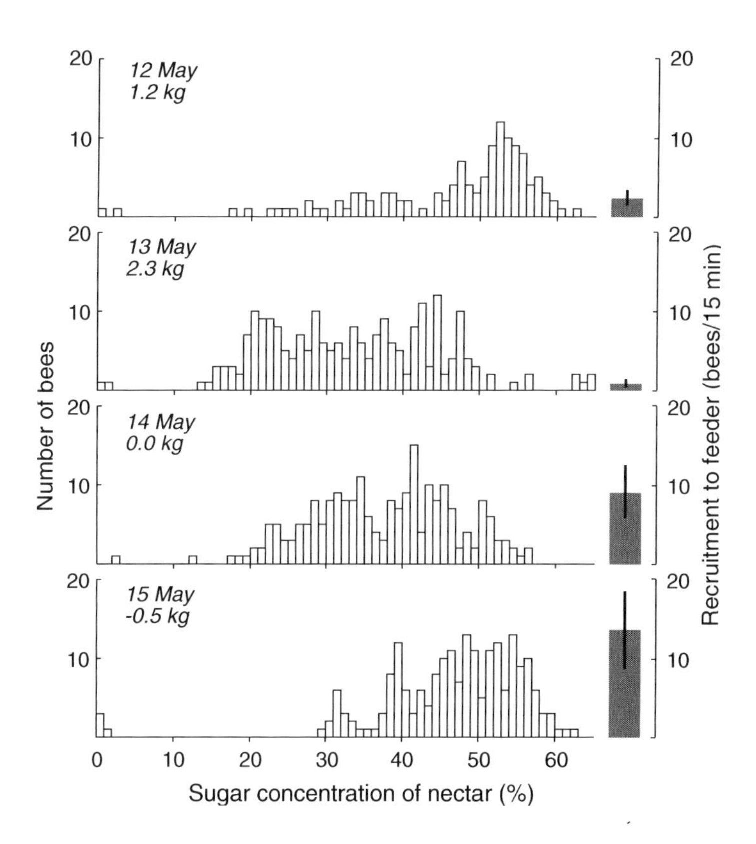

Figure 5.15 Daily patterns of nectar quality during a transition from abundant to sparse forage. Randomly selected foragers were captured upon arrival at their hive and the sugar concentration of each bee's nectar load was measured. The colony under study consisted of approximately 20,000 bees in a full-size hive. Data were collected at the end of the dandelion bloom in May 1985. The number below each date denotes the daily weight gain (or loss) of the colony, which reflects mainly how much nectar it gathered that day; the weight loss of 0.5 kg on 15 May indicates that the colony gathered essentially no nectar this day. The mean dance threshold of the colony's foragers was indirectly assayed each day by letting 30 bees visit a 2.0-mol/L sucrose solution feeder 500 m from the hive, capturing the bees they recruited to the feeder, and recording the rate of recruit captures. On all 4 days the weather was warm and sunny, hence ideal for foraging. The mean recruitment rates for 12–13 May and 14–15 May are significantly different; $P < 0.005$. Based on data partially published in Seeley 1986.

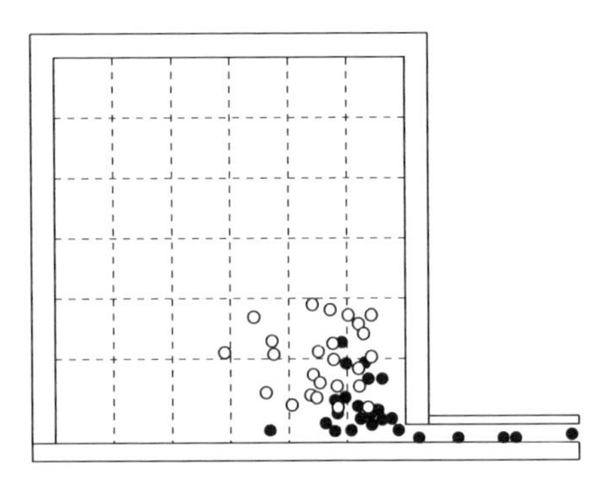

Figure 5.16 Spatial distribution of nectar unloadings in relation to a colony's rate of nectar collection. On 16 July 1992, a colony in an observation hive at the Cranberry Lake Biological Station was provided with two feeding places, one 550 m and the other 250 m from the hive. In the afternoon, at 3:45, 20 and 40 bees were foraging at the far and near feeders, respectively, and the total rate of nectar forager arrivals at the hive was 8.2 bees/min. A total of 25 unloading sites for bees from the far feeder were recorded, as shown, in the period 3:45–4:00. Immediately thereafter, the near feeder was shut off, causing the forager arrival rate to drop to 2.7 bees/min. Then another 25 unloading sites for bees from the far feeder were recorded, in the period 4:05–4:20. Note that when the forager influx dropped, the unloading sites shifted toward the hive entrance (the two distributions differ significantly in the mean distance from the entrance opening; $P < 0.001$). Based on unpublished data of T. D. Seeley.

feeders, he observed that each forager required 45–90 sec to unload. He also reported that several other aspects of a forager's unloading experience likewise covary with the colony's rate of nectar collection, including the distance inside the hive of the unloading, the number of food storers simultaneously unloading each forager, and the seeming eagerness of the food storers to obtain a forager's nectar.

My own observations agree fully with those of Lindauer. As already shown (Figure 5.14), the time spent searching for a food-storer bee is correlated with the colony's rate of nectar intake. In addition, as Figure 5.16 indicates, the location of unloading is strongly influenced by the nectar influx. Initially, when the forager return rate was relatively high, each bee had to crawl 10 cm or more into the hive before contacting a food-storer bee, but once the influx of foragers was lowered, each forager found a willing unloader just inside the hive, if not out in the entrance tunnel. Moreover, the maximum number of bees simultaneously unloading each forager differed markedly between the two foraging conditions depicted in Figure 5.16: 1.2 ± 0.3 (range 1–2 bees) versus 2.5 ± 0.6 (range 1–4 bees) simultaneous unloaders per forager, for the periods of high and low forager influx, respectively.

A returning forager could use any one, or some combination, of these variables of the unloading experience as an indicator of her colony's success in nectar collection. There are, however, at least two other possible indicators. The level of floral odor inside the hive might rise with increasing nectar influx, and the bees might monitor this as a cue of their colony's foraging status. Beekeepers can sometimes tell when their colonies are gathering copious nectar by a strong floral aroma wafting from their hives. Another possibility is that the returning foragers might monitor the level of bee traffic in the hive entrance, which would provide a rather direct indication of the colony's foraging rate. The unloading-experience hypothesis is resolved from the nectar-odor and bee-traffic hypotheses by an experiment in which I increased the search times experienced by nectar foragers (and altered the associated variables of unloading location, number of unloaders, and so forth), but I did not increase the colony's rate of nectar collection and hence did not increase the level of nectar odor or bee traffic (described in detail in Seeley 1989a). This trick was accomplished by *removing most of the colony's food-storer bees.* The unloading-experience hypothesis predicts that when the food storers are removed and the foragers experience increased search times, the foragers will decrease their recruitment to a standard feeder, because it should seem to the foragers that the colony has markedly boosted its nectar collection. In contrast, the nectar-odor hypothesis and the bee-traffic hypothesis both predict that under the same circumstances the colony's foragers will not decrease their recruitment because in this experiment there will be no increase in nectar odor and no increase in bee traffic, and hence it should seem to the foragers that the colony has not boosted it nectar collection. So what do the nectar foragers actually do?

At the Cranberry Lake Biological Station, I set up two observation hives side by side and on 13 July 1989 established one feeding station for each colony 420 m south of the hives, with the two feeders separated by 300 m (Figure 5.17). Each colony's feeder was visited by 30 foragers and was filled with a rich, 2.0-mol/L sucrose solution to elicit dancing. All recruits to the feeders were captured.

On 14 July, a hot, sunny day, the rate of recruitment to each feeder was recorded and for one colony (the experimental) the behavior of its foragers upon return to the hive was monitored. Both colonies' foragers recruited numerous bees to their feeder, and, not surprisingly, the observations of the experimental colony's foragers revealed that they searched only briefly to find an unloader (11.1 ± 4.6 sec) and danced strongly (73% danced, producing 4.8 ± 5.1 waggle runs/dance). During the next 2 days, most of the food-storer bees were removed from the experimental colony by daubing paint on the back of each bee seen unloading a forager from the feeder and then plucking all the painted bees off the combs at the end of each day. This procedure removed some 20% of the experimental colony's bees. On the morning of 17 July, the measurements of recruitment, search times, and dancing were resumed. The effects of removing most of the food storers from the experimental colony were striking. This colony's foragers searched nearly twice as long as before to locate an unloader (21.4 ± 13.6 sec), and, most important, they danced weakly (only 7% danced, producing just 0.3 ± 1.2 waggle runs/dance) and recruited few bees to the feeder (Figure 5.17). This drop in dancing and recruitment could not be attributed to poor weather, for this day was again hot and sunny; moreover, the recruitment to the control colony's feeder had not declined at all. Evidently the foragers in the experimental colony had noted the change in unloading experience and had accordingly reduced their dancing, but had ignored the fact that neither the forage-odor level nor the bee-traffic level had in-

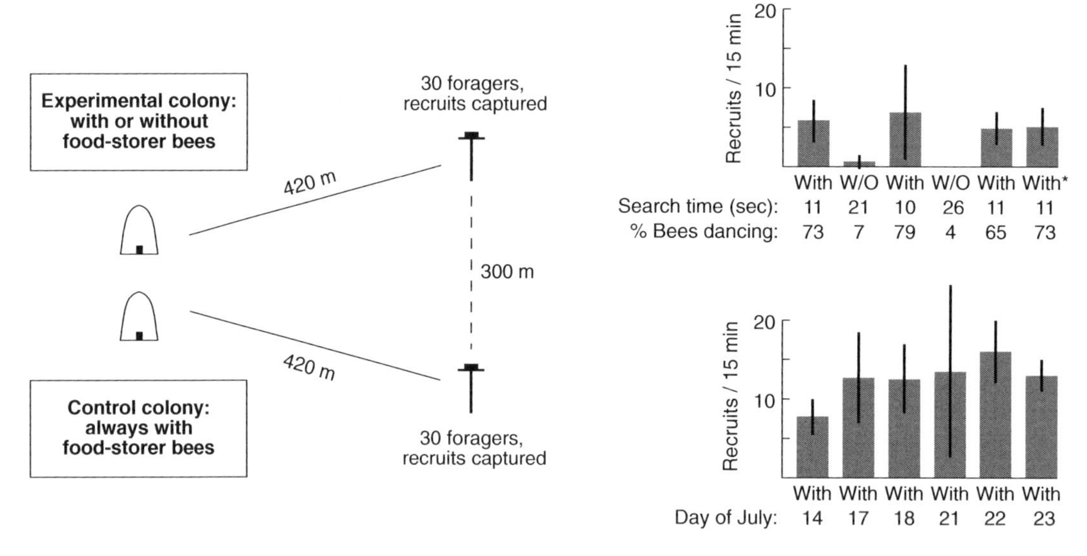

Figure 5.17 Experimental array and results of the test of how nectar foragers acquire information about their colony's rate of nectar collection. The experimental colony was with or without its food-storer bees whereas the control colony always had these bees. The "With*" for the experimental colony on 23 July denotes the condition that this colony still possessed its food-storer bees, but that a different 20% of the colony's population had been removed for this day. After Seeley 1989a.

creased. These results were confirmed when the experiment was repeated over the next several days. Also, on 22 and 23 July, an experiment was performed to control for any general effects of removing some 20% of the bees from a colony. At the end of 22 July, 20% of the experimental colony's bees (but not any of its labeled food-storer bees) were plucked off the combs, but observations on 23 July indicated no change in the unloading experience of this colony's foragers and no change in their dancing and recruitment (Figure 5.17).

Thus it is clear that some variable feature of the unloading process informs a nectar forager of her colony's rate of nectar collection. To the human observer, the feature of the unloading process that appears to be most closely and strikingly correlated with the variation in nectar influx is the time that a newly returned forager spends searching for someone to unload her. Furthermore, the time spent searching for an unloader is probably a variable that is easily measured by bees. In the next section, I will consider why search time reliably varies with a colony's nectar influx, but it should be remembered that other variables of the unloading experience may also contribute to a forager's sensitivity to her colony's nectar influx.

5.7.4. WHY SEARCH TIME RELIABLY VARIES WITH A COLONY'S NECTAR INFLUX

The reason search time varies reliably with nectar influx can be established with the aid of a simple probability model that relates the two variables of interest—the rate at which nectar foragers arrive at a hive and the average time that a nectar forager spends searching for a food-storer bee—under the condition that there is no change in the colony's capacity for nectar processing (see Seeley and Tovey 1994). (As will be shown in Section 6.3, a colony can adjust its nectar processing capacity, but does so only if there is a large discrepancy between it and the nectar collecting rate.) The starting assumption for this model is that when a nectar forager returns to the hive and searchs for a food storer among the bees standing in the unloading area (Figure 5.16), she searchs at random, sampling one bee at a time to determine whether or not it is a food-storer bee ready to accept her nectar load. I also assume that the food-storer bees are distributed at random among the other bees in the unloading area and that their total number in the colony does not change. The nectar forager's sampling of bees in the unloading area is therefore analogous to an individual's sampling (with replacement) of marbles in an urn, with the marbles of two colors, one corresponding to food-storer bees and the other corresponding to non-food-storer bees (other foragers, guards, and so on). The parameters needed to model this sampling process are defined in Table 5.2.

Table 5.2. Definitions of the variables used in the urn model. After Seeley and Tovey 1994.

Variable	Definition (units)
C	Average time for a food-storer bee's work cycle, which consists of loading with nectar, leaving the unloading area, storing nectar in honeycombs or distributing it to nestmates, and returning to the unloading area (min)
F	Total number of food-storer bees (bees)
N	Number of bees in the unloading area (bees)
R	Rate of nectar forager arrival at the hive (bees/min)
S	Average search time, that is, the time a forager spends searching for a food-storer bee (it does not include the time spent unloading nectar to the food-storer bee) (min)
T	Average time for a returning forager to sample one bee in the unloading area and determine whether or not it is a food-storer bee willing to take her nectar (min)

At equilibrium, there will be CR food-storer bees processing nectar, hence outside the unloading area and not available to unload foragers, and $F - CR$ food-storer bees within the unloading area, ready to unload foragers. The probability of a "successful" sampling of a bee in the unloading area by a nectar forager is therefore simply $(F - CR)/N$. This assumes that N is so large that we may neglect the change from N to $(N + HR - CR)$, where H is the average time spent in the unloading area by a forager upon return to the hive.

By the properties of the geometric distribution, the expected number of samples until (and including) a successful sampling is $1/p$, where p is the probability of a successful sampling, which in this case is $(F - CR)/N$. Hence the expected number of samples is $N/(F - CR)$. Therefore

$$S = T\frac{N}{(F - CR)} \tag{5.1}$$

Rearranging this, we see that

$$\frac{1}{S} = \frac{F}{TN} - \frac{C}{TN}R \tag{5.2}$$

To express this in words, this model predicts that the reciprocal of the average search time $(1/S)$ is a negative linear function of the rate of arrival of nectar foragers (R).

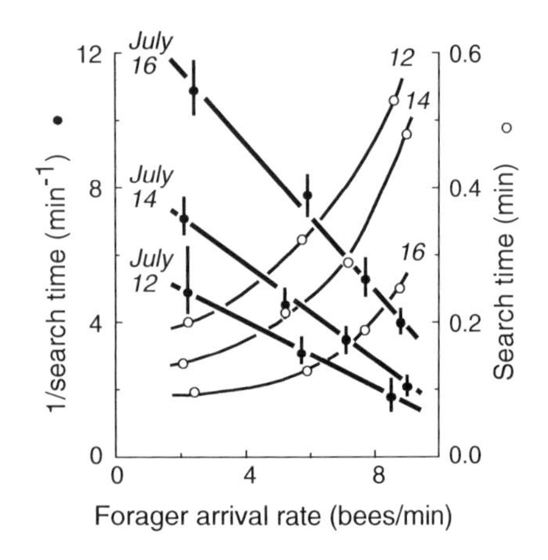

Figure 5.18 The mean (± SE) time that a nectar-forager bee spends searching for a food-storer bee to unload her nectar upon return to the hive ("search time," *open circles*), and the reciprocal of this search time *(filled circles)*, both in relation to the arrival rate of nectar foragers at the hive (July 1991). After Seeley and Tovey 1994.

This prediction was checked, and thus the model was tested, by conducting an experiment in which Craig Tovey and I could fix a colony's rate of forager arrival at different levels and measure the average search time of its foragers at each level (Seeley and Tovey 1994). The setup was essentially identical to the one depicted in Figure 5.14, in which a colony at the Cranberry Lake Biological Station was provided with two sucrose solution feeders in opposite directions, the colony's foraging rate was adjusted by changing the number of bees visiting one of the feeders, and behavioral observations were made on the bees visiting the other feeder. Three trials of the experiment were performed between 12 and 16 July 1991. All three revealed the same striking pattern of a negative linear relationship between forager arrival rate and the reciprocal of mean search time (Figure 5.18), thereby confirming the critical prediction of the model. Evidently this simple model accurately describes the dynamics between foragers and food storers, and so clarifies why search time is a good indicator of a colony's rate of nectar collection. In short, we can visualize the search for a food storer by a forager as a random sampling process in which the per-trial probability of success decreases as the arrival rate of foragers increases. Accordingly, there is an automatic increase in the expected number of samples that a forager must make, and hence by the unbreakable rules of probability, the mean search time grows longer as the colony's rate of nectar collection increases.

One unanticipated feature of the results is the different arrival rate–search time relationships recorded on different days. Across the 5-day period of data collection, the mean search time associated with any given arrival rate showed a steady and dramatic decline. This downward shifting of the search time curves most probably arose because the colony was gradually increasing its nectar processing capacity over the 5-day period. I will consider this further in the next chapter, where I address the issue of how a colony keeps its nectar processing capacity in balance with its rate of nectar collection.

5.8. How a Forager Determines the Profitability of a Nectar Source

To perform a dance of proper duration, and thereby present an accurate report on the profitability of her food source, a bee must first know the profitability of her source. How does a forager acquire this bit of knowledge? More specifically, how does a nectar forager know whether the flower patch she has just visited represents a food source of low, medium, or high quality? Of the three hypotheses presented in Figure 5.19, the first and simplest—H1: she compares her current patch with previously experienced patches—is evidently not the right one. This hypothesis predicts that a novice forager will not perform dances, since a forager just starting out cannot have experienced a wide range of food sources, and hence she cannot be adequately calibrated to make correct evaluations and so should not be able to determine whether a food source deserves to be advertised with a dance. Lindauer (1952) observed, however, that fully 47 out of 91 novice foragers performed a dance upon completing their first foraging trip. Assuming that these bees were producing dances appropriately, we see that foragers with minimal foraging experience know whether or not a given flower patch merits reporting in the hive. How is this possible?

H1. **Direct comparisons:** personal global knowledge
H2. **Indirect comparisons:** global knowledge by supervisor (S)
H3. **No comparisons:** local knowledge + shared scale of profitability

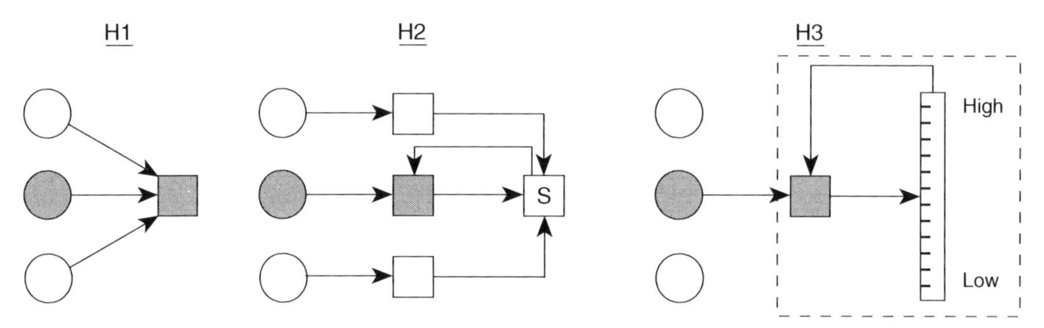

Figure 5.19 Three hypotheses for how a nectar forager can know whether the profitability of her current flower patch is low, high, or some point in between. In the schematic drawings, a circle represents a flower patch, an arrow denotes a flow of information about profitability, and a square represents a bee (in each case, the *shaded circle and square* denote the focal flower patch and focal forager). The three hypotheses are distinguished by the amount of information possessed by the focal forager (local or global knowledge) and by the locus of the assessment process (forager or food-storer bee). In H1, the forager possesses broad knowledge of flower patch quality and so can evaluate her current patch in comparison with other patches. In H2, a supervisor (food-storer bee) acquires broad knowledge about patch quality from multiple foragers, makes comparisons, and provides feedback to each forager she unloads regarding the quality of her patch. In H3, the forager uses only knowledge of her current patch and is able to assess its profitability through an internal scale of quality (*dashed lines* represent the boundaries of the forager's body). Presumably this internal scale is built into the bee's nervous system during development and is similarly calibrated for different foragers.

Until recently, most biologists studying honey bees believed that this puzzle is solved by shifting the locus of the assessment process from the nectar foragers to the food storers (Figure 5.19, H2). According to this hypothesis, the food-storer bees sample the various nectar loads brought into the hive and preferentially accept those with the highest sugar concentration. The speed of unloading by the food storers supposedly informs each forager of the relative quality of her nectar. Hence it was believed that the food-storer bees function, in effect, as supervisors of the nectar foragers. Lindauer, for example, states, "day by day, throughout the year, the different groups of foragers are informed who among them has discovered the best nectar sources. Those foragers that have collected their food from low-quality food sources have difficulty unloading their food in the hive [hence they refrain from dancing], and this has the effect that the majority of the unemployed foragers are ultimately recruited to the foraging sites which at the time offer the most concentrated nectar" (Lindauer 1975, p. 30; my translation). This intriguing hypothesis grew out of Lindauer's (1954) pioneering study of the regulation of water collection in honey bee colonies, in which he observed that when a colony has an emergency need for water, the food storers seem to preferentially unload foragers returning with water or dilute nectar. He presumed that under normal conditions, when a colony needs mainly food rather than water, the food storers preferentially unload foragers returning with the most concentrated nectar because, all else being equal, this is the best food. This idea seemed to be beautifully confirmed two years later by the results of an experiment by Boch (1956). He provided a colony with two feeding places, A and B, both of which were initially filled with 3/8-mol/L sucrose solution and visited by 40 bees (Figure 5.20). At first, the bees visiting feeder B performed dances at a moderate rate, about 25 dances every 5 min. But when feeder A was reloaded with 2-mol/L sucrose solution, the bees from feeder B essentially ceased dancing. Boch (1956) explained this change in the behavior of the bees from feeder B as a result of the bees from feeder A returning home with superior food, thus drawing to themselves the food storers and so causing the bees from feeder B to experience slow unloadings in the hive. In short, Boch argued that the food-storer bees had informed the bees from feeder B that their forage had become second-rate when better food began to flow in from feeder A.

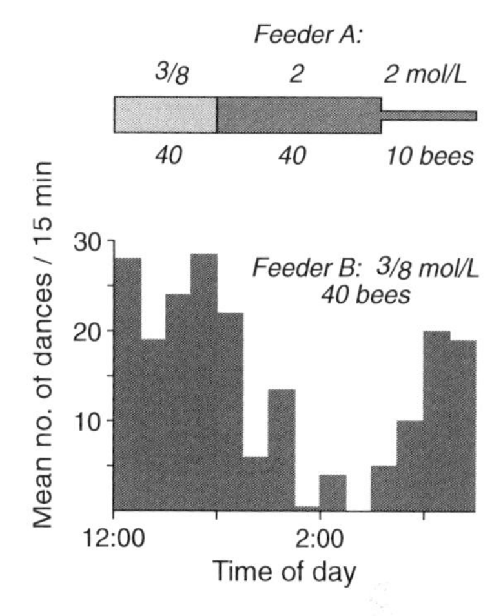

Figure 5.20 The influence of changes at one nectar source on the behavior of bees foraging at another nectar source. When the bees from feeder A began collecting a richer sugar solution, the bees from feeder B dramatically reduced their dancing. Evidently, this change in behavior was caused by an increase in the quantity of sugar solution flowing into the hive (because the bees from feeder A boosted their foraging tempo in response to the richer food), not by an increase in the quality of the incoming sugar solution. This is indicated by the resumption of dancing by the feeder-B bees when the number of feeder-A bees was trimmed from 40 to 10. Based on data in Boch 1956.

Although appealing when originally proposed, the idea that food storers inform foragers of the quality of their nectar sources by selectively unloading those with the sweetest nectar is now known to be mistaken. The evidence contradicting this hypothesis is both conceptual and empirical. The conceptual failing of the hypothesis is that it assumes that the quality of a nectar source can be judged simply from the sugar concentration of its nectar. This is surely wrong. Many other variables influence the energetic profitability of a nectar source, including distance from the hive, abundance of nectar, and the spacing of the flowers (Section 5.5). The correct hypothesis for how bees evaluate nectar sources must include mechanisms for factoring in all variables that influence the energy gains and costs of foraging.

I have already examined (Section 5.7) one of the principal pieces of empirical evidence contradicting the hypothesis that a forager is informed about nectar-source quality by how long she must search to find an unloader. This is the fact that a nectar forager's search time is a function of the quantity, not the quality, of the nectar flowing into the hive. Why, then, did the bees from feeder B in Boch's experiment cease dancing when the quality of the food at feeder A increased from 0.375 to 2.0 mol/L? The most likely explanation is that the rise in food quality at feeder A stimulated the 40 bees foraging there to work faster, thereby boosting the colony's rate of nectar influx and so lengthening the search times experienced by the feeder-B bees. This is not mere speculation. On 11 July 1987, I established a feeding place for 30 bees from an observation hive, and over the course of the day I observed their behavior at four levels of sugar concentration, from 0.5 to 2.0 mol/L. When the 0.5-mol/L solution was offered, the bees foraged slowly, spending 7.7 min on average per round trip to the feeder. But when the higher concentration solutions were offered, the bees foraged at a much higher tempo, needing on average only 4.9 min per round trip (data from table 1 in Seeley, Camazine, and Sneyd 1991). Similar effects of food-source quality on foraging tempo have been reported by Schmid (1964), Núñez (1966, 1970), and Waddington (1990). Indeed, Boch's own experimental results strongly suggest that the filling of feeder A with a 2.0-mol/L solution triggered reduced dancing by bees from feeder B because of a rise in the quantity rather than in the quality of the sugar solution brought into the hive. This is indicated at the end of his experiment, when he lowered the colony's nectar influx by removing all but 10 bees from feeder A, and observed that the bees from feeder B quickly resumed their dancing (Figure 5.20).

The most direct empirical evidence against the hypothesis that food-storer bees inform foragers about nectar-source profitability comes from an experiment designed specifically to determine whether it is the foragers or the food storers who assess nectar sources (described in detail in Seeley, Camazine, and Sneyd 1991). On 17 August 1989 two feeding places were established, one 50 m from an observation hive and loaded with 0.75-mol/L sucrose solution, the other 1250 m from the hive but loaded with a richer, 1.00-mol/L solution (Figure 5.21). Fifteen bees from the observation hive were trained to visit each feeder, and any recruits to the feeders were captured to stabilize the conditions at each feeder. The critical feature of this experimental layout is that it presents the bees with a situation in which the food source with the highest sugar concentration is not the one with the highest overall energetic profitability. Thus, on the one hand, the hypothesis that *food storers* assess nectar-source quality predicts that the bees from the 1250-m, 1.00-mol/L feeder will dance more strongly, because their nectar is more concentrated and so should be favored by the food storers. On the other hand, the hypothesis that

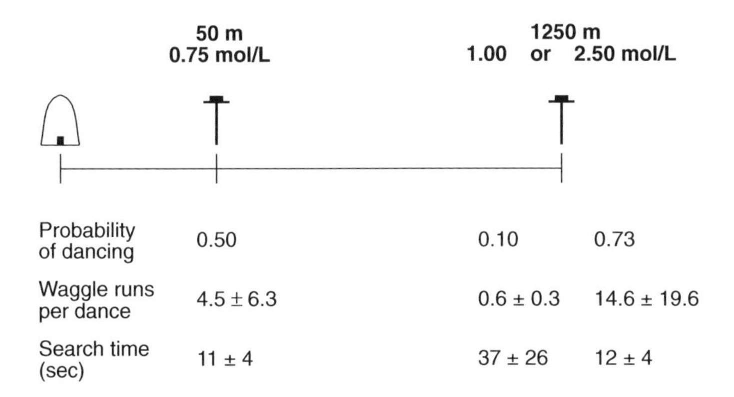

Figure 5.21 Experimental design and results of the test for which bees evaluate a colony's nectar sources: foragers or food storers. After Seeley, Camazine, and Sneyd 1991.

foragers assess nectar-source quality predicts that the bees from the 50-m, 0.75-mol/L feeder will dance more strongly, because its shorter distance from the hive more than compensates for its lower sugar concentration in terms of overall energetic profitability. Which prediction is correct? As shown in Figure 5.21, the bees from the near feeder danced much more strongly than those from the far feeder. Clearly, it must be the foragers who assess nectar-source quality.

This conclusion is reinforced by the average search times for the two forager groups. First, note that the bees from the 1250-m, 1.00-mol/L feeder took more time to start unloading than did the bees from the 50-m, 0.75-mol/L feeder. This runs contrary to the hypothesis of assessment by food storers, which assumes that food-storer bees preferentially unload foragers with more concentrated nectar. This difference in search times evidently arose because the foragers from the far feeder concluded that their feeder's profitability was quite low and so they slowed their foraging tempo, in part by not seeking to unload their nectar immediately upon entering the hive. Second, note that foragers from the 50-m, 0.75-mol/L feeder and the 1250-m feeder, when refilled with a 2.50-mol/L sucrose solution, began unloading equally quickly, despite an enormous difference in concentration between their sugar solutions. This too contradicts the basic assumption of the hypothesis of assessment by food storers. This equality in mean search times for the two forager groups no doubt occurred because foragers from both near and far feeders judged that their feeder was highly profitable and so they decided to forage with a high tempo, which included seeking to unload nectar immediately upon entering the hive.

All the evidence at hand now strongly favors the idea that each nectar forager independently assesses the profitability of her food source and scales her dance output accordingly (Figure 5.19, H3). This information-processing task evidently has the following general form (Figure 5.22). While gathering nectar from a patch of flowers, a bee takes in information about the energetics of her particular forage site—determined by such variables as distance from the hive, sugar concentration of the nectar, and nectar abundance—and by the end of her foraging trip she has integrated this information to give her a sense of the overall energetic profitability of her nectar source. Each

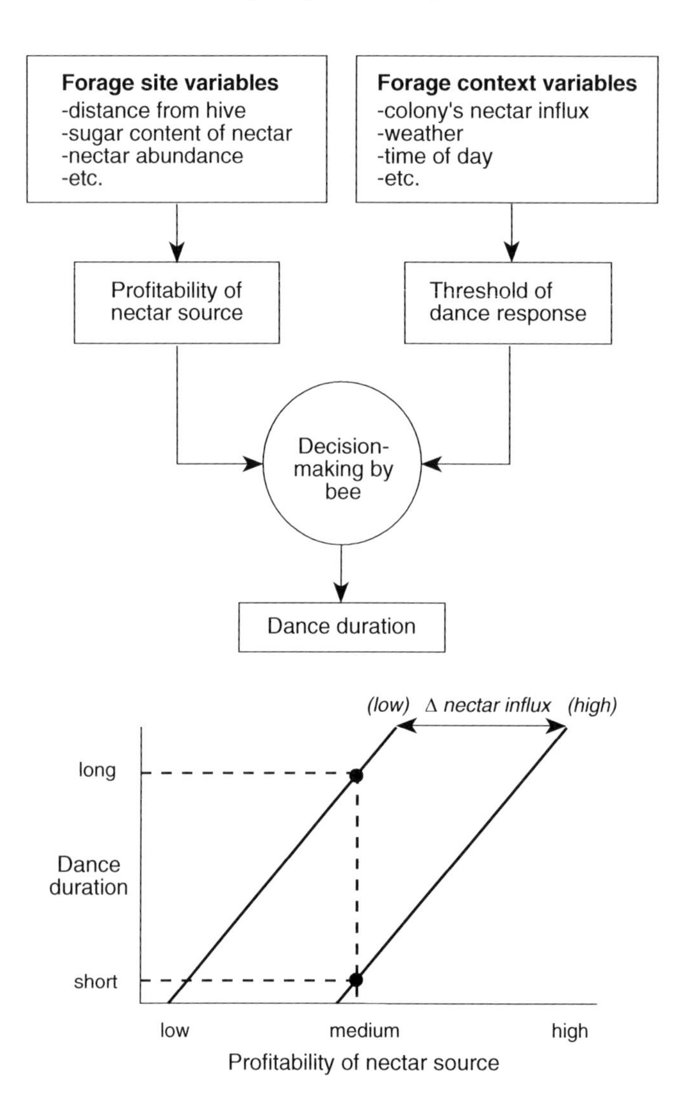

Figure 5.22 Flow diagram illustrating the information processing performed by a nectar forager in deciding how long to dance upon return to the hive. The bee integrates information about her particular flower patch to estimate its energetic profitability, and she integrates information about the general foraging conditions to set a dance threshold. The final decision of how long to dance takes into account both the level of nectar-source profitability and the threshold of dance response. Thus, for example, a medium-quality nectar source can arouse short or long dances, depending on the state of the variables affecting the dance threshold, such as the colony's success in nectar collection.

bee's nervous system appears to be calibrated during development so that she knows, even at the start of her foraging career, where any given level of profitability falls within the range of profitabilities experienced by bees: low, high, or some point in between. The forager also takes in information about the foraging conditions in general—determined by such variables as the colony's nectar influx and the weather—and integrates this information to set a level of profitability that is the threshold for dancing. Finally, the bee's nervous system combines her sense of the "goodness" of her nectar source and her sense of the proper threshold for dancing, to make a decision on how long she should dance.

Summary

1. Each forager bee returning to the hive from a flower patch brings home information about a food source. In a typical colony, there will be several thousand foragers, creating a strong influx of information (Figure 5.2). Most foragers are exploiters of old sources and so provide updates on previous finds, while a few are explorers (or scouts) for new sources and so bring back news of fresh discoveries (Figure 5.3). The scouts are unemployed foragers that seek new food sources by independent searching rather than by following recruitment dances. Approximately 10% of the unemployed foragers go scouting, which means that a colony typically fields some 100 scouts per day. Little is known about the searching behavior of scout bees.

2. Although all foragers bring home information about food-source location and profitability, only bees returning from highly profitable sources perform dances and so share their information. Thus the pool of shared information within a hive consists almost exclusively of information about rich food sources.

3. Foragers share their information in a small region of the hive located just inside the entrance, called the dance floor (Figure 5.4). Within the dance floor, dances for different forage sites are mixed together (Figure 5.5); hence a bee standing in the dance floor is exposed to information about multiple food sources.

4. A forager reports on the profitability of her food source by adjusting the strength of her dance. In principle, this could involve modulation of signal duration (number of waggle runs per dance) or signal intensity (vigor of each waggle run), or both. The bees rely primarily

upon modulation of signal duration (Figure 5.7). Dance duration varies by two orders of magnitude, from 1 to about 100 waggle runs per dance in accordance with food-source profitability (Figure 5.8). The duration of any one dance, however, does not provide precise information about the profitability of the food source it represents. The bees rely at most only slightly upon modulation of signal intensity to code information about food-source profitability. Threshold-level sources evoke only feeble waggle runs, whereas suprathreshold sources elicit vigorous waggle runs. The waggle runs for food sources that are above the dance threshold evidently do not differ in strength or effectiveness, as is indicated by the experimental result that the proportion of a colony's recruits to two equidistant feeders with different profitabilities is accurately predicted by the proportion of waggle runs for each feeder (Figure 5.9).

5. Bees must have a yardstick by which they measure the profitability of a nectar source. The fact that two sucrose solution feeders will elicit dances with equivalent mean durations when the bees exploiting them experience equal energetic efficiencies, but not equal rates of energy gain, suggests that bees use energetic efficiency as the criterion of profitability of a nectar source (Table 5.1). Using this criterion will help a colony maximize the lifetime energy collection of each forager, assuming that a bee's foraging performance is limited by energy expenditure rather than time. It is possible, however, that nectar foraging bees use different criteria of nectar-source profitability depending on the time of the year and the state of the colony.

6. The relationship between the stimulus of nectar-source profitability and the response of waggle dance duration is characterized by several features. First is *linearity of the stimulus-response function* (Figure 5.10). Because the range of stimuli that elicit responses is far smaller than the range of responses, a linear function is sufficient for reporting on the full range of suprathreshold stimuli while providing good resolution of low-level stimuli. The second feature is *no adaptation in the response* (Figure 5.11). Each bee functions as a "tonic" sensory unit, reporting on the current level of profitability of her nectar source rather than on just the changes in its profitability. The third feature is *strong variation among bees in the response*. Individual bees vary greatly in their dance thresholds. This fact may enable a colony to respond in a graded fashion over a broader range of stimuli than if all bees had identical thresholds.

7. Nectar foragers adaptively tune their dance thresholds in relation

to their colony's foraging status. Each bee's dance threshold is lower when the colony's nectar influx is low than when it is high (Figure 5.12). This adaptation enables a colony to exploit a wide range of nectar sources when forage becomes sparse, thereby maintaining an energy influx into the hive, and to exploit only highly profitable sources when forage becomes abundant, thereby enhancing the efficiency of energy acquisition. Bees adjust their thresholds in response to changes in the quantity of nectar coming into the hive, a fact demonstrated by changes in dancing when the quantity, but not the quality, of the colony's nectar influx was experimentally manipulated (Figure 5.14). A forager detects changes in her colony's rate of nectar collection by taking note of some variable in the unloading process (Figures 5.16–18). Probably this is the amount of time a forager spends searching for a food-storer bee to unload her, since this is a variable that is easily perceived by bees and changes reliably with the colony's nectar influx. This reliability is guaranteed by the rules of probability, because a forager's search for a food storer among the bees in the unloading area is evidently a random sampling process in which the per-trial probability of success decreases as the arrival rate of foragers increases.

8. A forager must have some means of knowing whether the flower patch she has just visited represents a source of nectar of low, medium, or high quality. Although it had previously been widely believed that the food-storer bees inform the foragers which among them have visited the best nectar sources, by selectively unloading those with the sweetest nectar, this idea is now known to be mistaken. The evidence against this hypothesis is both conceptual and empirical. Conceptually, the hypothesis is wrong because the quality of a nectar source cannot be assessed solely by the sugar concentration of its nectar. Empirically, the hypothesis is contradicted by an experiment in which bees bringing in the less concentrated of two sucrose solutions began unloading more quickly and danced more vigorously (Figure 5.21). It is now clear that each nectar forager independently assesses the profitability of her flower patch by integrating information about the energetics of foraging at her particular patch. The bee also integrates information about foraging conditions in general (colony's nectar influx, weather, and so on) to set a level of profitability that is the threshold for dancing. Finally, the bee's nervous system combines the inputs on nectar-source profitability with those on dance threshold to form a decision about her behavioral output: how long the bee should dance (Figure 5.22).

How a Colony Acts on Information about Food Sources

5.9. Employed Foragers versus Unemployed Foragers

An important first step toward understanding how a colony responds to its information about food sources is to again draw the distinction between employed and unemployed foragers, that is, between those forager bees that are and are not currently engaged in exploiting a patch of flowers. This distinction is crucial because the members of these two groups draw upon the colony's pool of information about food sources in completely different ways. As a rule, an employed forager exploits only the small body of information about her own food source that she carries within her nervous system, whereas an unemployed forager (if she is a recruit rather than a scout; see Figure 5.3) draws upon the large body of information about food sources that her nestmates present on the dance floor.

What is the evidence that employed foragers actually ignore the information on the dance floor? In the summer of 1990, I performed an experiment which required that I closely watch forager bees inside an observation hive, one at a time, from time of arrival at the hive to time of departure (see table 2 in Seeley and Towne 1992). Each bee was engaged in exploiting a feeder positioned 400 m from the hive and loaded with a sucrose solution in the range of 1.0–2.5 mol/L. My principal aim was to measure the duration of dancing by each bee, but the observation protocol also allowed me to assess the extent of dance following by these bees. Over a 24-day period (5–29 July), I watched 1712 instances when a forager returned to the hive, unloaded her sucrose solution, sometimes performed a dance, and then left the hive to return to the feeder. Not even once did I see a bee orient to a dancer, much less follow one closely. This result is made all the more remarkable by the fact that the nectar unloading area is essentially congruent with the dance floor (compare Figures 5.4 and 5.16), and hence the bees that I watched routinely stood within a few centimeters of other foragers performing dances. On 23 June 1992, I looked again for dance following by employed foragers. This time I established a feeding place 350 m from an observation hive at the Cranberry Lake Biological Station and watched, one by one, each of the 30 bees from the feeder when they returned to the hive. Over the day, the feeder was set at four different levels of sugar concentration (0.5, 1.0, 1.5, and 2.0 mol/L), and at each level each of the 30 bees was followed throughout two returns to the hive, for a total of 240 observations. All four sugar concentration settings yielded the same result: none of these employed foragers showed any interest in the dances performed all around them. Núñez (1970) reports a similar finding based on detailed observations of the in-hive behavior of employed foragers: "It has not been observed that bees regularly visiting the automatic feeder follow other dancers 'advertising' the [only other available] feeder." Given that Núñez and I have observed approximately 2000 returns to the hive by bees actively exploiting food sources, and have not witnessed any dance following by these bees, it seems reasonable to conclude that employed foragers rarely, if ever, follow dances.

One sees precisely the opposite response to dances by unemployed foragers. Consider, for example, the case of foragers that begin their day unemployed because the flowers where they foraged previously have not yet begun to bear nectar or pollen. These bees position themselves on the dance floor and follow dancer after dancer for a waggle run or two, apparently scanning the morning's forage reports for news regarding yesterday's flower patch. This situation is easily recreated experimentally. On the morning of 3 September 1981, 40 labeled bees from an observation hive were trained to visit a sugar water feeder 200 m away, and were allowed to forage there until 5:30 that evening, when the food was taken away. The next morning, before refilling the feeder, I recorded the positions of the labeled bees inside the observation hive and noted all instances of dance following by these bees. During the first half hour of observation, 6:00–6:30, the bees were scattered thoughout the hive (Figure 5.23). Gradually, over the next hour, the bees drifted down to the dance floor, so that by 7:30 nearly all the labeled bees were assembled as a dense throng on the dance floor, where dances were under way for various natural food sources. During the time interval 7:30–8:00, 27 (68%) of the 40 bees oriented to at least one dancer reporting on a flower patch, though each labeled bee turned away after following just 1 or 2 waggle runs, as if quickly realizing, "She is not advertising my forage site." Meanwhile, several of the labeled bees were making occasional reconnaissance visits to the feeder, so that when I refilled it with sugar solution at 8:30, a bee appeared there just 7 min later. She loaded up, flew back to the hive, and announced her discovery with a vigorous dance. By 9:30, 32 of the remaining 38 bees (one bee had disappeared) had followed a dance advertising the feeder and was observed for-

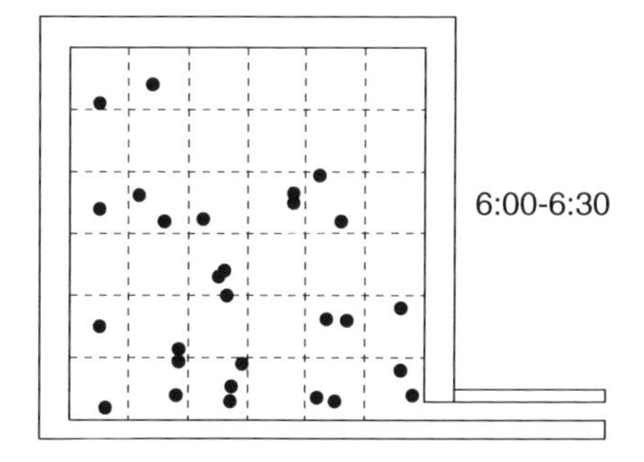

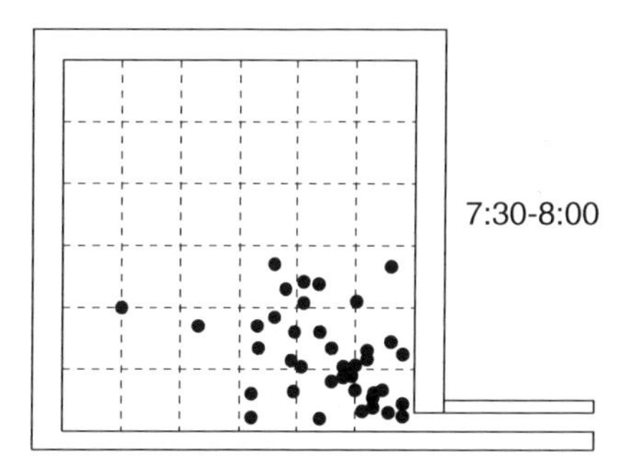

Figure 5.23 The positions of a group of bees within a two-frame observation hive (see Figure 4.2) on the morning of 4 September 1981. All 40 bees in the group had been trained to a sugar water feeder on the previous day, and all 40 were labeled for individual identification. In the figures, each dot depicts the location where a bee was first sighted during the time period shown on the morning of 4 September; not all bees were visible. Initially, many of the bees were resting in quiet places away from the dance floor, but by 7:30 they had assembled from all directions on the dance floor, where they followed dances for news of the renewal of their food source. Based on unpublished data of T. D. Seeley.

aging at the feeder. Clearly, these unemployed foragers showed great interest in the dance reports of their nestmates. Such behavior is evidently typical, for the bees repeated this performance on the following 2 days (Seeley, unpublished results); moreover, other investigators (Körner 1939; von Frisch 1967) describe similar behavior patterns by temporarily unemployed foragers at the start of the day.

Many unemployed foragers, however, do not move so easily into the ranks of the employed foragers, for frequently they face the challenge of locating a new forage site rather than simply resuming work at a familiar site. This situation arises either when the forager is experienced but her previous flower patch has greatly deteriorated, or when she is inexperienced and therefore has no previous work site to which she can return. The novice foragers do constitute a sizable fraction of a colony's forager force. A forager bee typically lives only about 10 days, and hence some 10% of a colony's foragers die each day (Sakagami and Fukuda 1968; Dukas and Visscher 1994). These bees must be replaced lest the forager force dwindle, which implies that on any given day approximately 10% of the foragers within a colony are just beginning their foraging careers.

An unemployed forager can locate a new food source either by searching on her own or by taking advantage of the information available on the dance floor to guide her to a profitable source (Figure 5.3). As previously seen (Section 5.1), when foragers, both experienced and inexperienced, need to find new food sources, the vast majority do so by following the dances of their nestmates.

In summary, we can now see a remarkable feature of the organization of a honey bee colony's foraging: Each employed forager knows only about her particular patch of flowers, and hence she functions with extremely limited knowledge of the current foraging opportunities. But unemployed foragers have the opportunity to acquire broad knowledge of the foraging opportunities, by reading the information displayed on the dance floor.

5.10. How Unemployed Foragers Read the Information on the Dance Floor

An unemployed forager has an opportunity to become well informed about the various food sources exploited by her colony whenever she goes onto the dance floor to follow dances to locate a new rich food source. Certainly, the dance information is arrayed in the hive in a manner that should make it possible for a bee to acquire an overview of the foraging opportunities. As shown earlier, throughout the day dances representing multiple flower patches are performed close together in the hive (Figures 5.4 and 5.5). Also, nectar-source profitability is coded in the dances—in dance duration—so there is the possibility that dances for richer and poorer nectar sources are distinguished by the bees (Figure 5.7). Furthermore, as noted previously, among dances for natural food sources, the strongest ones are more than 100 times longer than the weakest ones, which implies that the profitability of natural food sources varies greatly (Figure 5.8). Hence it is conceivable that an unemployed forager could follow numerous dances on the dance floor and acquire broad knowledge of the available food sources, including their relative profitabilities. Having acquired this information, she could then compare the various food sources and choose the best one as her next work site. Do unemployed foragers do so, or do they actually adopt a new forage site based on much more limited information?

5.10.1. THE BEHAVIOR OF DANCE-FOLLOWING BEES

A logical first step to answering this question is simply to observe the dance-following behavior of unemployed foragers. This is easily accomplished. One trains a small group of foragers from an observation hive to a feeder, labels them with paint marks for individual identification, then shuts off the feeder and observes the labeled bees in the observation hive as they follow dances to find a new food source. In September and October of 1990, I performed this procedure three times, each time letting 10 bees forage for 2 days from a sugar solution feeder, then removing the feeder on the morning of the third day and watching the labeled bees in the hive. I observed 21 of the 30 bees follow dances to find their next food source. Although the behaviors of these bees varied considerably, they all followed the general pattern that is shown in Figure 5.24 and is summarized in the following composite description of a dance-following bee's behavior (described in greater detail in Seeley and Towne 1992).

For the first several hours, the bee does not follow any dances even if there are numerous dances in the hive advertising rich flower patches. Instead, during this time she makes several trips outside the hive to inspect the feeder (now empty). Each trip is brief, generally lasting less than 5 min. Eventually, however, she starts to follow dances. This begins with a period, of about 3 hours, when the unemployed forager typically follows several dancers, but each one for just 1–2 waggle runs and without a noticeable change in her level of arousal. This period of desultory dance following ends suddenly when the bee begins following one dancer with great enthusiasm. She excitedly tracks this dancer through numerous waggle runs, rarely less than 6 and sometimes 12 or more, twisting along close behind the dancer, until she abruptly breaks away from the dancer and scrambles out of the hive, presumably to search for the food source indicated by the dance that she has just followed.

If the bee locates this food source, she will return in about an hour with a load of pollen or nectar. More probably, however, she will crawl back into the hive after only about 15 min outside and will hungrily beg for food, having failed to locate the recruitment target. During the next hour or so, she may again pay cursory attention to one or two dancing bees before again springing into action, tripping closely along behind a dancer for about half a dozen waggle runs, and then scurrying across the comb and out of the hive. This search, like the first, is likely to be unsuccessful so that some 15 min later the bee again crawls back in the hive empty-handed. On average, a bee will need to make approximately four tries, that is, conduct some four dance-guided searches, to locate a flower patch advertised by a dancer. Evidently it is not a trivial matter for a bee to find the flower patch that a dancing bee has advertised.

Figures 5.24 and 5.25 both illustrate another characteristic feature of the behavior of dance-following bees, which is that a bee's con-

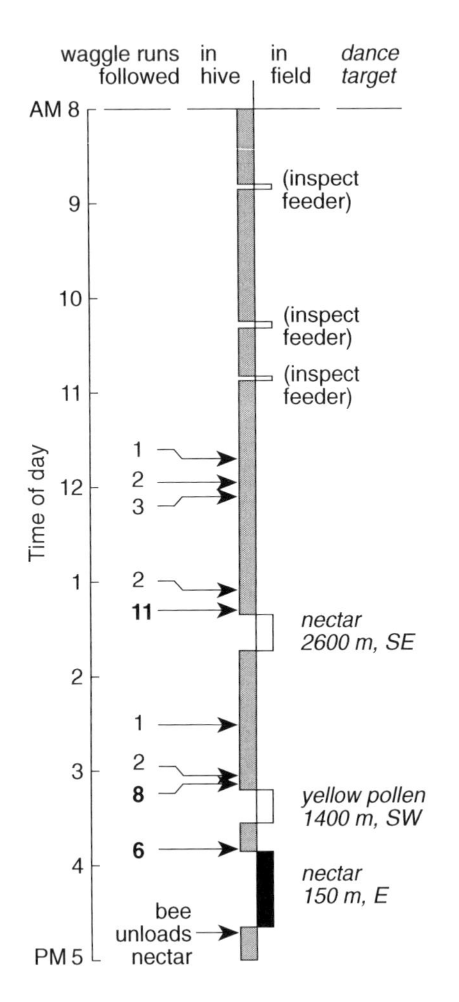

Figure 5.24 Record of dance following by an unemployed forager trying to locate a new food source. Initially, she did not follow any dances and instead inspected her previous food source, a feeder that was not refilled on the day of observations (16 September 1990). Then from 11:30 to 1:00 she followed several dances, but each only very briefly. At 1:21 she followed extensively (for 11 waggle runs) a dancer advertising a source of nectar located 2600 m to the SE, and immediately thereafter she left the hive to search for these flowers, but evidently was unsuccessful, for she returned 22 min later without food. At 3:09 she again followed one pollen dancer closely (for 8 waggle runs), promptly left the hive, and spent some 20 min outside, but again returned without food. Finally, at 3:50 she followed closely (for 6 waggle runs) a dancer advertising a nearby nectar source, scurried out of the hive, and 49 min later crawled into the hive with a large load of nectar. The general pattern of this bee's behavior is typical, for 20 other bees showed a similar pattern of initially following dances only briefly, then following just one dancer closely before each departure from the hive. These 20 other bees also made several dance-guided searches outside the hive before finally finding a new food source (see Figure 5.25). Based on data in Seeley and Towne 1992.

secutive searches are frequently directed toward different recruitment targets. A bee may even alternate between two different recruitment targets in the course of a series of search attempts (see the bottom record in Figure 5.25). Clearly, an unemployed forager does not restrict her search attempts to a single, best food source, or even to a single type of forage (nectar or pollen).

These observations allow us to draw two important conclusions about how unemployed foragers read the information on the dance floor. First, they do not acquire information from a large number of dancers before setting out to search for a new food source. Indeed, each such search is preceded by just a single bout of following a dancer for numerous (more than 1 or 2) waggle runs. It is clear, therefore, that a dance-following bee does not conduct a thorough survey of the information available on the dance floor before leaving the hive. Second, it is exceedingly rare for an unemployed forager to follow a dance from start to finish. This fact suggests that these bees do not acquire information about food-source profitability in the course of dance following since, as shown in Section 5.4, information about profitability is strongly expressed only in dance duration. That bees do not attempt to measure dance duration is not so surprising. After all, few bees will be near a dancer when she begins dancing; so only a small fraction of the bees following a given bee's dance will be able to measure the full duration of her dance. Moreover, since the coding of food-source profitability in dance duration is extremely noisy (see Figure 5.7), a dance-following bee that did manage to accurately measure the duration of a dance would still not have precise information about the profitability of the food source represented by the dance. In summary, the typical behavior pattern of a bee following dances suggests strongly that unemployed foragers do not acquire broad knowledge of the available food sources, compare them in terms of profitability, and select the richest one for their next work site. Proof of this, however, required performing an experiment in which the information on the dance floor was well controlled so that different hypotheses for how bees follow dances could be precisely tested.

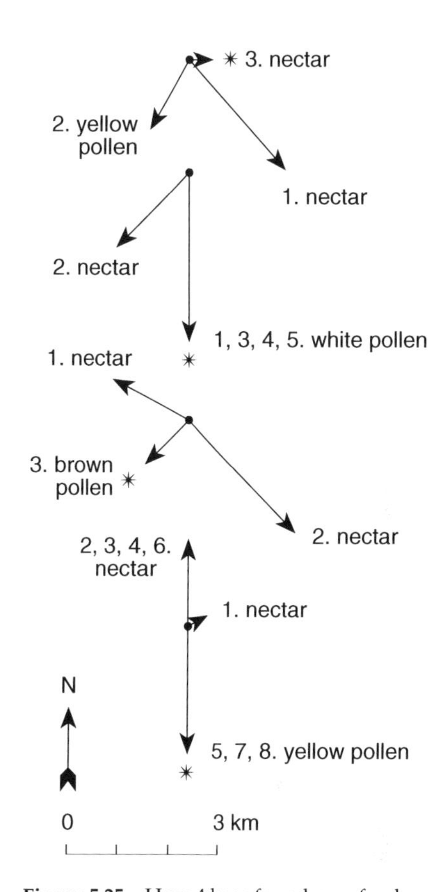

Figure 5.25 How 4 bees found new food sources by following dances, from data collected in September 1990. Each arrow denotes the flower patch indicated by the dance that the bee followed just before leaving the hive. The direction and length of the arrow give to scale the location of the reported patch of flowers. Bees typically make several dance-guided searches before finding a new food source. The order of the searches is indicated by the numbers beside each arrow; a star beside an arrow specifies the food source that finally was found. Note that consecutive searches are frequently to recruitment targets that differ in both location and forage type (nectar or pollen). The top diagram depicts the pattern of searches performed by the bee represented in Figure 5.24. After Seeley and Towne 1992.

5.10.2. EXPERIMENTAL ANALYSIS OF HOW BEES GATHER INFORMATION ON THE DANCE FLOOR

To test whether or not each unemployed forager follows multiple dances and selectively responds to the strongest one, I created a situation in which the vast majority of the dances in an observation hive

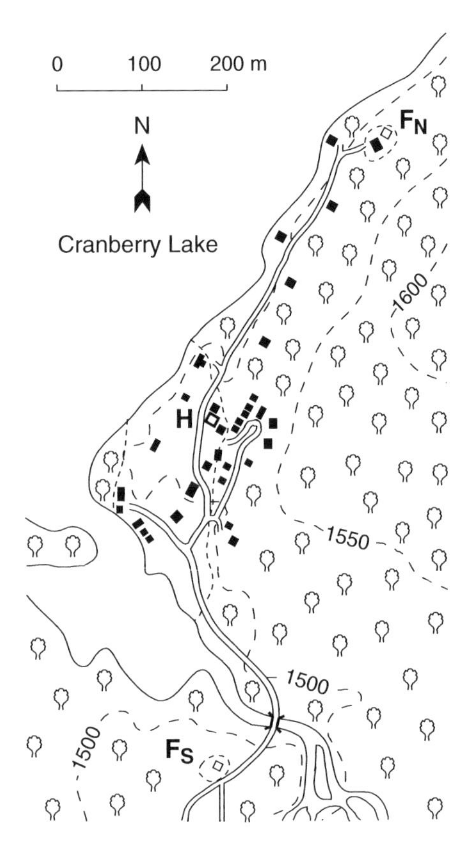

Figure 5.26 Sketch of the experimental layout at the Cranberry Lake Biological Station in July 1990: *H*, observation hive; F_N, feeding station 400 m north of the hive; F_S, feeding station 400 m south. Both feeders were positioned in clearings approximately 30 m in diameter. Contour lines indicate feet above sea level.

signaled just two feeders, one highly profitable and eliciting strong dancing, the other less profitable and eliciting weaker dancing. The two feeders were equidistant from the hive, but in opposite directions, and each was visited by 30 labeled foragers. I measured the total amount of dancing (number of waggle runs) and the total amount of recruitment (number of bees) for each feeder. The logic of the test runs as follows. If each unemployed forager follows multiple dances, compares them, and responds only to the dance advertising the best food source, then dances for the more profitable feeder will be disproportionately effective per waggle run; that is, the proportion of recruits to the richer feeder will exceed the proportion of waggle runs for that feeder. If, however, each forager follows just one dance, chosen more or less at random, then the proportion of recruits to each feeder will match the proportion of waggle runs for that feeder. It should be noted that these predictions depend critically on the assumption that there is no difference between richer and poorer feeders in attractiveness or arousal effectiveness of a waggle run (see Section 5.4). Hence this experiment addressed simultaneously the question of whether richer food sources elicit livelier, not just longer, dances, and the question of how dance-following bees sample the information on the dance floor.

In July 1990, two feeding stations were established 400 m north and south of an observation hive located in the center of the Cranberry Lake Biological Station (Figures 5.26 and 5.27). Thirty individually labeled bees were trained to each feeder, and all other bees arriving at each feeder were captured. Each trial of the experiment lasted 1 day and was started by loading the two feeders with sucrose solutions that differed in concentration but contained the same scent (anise). The sugar concentration for each feeder was carefully adjusted at the start of each trial so that the richer feeder elicited dancing that was strong—but not so strong that recruits arrived faster than they could be easily captured—and so that the poorer feeder elicited waggle runs at a rate roughly 2/3, 1/3, or 1/10 that for the richer feeder. Then, for the next 3–5 hr, my assistants and I recorded the dances for each feeder and captured the recruits to each feeder.

Eleven trials of this experiment were performed, and in each we found that the proportion of waggle runs for each feeder accurately predicted the proportion of recruits arriving at the feeder. For example, on 13 July 1990, the north and south feeders were loaded with 1.75 and 1.25 mol/L sucrose solutions, and between 9:00 and 1:00 the

Figure 5.27 Aerial photograph of the Cranberry Lake Biological Station, looking south. The observation hive is placed in the large clearing, and the feeders are positioned in smaller clearings outside the bounds of this photograph, as depicted in Figure 5.26. Photograph by W. M. Shields.

30 bees from each feeder performed a total of 8722 waggle runs, of which 90.4% were for the richer, north feeder. These dances aroused 153 recruits to the feeders, of which 135, or 88.2%, arrived at the north feeder (see also Figure 5.9). Ten more trials of this experiment yielded the same pattern of recruitment to the feeders, one strictly propor-

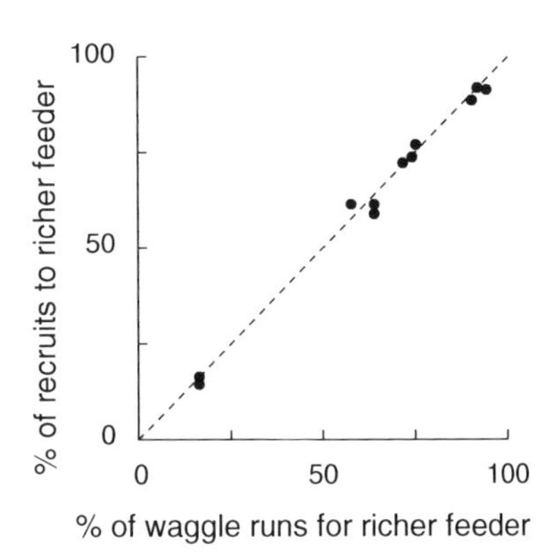

Figure 5.28 Results of the experiment designed to test whether or not bees compare dances and preferentially respond to those representing better food sources. Based on data reported in Table 2 of Seeley and Towne 1992.

tional to the number of waggle runs for each feeder (Figure 5.28). Certainly, the most convincing results in this experiment came from the trials of 22 and 29 July, when the number of bees visiting the richer feeder was reduced from 30 to 5. On these days, therefore, there was more dancing *per bee* for the richer feeder, but more dancing *total* for the poorer feeder, because there were many more bees (30) visiting it. Under these conditions, if each dance-following bee does not preferentially respond to waggle runs for the richer feeder, then the colony will make a mistake—it will devote more recruits to the poorer feeder. This is precisely what we found! On both 22 and 29 July, the dances for the richer feeder contained twice as many waggle runs, on average, as those for the poorer feeder, but only 17% of the total waggle runs were for the richer feeder, and only 15–16% of the recruits arrived at the richer feeder (see Figure 5.28).

Thus it is unequivocally clear that unemployed foragers do not follow multiple dances and selectively respond to those advertising the best food source. The results indicate instead that each bee sampled just one dance before exiting the hive to search for a new food source. This, in turn, implies that the brief episodes of dance following that were observed (Figure 5.24) were not part of a tactic for surveying the available recruitment options. Instead, they were probably low-cost attempts to gain information about the previous day's food source, which might again provide rich food. The experimental results also demonstrate, as previously noted (Section 5.4), that the individual waggle runs for richer and poorer food sources are, on average, equally attractive to dance-following bees.

Recent research by Oldroyd, Rinderer, and Buco (1991) suggests that a bee's sample of the dances is not totally random, but is biased somewhat as a function of her genotype. When they created colonies consisting of two genetically labeled patrilines, distinguishable by their color (dark versus light brown), installed them in observation hives and observed the dances occurring in these colonies, they found that bees of a particular patriline sometimes showed a statistically significant bias toward following dances performed by members of their own patriline. The authors point out that this bias does not necessarily reflect patriline recognition per se, for it can be explained, at least in part, by differences between the two patrilines in foraging behavior. For example, their light-brown bees preferentially followed dancers bearing pollen, and may have also preferentially performed dances for pollen sources. These two tendencies, acting together, would produce the stronger than expected association between dancers and dance followers of the same patriline. A follow-up study by Oldroyd and his colleagues (1993), which looked for patriline differences in foraging distance, likewise found that bees followed dances performed by members of their own patriline more often than expected by chance, though this bias was sometimes extremely weak or missing altogether. As in the previous study, the higher than expected association between full sisters on the dance floor evidently arose partly as a result of differences between the two patrilines in producing and following dances for food sources differing in some variable, in this case distance from the hive.

In nature, such biases toward following dances by members of the same patriline are likely to be far weaker than in the two-subfamily colonies studied by Oldroyd and his colleagues, since natural colonies generally consist of 10 or more patrilines and hence the vast majority of the dancers encountered by any given bee will be from patrilines other than her own. Nevertheless, the finding that bees of different patrilines sometimes show preferential following of particular types of dances, such as pollen dances versus nectar dances, probably does pertain to natural colonies. It should be noted, however, that these preferences are certainly not absolute and indeed are evidently quite weak, for as noted earlier (Figures 5.24 and 5.25), a bee will typically follow dances representing extremely diverse food sources, including ones located at different distances from the hive and bearing different types of forage (pollen or nectar). When Lindauer (1952) witnessed 30 cases in which a forager followed a nestmate's dance to switch from one forage site to another, he observed that in 14 instances the bee switched from pollen foraging to nectar foraging, or vice versa. Clearly a dance-following bee does not, as a rule, show an overwhelming preference for a particular type of dance.

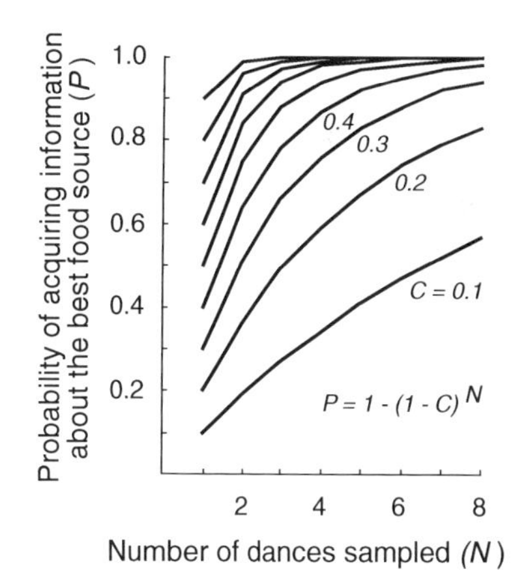

Figure 5.29 The probability (P) that a dance-following bee will learn about the best food source being advertised on the dance floor, as a function of the number of dances she samples (N) and the fraction of the total dance circuits that are for the best food source (C). Except when C is low (< 0.30), if a bee samples several dances, her probability of acquiring information about the best food source will be close to 1.00. After Seeley and Towne 1992.

5.10.3. Why Bees Do Not Broadly Sample the Information on the Dance Floor

At first thought, it seems grossly maladaptive for an unemployed forager not to follow multiple dances in the hive and selectively respond to the strongest one encountered. After all, by failing to survey broadly the foraging opportunities, a forager will often fail to learn about the best food source. The likelihood of this happening is shown in Figure 5.29, where the probability of acquiring information about the best food source is shown as a function of the number of dances a bee samples and the fraction of the waggle runs in the hive that are for the best food source. A bee sampling just one dance has a low probability of learning about the best source unless the fraction of waggle runs for this source happens to be high. But if a bee samples several dances rather than just one, she can raise her probability of learning about the best source to nearly 1.00.

So why don't bees sample broadly and thereby give themselves a high probability of gaining information about the best food source? One might argue that they don't because information about food-source profitability is not expressed in dances in a way that is easily extracted by the dance followers, but this answer begs the question of why this information is not more accessible to the dance followers. I suggest that bees don't sample broadly because, all things considered, it is better for the colony if the dance followers are not choosy. By responding to all dances in the hive, foragers are dispatched to an array of different food sources, each of which has a known, relatively high profitability (indicated by the fact that it elicited dancing), but an unknown size and largely unpredictable future. And by distributing themselves among all reasonably profitable forage sites, the colony's foragers are able to expand and contract their efforts appropriately over the ever changing food-source array. Following only the dances for the one, best food source, by contrast, would lead to an all-or-none approach that could leave the colony overinvested (in a highly profitable but small or short-lived source, for example) and underinformed should foraging conditions change. This argument is further developed below (Section 5.14), in a theoretical study that reveals that a colony achieves high foraging success by allocating foragers simply in proportion to the amount of dancing for each source.

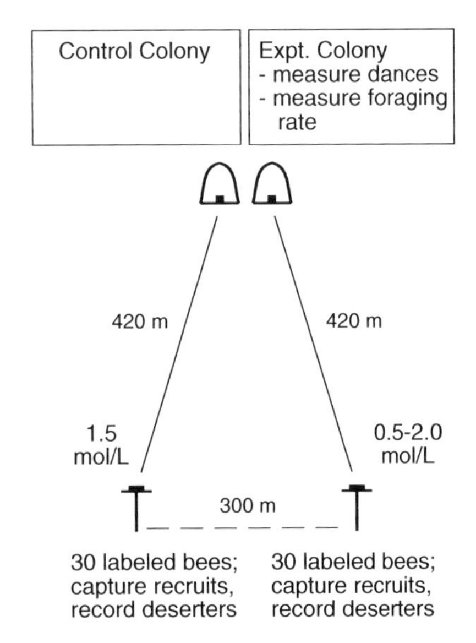

Figure 5.30 Experimental array used to investigate how nectar foragers modulate their behavior in accordance with nectar-source profitability.

5.11. How Employed Foragers Respond to Information about Food-Source Profitability

Once a bee has located a nectar source, she carefully adjusts multiple variables of her behavior in accordance with the profitability of the source, and it is these adjustments that underlie much of a colony's ability to wisely distribute its foraging efforts among the various sources of nectar. To obtain a precise picture of this behavior modulation in relation to nectar-source profitability, I needed to record the behavior of foragers in an experimental setting where I could precisely vary the quality of a nectar source and hold constant all other factors influencing a nectar forager's behavior, such as her colony's nectar influx and the fullness of its combs (Lindauer 1948; Seeley 1989a). This was accomplished in July 1987 by establishing two observation hives, one experimental and one control, at the Cranberry Lake Biological Station, and setting up a sugar water feeder for each hive 420 m from the hives (the geometry of the experimental layout is shown in Figure 5.30). Each colony's feeder was visited by 30 labeled bees, and all recruits to and deserters from each feeder were recorded. All recruits were captured to stabilize conditions at both feeders. Capturing the recruits also ensured that the nectar influx from each feeder remained at a steady, low level, thereby controlling the variables of nectar influx and amount of empty comb in each colony. Also, the counts of recruit captures provided another measure of the behavior of the labeled bees. The scarcity of natural forage at Cranberry Lake guaranteed that each colony's food collection from sources besides the feeders would remain low and steady throughout each trial of the experiment. Each trial lasted 5–7 hr. During this time, the profitability of the experimental colony's feeder was systematically changed by altering the concentration of its sugar solution (0.5–2.0 mol/L) and the behaviors of its foragers were recorded. Meanwhile, the sugar concentration at the control colony's feeder was left unchanged (at 1.5 mol/L) and the behaviors of its foragers were recorded to double-check for possible confounding changes in such influential variables as air temperature, weather conditions, and nectar availability.

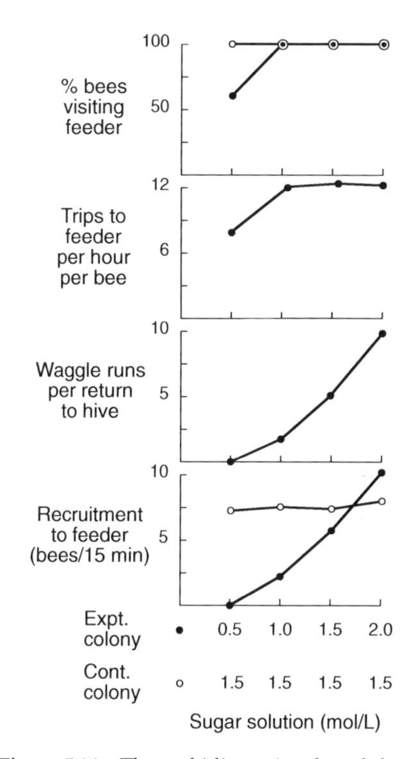

Figure 5.31 The multidimensional modulation of foraging behavior in relation to nectar-source profitability. The concentration of sucrose solution in the experimental colony's feeder was raised in 0.5-mol/L steps over the course of a day, 10 July 1987, and the behavior of 30 bees visiting the feeder was monitored, both at the feeder and inside an observation hive. The behavioral changes shown by bees from the experimental colony were due to changes in food-source profitability, and not other factors (weather, for instance) as is indicated by the lack of change in data gathered simultaneously on bees from a control colony whose feeder contained a constant, 1.5-mol/L, sugar solution. Based on data in table 1 of Seeley, Camazine, and Sneyd 1991.

Figure 5.31 illustrates how the bees finely adjusted several components of their foraging behavior in accordance with food-source quality on 10 July 1987. When the quality was high, as when the experimental colony's feeder was loaded with a 2.0-mol/L solution, all 30 bees continued visiting the feeder, worked quickly, and danced strongly, so that additional nestmates appeared at their feeder at a high rate. When the quality was low (0.5-mol/L solution), by contrast, nearly half the bees stopped visiting the feeder, and those that persisted foraged only slowly, spending more and more time inside the hive between trips to the feeder, and they did not perform recruitment dances. This set of responses resulted in a drop in the number of bees visiting the feeder. Setting the feeder at intermediate levels of quality elicited behavioral responses of intermediate strength, with correspondingly intermediate rates of recruitment to the feeder. It should be noted that data collected over the course of the day at the control colony's feeder showed no significant changes in number of bees visiting the feeder or in recruitment rate, which implies that the ambient conditions were stable during the trial. Hence it seems clear that the behavioral changes recorded for the experimental colony's bees were strictly in response to changes in food-source profitability.

Such multidimensional modulation of foraging behavior in relation to nectar-source profitability is typical. Not only did I witness it in all three trials of the experiment just described (Seeley, Camazine, and Sneyd 1991), but other investigators (von Frisch 1967, p. 45; Núñez 1966, 1970, 1982) have analyzed one or more behavioral variables using similar experimental procedures and have likewise reported modulation of dance strength, foraging tempo, and abandonment of the food source in accordance with nectar-source profitability. Moreover, Waddington (1990) observed that when bees were given a 60% (2.25-mol/L) sucrose solution they maintained a higher thoracic temperature and foraged more briskly than when they were given a 20% (0.6-mol/L) solution. Núñez (1966, 1970, 1982) also reports that bees adjusted the size of the nectar load in relation to food-source profitability, especially when profitability was varied by changing the availability, rather than the concentration, of the sugar solution. The behavioral modulation of nectar foragers is therefore summarized by the following general rule: as food-source profitability increases, the tempo of foraging increases, the volume of nectar loaded increases, the duration of dances increases, and the probability of abandoning the food source decreases.

5.12. The Correct Distribution of Foragers among Nectar Sources

We have seen that an *employed* forager's knowledge of the array of food sources exploited by her colony is limited to her own particular source (Section 5.9), that she makes her own independent assessment of this source (Section 5.8), and that based on this assessment plus information about the colony's foraging status (Section 5.7) she chooses an appropriate foraging response. This involves deciding whether to continue foraging at the source, and if so, deciding how strongly to advertise the source by dancing (Section 5.11). We have also seen that an *unemployed* forager—unless she is a scout bee—follows just one dance, chosen more or less at random, each time she attempts to locate a new food source (Section 5.10). Thus we now know that each forager within a honey bee colony possesses only extremely limited knowledge of her colony's foraging oppportunities. But at the same time we know that the colony as a whole responds in a way that takes account of the full array of foraging opportunities outside the hive (Section 3.4). How can we reconcile these two seemingly contradictory facts?

More precisely, how can we account for the ability of a colony to act wisely in terms of the actions of poorly informed foragers? An answer is suggested by the experiment depicted in Figure 5.32. On 8 June 1989, a colony consisting of some 4000 bees labeled for individ-

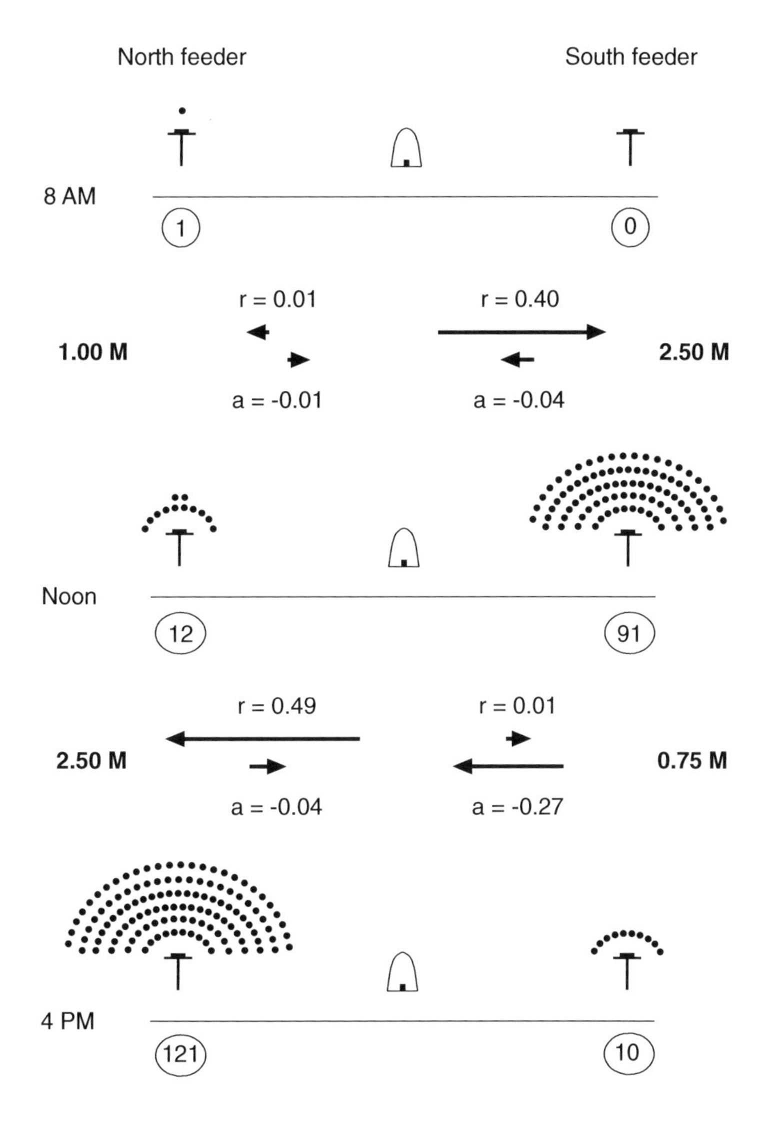

Figure 5.32 Preferential exploitation of the richer of two food sources, as observed on 19 June 1989, together with data on the pattern of recruitment (*r*) and abandonment (*a*) for each feeder. The two feeders were located 400 m from the hive and were identical except for the concentrations of their sugar solutions (*M* = mol/L). The sugar solutions were changed at noon to reverse the relative qualities of the two feeders. The number of dots above each feeder denotes the number of different bees that visited the feeder in the half hour preceding the time shown on the left. The variables *r* and *a* denote the average per capita rates of recruitment to or abandonment of a feeder, measured in recruits (or deserters) per 30 min per bee visiting the feeder. On the day before the observations, 12 and 15 bees had experience at the north and south feeders, respectively, and so provided the initial link between the colony and each feeder. These bees returned to their respective feeders on the morning of 19 June, but were not counted as recruits. After Seeley, Camazine, and Sneyd 1991.

ual identification was established at the Cranberry Lake Biological Station in an observation hive, and was provided with two sucrose solution feeders 400 m north and south of the hive (layout identical to that shown in Figure 5.26.) Rainy weather prevailed for the next 10 days, but by the end of the day on 18 June my assistants and I had 12 and 15 different bees visiting the north and south feeders, each of which contained a 2.0-mol/L sucrose solution. Then on 19 June (warm and sunny weather), we loaded the north and south feeders with 1.0-and 2.5-mol/L solutions in the morning, and with 2.5- and 0.75-mol/L solutions in the afternoon, and from 7:30 to 4:00 we recorded the identities of the bees visiting each feeder for each half-hour period. From these records we could determine how many different bees visited each feeder every 30 min, and we could calculate the per capita rates of recruitment and abandonment for each feeder over the course of the day.

Figure 5.32 shows that one can think of the process of allocating foragers among nectar sources as a process akin to natural selection, that is, a process in which the distribution of individuals among alternative states (food sources) is determined not by some high-level, well-informed supervisor, but simply by the differential "survival" (persistence at a food source) and "reproduction" (recruitment to a food source) of the individuals, each responding to its own, immediate set of circumstances. For example, on the afternoon of 19 June, after the relative qualities of the north and south feeders had been reversed, the colony produced an appropriate redistribution of its foragers simply by having each employed forager respond to the changed conditions at her own feeder. The south-feeder bees responded to the deterioration of their feeder by lowering their recruitment rate and increasing their abandonment rate, while the north-feeder bees responded to the improvement of their feeder by increasing their recruitment rate (their abandonment rate remained low). The net result was that the percentage of the colony's forager population associated with the richer food source automatically increased while that for the poorer food source inevitably decreased. In short, through a process analogous to natural selection, the colony built a globally correct response out of the locally controlled actions of its members.

5.12.1. A MODEL OF COLLECTIVE WISDOM IN FORAGER ALLOCATION

The idea that a colony's skill in allocating foragers among nectar sources can be explained as a process of natural selection among poorly informed foragers can be rigorously tested by means of a mathematical model which endows each forager with strictly limited information (see Seeley, Camazine, and Sneyd 1991). This model, therefore, expresses in detail a hypothesis of how the allocation process works. I will assess the model's relevance to what actually happens inside a beehive by comparing the pattern of exploitation of two unequal nectar sources that is predicted by this model with the pattern of exploitation that was observed during the experiment illustrated in Figure 5.32.

The model deals with the situation of a colony choosing between two nectar sources, A and B, under a fixed set of foraging conditions (no changes in weather, nectar abundance, and so on). In this model (Figure 5.33), each forager is in one of seven distinct compartments, each of which is characterized by an activity:

A: foraging at nectar source A

B: foraging at nectar source B

D_A: dancing for nectar source A

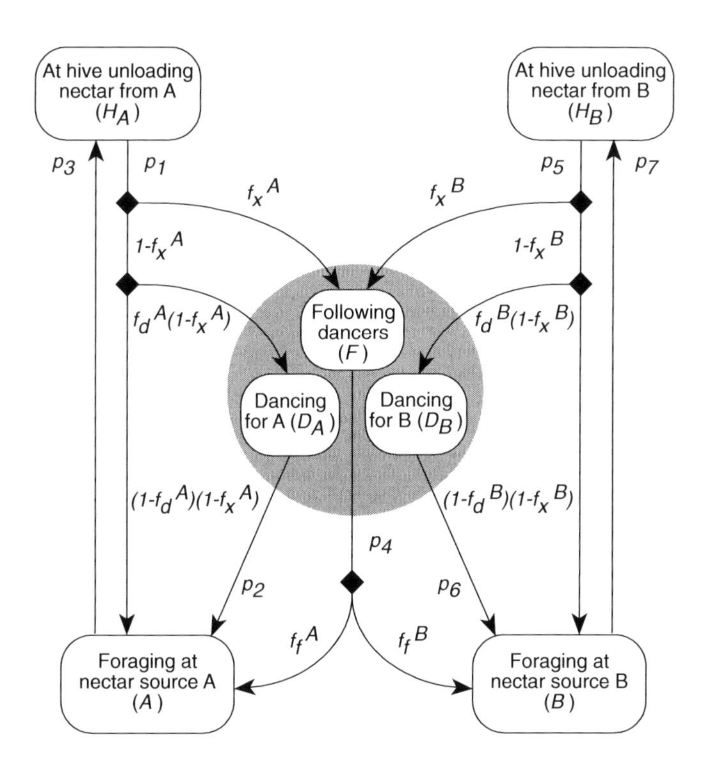

Figure 5.33 Flow diagram representing the mathematical model of how honey bee colonies allocate foragers between two nectar sources (A and B). At any moment each forager can be in one of the seven compartments shown (H_A, H_B, D_A, D_B, A, B, F denote the compartments as well as the number of foragers in the compartments). The rate at which bees leave each compartment is indicated by p_1-p_7. The functions f_x^A, f_x^B, f_d^A, f_d^B, and so on, indicate the probability of taking one or the other fork at each of the five branch points *(black diamonds)*. After Seeley, Camazine, and Sneyd 1991.

D_B: dancing for nectar source B

F: following a dancer (unemployed forager)

H_A: unloading nectar from nectar source A

H_B: unloading nectar from nectar source B

More precisely, a bee is in one of these seven compartments until the moment she enters her next compartment. Thus, for example, the time spent in D_A includes both the time spent dancing for and returning to nectar source A. Note that the dance floor (shaded area in Figure 5.33) contains three separate compartments: those bees dancing for source A, those bees dancing for source B, and those bees following a dancer. Note too that Figure 5.33 consists of two separate cycles, one for each food source, with the follower compartment (F) the only intersection point for the two cyles. Thus bees from one nectar source can switch over to the other source only by passing through the dance floor and following a dancer for the other nectar source.

Two factors affect the proportion of the total forager force in each of the seven compartments: (1) the rate at which a bee moves from one compartment to another and (2) the probability that a bee takes one or the other fork at the five branch points (black diamonds) in Figure 5.33. The fraction of bees leaving a compartment in a given time interval is denoted by the appropriate rate constant p_i, with units min^{-1}. For example, the rate constant for bees leaving compartment A is p_3 and the fraction of the bees at nectar source A that leave in time interval Δt is equal to $p_3 \Delta t$. The values of the rate constants p_i for a particular experimental situation will be presented in a later section.

Now let us consider what determines the probabilities of the different behaviors at each of the five branch points in Figure 5.33. The first branch point is encountered after a bee has unloaded her nectar in the hive. Here, she may abandon the nectar source and return to the dance floor to follow another dancer. The probability that a bee does so is denoted by the abandoning function, f_x. Its value will depend on the profitability of the nectar source, thus f_x^A denotes the probability that a bee leaving H_A will abandon nectar source A and become a follower bee (F). Of course, abandonment diminishes the number of bees committed to a nectar source and provides a pool of unemployed foragers that will follow a dance for one source or another.

The second branch point applies to bees that did not abandon their nectar source. It determines what proportion of these bees will dance for their nectar source. Although at this branch point there is no filtering of bees away from the nectar source to which they are committed, this branch point affects the probability with which a follower bee follows dances for one or the other nectar source. The probability that a bee performs a dance for her nectar source is denoted by the dancing function, f_d. As with the abandoning function, its value depends on the profitability of the nectar source, with f_d^A denoting the probability that a bee foraging at nectar source A performs a dance.

The third branch point is encountered on the dance floor when bees follow dancers for one or another nectar source. The probability of a follower bee following dances for nectar source A and then leaving the dance floor to go to this nectar source is denoted by the following function f_f^A. As we have seen (Section 5.10), each follower bee follows just one dancer, chosen essentially at random, before leaving the hive to search for a new food source. Hence in the situation of just two nectar sources, A and B, the probability of following a dancer for nectar source A (f_f^A) can roughly estimated by $D_A/(D_A + D_B)$. However, since only a portion of a bee's time in the dance area is actually spent dancing, it is necessary to weight D_A and D_B in the above expression by the proportion of time that the foragers actually dance. These fractions are denoted by d_A and d_B. Thus

$$f_f^A = \frac{D_A d_A}{D_A d_A + D_B d_B} \tag{5.3}$$

and

$$f_f^B = \frac{D_B d_B}{D_A d_A + D_B d_B} \tag{5.4}$$

Each function takes into account the number of dancers for each food source as well as the time spent dancing, and so indicates the proportion of the total dancing for each nectar source.

For simplicity in making calculations with the model, we make two further assumptions: (1) all the foragers go to one of the two nectar sources and (2) the total number of foragers (employed and unemployed) is fixed.

5.12.2. THE MODEL'S EQUATIONS

From Figure 5.33 we can write down the following set of differential equations:

$$dA/dt = (1 - f_d^A)(1 - f_x^A)\,p_1H_A + p_2D_A + f_f^A p_4F - p_3A \tag{5.5}$$

$$dD_A/dt = f_d^A(1 - f_x^A)\,p_1H_A - p_2D_A \tag{5.6}$$

$$dH_A/dt = p_3A - p_1H_A \tag{5.7}$$

$$dB/dt = (1 - f_dB)(1 - f_xB)\,p_5H_B + p_6D_B + f_f^B p_4F - p_7B \tag{5.8}$$

$$dD_B/dt = f_d^B(1 - f_x^B)\,p_5H_B - p_6D_B \tag{5.9}$$

$$dH_B/dt = p_7B - p_5H_B \tag{5.10}$$

$$dF/dt = f_x^A p_1H_A + f_x^B p_5H_B - p_4F \tag{5.11}$$

A detailed derivation and discussion of these equations is given in Camazine and Sneyd (1991).

5.12.3. TESTING THE MODEL

We can test the model, and the hypothesis it embodies, by comparing its predictions with the pattern of forager allocation to two food sources that was observed in a particular experimental situation. It is important to note that correct predictions by the model are not guaranteed either by the assumptions underlying it or by the correct determination of the parameters (f_x^A, p_2, . . .). Only if both the model's parameters are accurately measured and its structure accurately represents the mechanisms underlying a colony's process of forager allocation will the model's predictions resemble what is actually observed in a real colony of bees.

We will examine the model's ability to predict the allocation dynamics for the situation shown in Figure 5.32, namely a colony choosing between two equidistant sucrose solution feeders that are identical, except that one contains a 0.75-mol/L solution and the other contains a 2.50-mol/L solution. This requires an estimate for each of the rate constants, p_i, $i = 1$ to 7. Each is equal to $1/T_i$, where each T_i is the time required to get from the relevant compartment to the next. Values of T_i appropriate to the foraging situation shown in Figure 5.32 are given in Table 5.3. Values of the abandoning function (f_x) and the dancing function (f_d) have also been measured for the situation shown in Figure 5.32, and are likewise shown in Table 5.3. Values of the following function (f_f) are calculated for the Figure 5.32 foraging situation with

Table 5.3. Parameter values for the model of a colony allocating foragers between two nectar sources, as shown in Figure 5.32. A and B correspond to the 2.50-mol/L and the 0.75-mol/L feeders, respectively. Based on table 2 of Seeley, Camazine, and Sneyd 1991.

Parameter: definition		Value
f_d^A	probability of dancing for A	1.00
f_d^B	probability of dancing for B	0.15
f_x^A	probability of abandoning A, per foraging trip	0.00
f_x^B	probability of abandoning B, per foraging trip	0.04
T_1	time from start of unloading to start of following, dancing, or foraging, A foragers	1.0 min
T_2	time from start of dancing to start of foraging, A foragers	1.5 min
T_3	time from start of foraging to start of unloading, A foragers	2.5 min
T_4	time from start of following dancers to start of foraging, A and B foragers	60 min
T_5	time from start of unloading to start of following, dancing, or foraging, B foragers	3.0 min
T_6	time from start of dancing to start of foraging, B foragers	2.0 min
T_7	time from start of foraging to start of unloading, B foragers	3.5 min

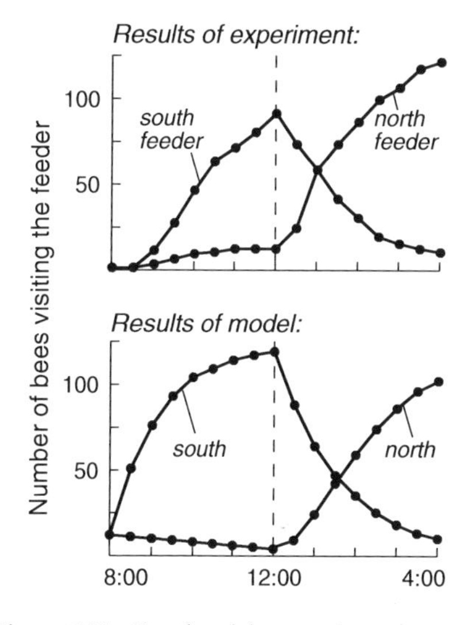

Figure 5.34 Results of the experiment conducted on 19 June 1989 *(top)* and of a mathematical model *(bottom)* analyzing how a colony selectively exploits the richer of two food sources. In the experiment, the number of bees visiting the feeder denotes the number of different individuals that visited a feeder during the previous half hour, as shown in Figure 5.32. In the simulation, the number of bees visiting the feeder is defined as the sum of the number of bees at the feeder, the number of bees at the hive unloading from that feeder, and the number of bees dancing for that feeder. The strong similarity between actual observations and computer simulation indicates that the mathematical model (Figure 5.33 and Equations 5.5–5.11) accurately describes the essence of the forager allocation process. After Seeley, Camazine, and Sneyd 1991.

the formulas shown in Equations 5.3 and 5.4, with $d_A = 0.38$ and $d_B = 0.02$ used as estimates for the proportion of time that bees in compartments *A* and *B* actually dance (see Seeley, Camazine, and Sneyd 1991 for a detailed discussion of how the model's parameters were estimated on the basis of empirical studies of bee foraging behavior).

Using the parameter values shown in Table 5.3, we can assess how well the model's predictions correspond with actual field observations. The top section of Figure 5.34 shows in detail the colony's response in the two-feeder choice test depicted in Figure 5.32, while the bottom section, for comparison, shows the computed solutions of the model. The starting conditions for the computer simulation were chosen to match the real-world example, where there were approximately 12 bees committed to each feeder, and during the course of the day a total of approximately 125 different bees visited the two feeders. A comparison of the top and bottom sections of Figure 5.34 shows that in the computed solutions of the model, as in reality, the colony exploits the most profitable nectar source and rapidly responds to changes in the location of the richer food source. In the experiment, the south feeder showed a rapid buildup of bees between 8:00 and 12:00 when it was loaded with a 2.5-mol/L sugar solution, followed by a decline in the number of bees visiting the feeder over the next 4 hr when the feeder was switched to a 0.75-mol/L solution. The computed solutions of the model show a similar pattern of rapid rise in bees at the south feeder (loaded with a 2.5-mol/L solution), and a marked decline 4 hr later when the feeder was switched to a 0.75 mol/L solution. As for the north feeder, in the experiment it initially contained a 1.0-mol/L sugar solution, rather than 0.75-mol/L solution, to prevent total abandonment of this feeder. Therefore, a slight build-up of bees was observed at the north feeder during the morning in the experiment. In the simulation, however, the north feeder initially contained a 0.75-mol/L solution (to keep the simulation simple) and it showed a slight decline in the number of bees. But during the afternoon, when for both the experiment and the simulation the north feeder was loaded with a 2.5-mol/L solution, in both settings the number of bees visiting the north feeder rose rapidly and with virtually identical trajectories.

These results show that the model captures the correct qualitative behavior, tracking the better of two nectar sources, and that it even provides a remarkably good quantitative match between simulation and observation. This demonstrates that the proper allocation of foragers among nectar sources can arise even if each bee has only extremely limited knowledge of the array of available nectar sources in the field. Indeed, the results of this simulation prove unequivocally that each employed forager needs only knowledge of the nectar source at which she is currently foraging and that each unemployed forager needs to follow only one randomly encountered dancer. Thus we can see that for a honey bee colony gathering its food, as for the natural world in general, the process of natural selection can generate a satisfactory solution to a problem without any of the participants having broad knowledge of the problem.

5.13. Cross Inhibition between Forager Groups

In the previous section, we treated the abandoning function (f_x) and the dancing function (f_d) as constants for a given level of nectar-source profitability (Table 5.3). We know, however, that the values of these two functions depend not only on the profitability of a bee's nectar source but also on several other variables, such as the colony's rate of nectar intake (see Section 5.7 and Figure 5.22). This fact is important to understanding how a colony's foragers become distributed among nectar sources, for the sensitivity of foragers to their colony's nectar influx endows the colony with a powerful mechanism of cross inhibition between groups of foragers gathering nectar from different patches of flowers. Such cross inhibition enhances a colony's ability to differentially exploit nectar sources with different levels of energetic rewards.

Two examples of this phenomenon are depicted in Figure 5.35, which shows how a colony that was exploiting two equally attractive food sources responded to an improvement in one both by increasing its exploitation of the richer source and by decreasing its exploitation of the poorer source. On 24 June 1991, a colony consisting of some 4000 bees labeled for individual identification was taken to the Cranberry Lake Biological Station, and was presented with two sucrose solution feeders, 400 m north and south of the hive (layout shown in Figure 5.26). On the morning of 28 June, the two feeders were filled with equivalent sucrose solutions, as is shown in Figure 5.35, and an assistant at each feeder recorded the identities of all bees visiting his or her feeder every half hour. Until 12:00, both feeders were visited by essentially the same, moderate number of bees. Then the south feeder was refilled with a richer sugar solution, and over the next 3 hr the number of bees visiting this feeder rose from some 25 bees to approximately 75 bees. Simultaneously, the numbers for the north feeder dwindled from 30 bees to about 15 bees, even though the conditions at this feeder were constant from morning to afternoon. Obviously, the stimulus of richer food in the south somehow led to an inhibition of the colony's response in the north. A similar pattern of response was found when the experiment was repeated 5 days later, on 2 July 1991.

How such cross inhibition arises is shown in Figure 5.36, which depicts the network of known inhibitory interactions between different groups of foragers inside a colony of bees. At the bottom we see two inhibitory pathways that operate in a rather simple, direct fashion. If, for example, the quality of nectar source A increases and the bees foraging there strengthen their dancing, the proportion of dances on the dance floor for nectar source B will decrease and the pool of unemployed foragers will shrink; hence the recruitment rate to nectar

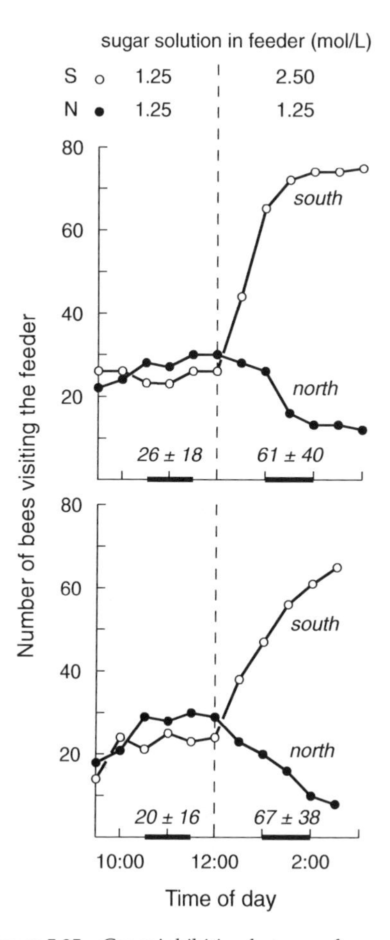

Figure 5.35 Cross inhibition between foragers gathering nectar from different sources. On 28 June 1991 *(top)* and 2 July 1991 *(bottom)*, a colony of 4000 bees labeled for individual identification was given two sugar water feeders, one 400 m north and the other 400 m south of the hive. When the concentration of sugar solution at the south feeder was raised at 12:00, the colony not only increased the number of bees visiting the south feeder, but also decreased the number of bees exploiting the north feeder. The numbers at the bottom of each plot indicate the mean (± SD) time to start of unloading by bees from the north feeder. The bar below each set of numbers indicates the time when data were gathered. Based on unpublished data of T. D. Seeley.

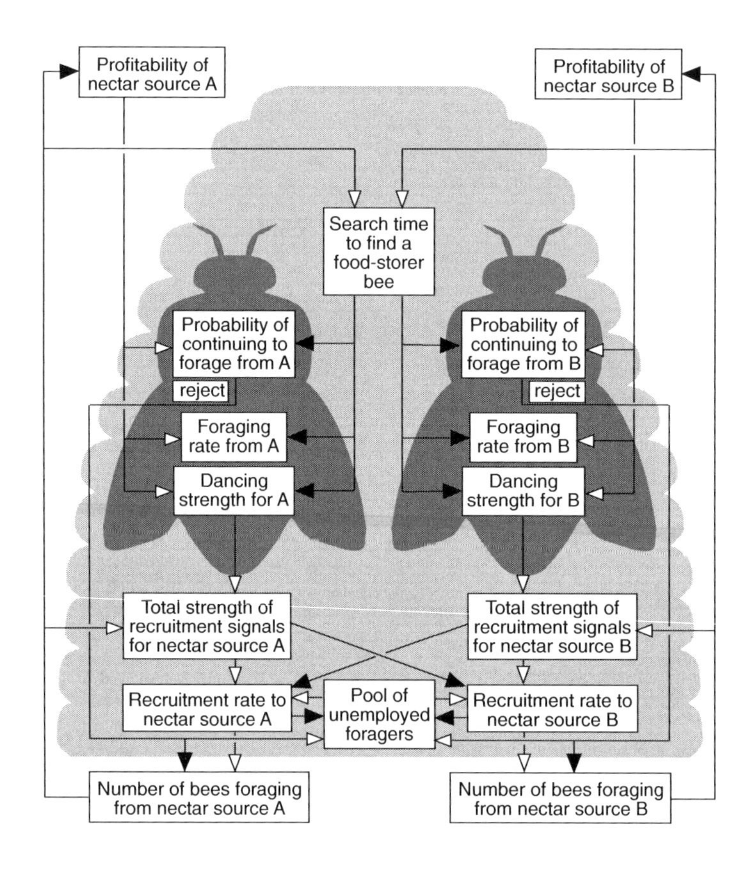

Figure 5.36 Flow diagram showing the pathways of excitatory and inhibitory influences among variables associated with nectar collection. White and black arrows denote pathways of excitation and inhibition, respectively. Note that a rise in the profitability of nectar source A, for example, has not only excitatory effects on the foraging rate, dance strength, and thus number of bees foraging from source A, but also inhibitory effects on the exploitation of a second source, B, whose profitability does not change. Such inhibition arises because an increase in the number of bees bringing in nectar from source A means that the bees from source B will have to search longer to find a food-storer bee, and so will reduce their dancing and their probability of continuing to forage at source B. Also, the dances for source B, if any, will become a smaller proportion of the total dances in the hive, and the pool of bees available for recruitment may shrink. All these changes will tend to increase the number of bees abandoning source B and to decrease the recruitment to source B. The net effect is that a rise in the profitability of nectar source A may trigger not only increased exploitation of nectar source A, but also decreased exploitation of nectar source B.

source B will be lowered. In short, stronger recruitment to one food source will *diminish the recruitment* to others through dilution of recruitment signals for them and through diminution of the colony's (finite) pool of recruits. Cross inhibition by this mechanism, however, cannot explain what is observed in Figure 5.35: namely a *rise in the abandonment* of another food source. This rise is caused by a more indirect means of cross inhibition. It is represented in the upper portion of Figure 5.36 and operates through a change in the colony's nectar influx, which alters the search time to find a food-storer bee by returning nectar foragers, and so causes these bees to adjust their acceptance and dance thresholds. Thus, for instance, if nectar source A becomes more profitable and is exploited more strongly, foragers from nectar source B will experience greater difficulty finding a bee to unload them upon return to the hive, and if nectar source B is only marginally profitable, the bees foraging there may decide to abandon it. This is almost certainly what happened in the two cases illustrated in Figure 5.35. In both instances, the time to start of unloading for the north-feeder bees increased by some 100 to 200% between morning and afternoon; hence the north-feeder bees must have markedly raised their acceptance thresholds. This fact, together with the fact that the north feeder contained only medium-quality food, makes it not so surprising that the number of north-feeder bees declined after rich food was supplied at the south feeder. Such cross inhibition via change in the colony's nectar influx probably also underlies the curious result reported by Boch (1956) of reduced dancing to one feeder when far superior food was suddenly provided at a second feeder (see Figure 5.20).

5.14. The Pattern and Effectiveness of Forager Allocation among Nectar Sources

The studies described above have revealed the mechanisms of forager allocation among nectar sources, but they tell us neither the allocation pattern that results nor its effectiveness for a colony of bees. These two topics have been investigated through the development of a system of differential equations that models the allocation process (Bartholdi, Seeley, Tovey, and Vande Vate 1993). It should be noted at the outset that this mathematical model is not a hypothesis about how the allocation process works; rather it is a quantitative description of the allocation process as revealed by empirical studies. Thus this model does not serve to analyze further the mechanisms of the allocation process, but instead helps us examine the logical implications of the bee's allocation process. Specifically, it sheds light on the distribution pattern of foragers among nectar sources and on the effects of this pattern on the colony's foraging success.

5.14.1. THE MODEL

Let F denote the set of nectar-bearing flower patches discovered by a colony, and for each patch x within the set F let n_x be the number of bees foraging at x (Table 5.4 summarizes the definitions of this and other variables used in the model). Let r_x denote the rate at which foragers return from patch x as a function of the number allocated to it (for simplicity of notation, we write r_x rather than the more cumbersome $r_x(n_x)$). These functions will differ at different patches. For example, r_x will depend on how long it takes a forager to fly from the hive to patch x and back. It will also depend on the distance between

Table 5.4. Definitions of variables used in the model of the allocation process. After Bartholdi, Seeley, Tovey, and Vande Vate 1993.

Variable	Definition (units)
f_x	Per capita rate of abandonment from patch x (abandonments/bee/hr)
g_x	Average duration of a dance for patch x (dance circuits/return to hive)
n_x	Number of bees foraging at patch x (bees)
N	Total number of active foragers (bees)
r_x	Rate at which foragers return from patch x (bees/hr)
T_x	Round-trip time to patch x: trip time (hr)

individual flowers at the patch, the amount of nectar per flower in the patch, and so forth. Generally, foragers working within the same patch will hinder each other's foraging. For example, the greater the number of foragers working at a patch, the greater the chance that a forager will land on a flower recently harvested by one of her nestmates. Thus we assume that r_x/n_x, the average number of trips a forager makes per unit time, is a nonincreasing function of n_x, the number of foragers allocated to the patch. Although foragers within a patch do hinder one another, we assume that this interference does not reduce the total rate of foraging at the patch. That is, we assume that the foraging rate r_x is a nondecreasing function of n_x, the number of foragers allocated to the patch.

When a forager returns from a patch, she decides whether to go back to the patch or abandon the patch. We let f_x denote the average rate at which each forager is diverted from patch x. The value of f_x will depend on n_x, though not because a bee assesses n_x directly, but because the foraging profitability that she experiences at the patch is a function of n_x. As with r_x, we write f_x rather than $f_x(n_x)$, for simplicity of notation. When a forager abandons one patch, she stops foraging for a while and then reenters the foraging process. In this model, we will assume that all foragers do so by following recruitment dances. We will also assume that foragers abandon patches (and so stop foraging) and get recruited to patches at the same average rate, so that the average number of active (employed) foragers remains constant.

We will denote the average duration of each dance for patch x as g_x, where g_x is a function of n_x but is represented simply as g_x. Therefore an allocation n leads to dancing for patch x at the rate $g_x r_x$, and to dancing for all patches at a combined rate of

$$GR = \Sigma g_x r_x \qquad x \text{ in } F \tag{5.12}$$

Thus, of all foragers reentering the foraging process, the fraction $g_x r_x / GR$ are recruited to patch x on average (see Section 5.10).

Since foragers abandon patch x at rate $f_x n_x$, to ensure that the number of employed foragers remains fairly constant, workers must reenter the foraging process at the average rate

$$fN = \Sigma f_x n_x \qquad x \text{ in } F \tag{5.13}$$

Thus, the system of differential equations

$$dn_x / dt = fN \frac{g_x r_x}{GR} - f_x n_x \qquad \text{for each } f_x \text{ in } F \tag{5.14}$$

describes the allocation process.

In a steady state, the average rates of recruitment and abandonment must balance, and so

$$\frac{n_x}{r_x} = g_x \frac{f}{f_x} \frac{N}{GR} \tag{5.15}$$

for each active patch x. Thus, in a steady state the average trip-time of each active patch x must be proportional to $v_x = g_x\,(f/f_x)$, that is,

$$T_x = v_x \frac{N}{GR} \tag{5.16}$$

for each patch x. The term v_x is a measure of the value of patch x, since as patch quality increases, g_x (the average dance duration) will rise and f and f_x (the per capita abandonment rate) will fall. In summary, over time a colony converges to the allocation with

$$\frac{v_x}{T_x} = \frac{GR}{N} \tag{5.17}$$

whereby each bee will accumulate value at the same rate, regardless of which patch she is visiting. We will call this the *equal value rate* allocation.

It is easy to see how this equalization can arise. If a patch starts out with too few foragers, its value will be disproportionately high (its foragers will have a higher than average dancing rate and a lower than average abandonment rate). This will result in an increase in the number of foragers working the patch. Conversely, if a patch has too many foragers, it will lose foragers due to a disproportionately low value (its foragers will have a lower than average dancing rate and a higher than average abandonment rate). Such gains and losses in the number of foragers will tend to bring the per capita rate of value intake from each exploited patch to the same, steady-state level.

5.14.2. TESTING THE MODEL

One can test the model by checking the prediction that in a steady state bees from different nectar sources will have equal rates of value accumulation. One can accomplish this by creating two artifical nectar sources which differ in distance from the hive or concentration of the sugar solution (or both), letting the colony's exploitation of the two feeders stabilize, and then measuring the average amount of dancing per return to the hive (g_x), the average abandonment rate (f_x), and the average trip-time (T_x) for the foragers from each feeder. If the model is correct, then from Equation 5.17 we can predict that the following relation will hold:

$$\frac{g_1}{f_1 T_1} = \frac{g_2}{f_2 T_2} \tag{5.18}$$

On 4 July 1991, Craig Tovey, two assistants, and I performed one test of this prediction (Bartholdi, Seeley, Tovey, and Vande Vate 1993). At this time we had a colony of some 4000 bees labeled for individual identification established in an observation hive at the Cranberry Lake Biological Station. We trained approximately 10 bees from this colony to each of two sucrose solution feeders which were identical in distance from the hive (350 m, one north and one south of the hive), but which differed in the concentration of the sugar solution (2.0 and 2.5 mol/L, respectively). These two feeders were essentially the only sources of nectar available to the colony. At 9:00 the feeders were loaded with their sugar solutions, and by 11:00 the colony had produced a stable distribution of foragers between the two feeders. This steady state was maintained for the next 4 hr: north feeder 20.9 ± 0.8 bees, south feeder 39.0 ± 2.2 (x ± SD). During this 4-hr period we measured the following: (1) the average trip-time of foragers, by recording the trip-times of 20 randomly chosen foraging trips to each feeder; (2) the total number of dance circuits performed for each feeder, by steadily monitoring the dancing in the hive; and (3) the number of abandonments from each feeder, by performing rolls calls every half hour of the bees visiting each feeder. A bee was judged to have abandoned a feeder if she was not recorded on two consecutive roll calls for the feeder.

The results of these measurements, together with the calculations of g_x, f_x, and g_x/f_xT_x, are summarized in Table 5.5. They reveal a close match between the model's prediction and reality. Although the two feeders differed markedly in initial profitability, so that the richer (south) feeder ended up with nearly twice as many foragers and received more than three times as many dance circuits as the poorer one, at steady state the two feeders differed only slightly (by a factor of 1.2) in their values of g_x/f_xT_x. Indeed, if there had been just one more abandonment of the south feeder (7 instead of 6), then the two feeders would have had identical values of g_x/f_xT_x.

5.14.3. EFFECTIVENESS OF THE ALLOCATION PATTERN

How effective is the equal value rate allocation pattern in terms of enabling a colony to successfully gather its nectar? This brings us again (see Section 5.5) to the question of what the proper currency is for measuring the foraging success of a colony. Natural selection among honey bee colonies probably does not favor simply efficiency *or* rate of energy collection. Rather, it is more likely that colonies that are able

Table 5.5. Test of the prediction that at equilibrium a colony's nectar foragers are allocated among nectar sources such that the rate of value accumulation is the same for all sources. After Bartholdi, Seeley, Tovey, and Vande Vate 1993.

	Feeder	
Variable	North	South
Sugar concentration (mol/L)	2.00	2.50
Number of foragers, n_x (bees)	20.9	39.0
Trip-time, T_x (hr)	0.165	0.140
Total trips to feeder in 4 hr, $n_x(4/T_x)$	507	1114
Total dance circuits in 4 hr	37	130
Dance duration, g_x (dance circuits/trip)	0.073	0.117
Number of abandonments in 4 hr (bees)	2	6
Per capita abandonment rate, f_x (abandonments/bee/hr	0.024	0.038
g_x/f_xT_x	18.4	22.0

to balance various foraging considerations, such as energetic efficiency *and* rate of energy intake, will have an evolutionary advantage, because ideally a colony will gather its food as quickly and efficiently as possible. Thus the best currency for evaluating the effectiveness of forager allocation patterns may be rather complex. It can be proven mathematically (Bartholdi, Seeley, Tovey, and Vande Vate 1993) that the equal value rate allocation is effective with respect to the currency:

$$VR = \Sigma v_x r_x \qquad x \text{ in } F \tag{5.19}$$

which we call the "rate of value accumulation." This is expressed more precisely as follows. Let n be any allocation, and let n^* be the equal value rate allocation. If at each patch x, $v_x r_x$ (the total rate of value accumulation from patch x) is a nondecreasing function of n_x, and $v_x r_x / n_x$ (the per capita rate of value accumulation from patch x) is a nonincreasing function of n_x, then

$$\frac{VR(n)}{VR(n^*)} \leq 2 \tag{5.20}$$

This equation states that no allocation is more than twice as effective in maximizing the colony's rate of value accumulation as the equal value rate accumulation.[3] Precisely what this implies about a colony's foraging success depends on the meaning of v_x. One plausible interpretation is that v_x is proportional to the energetic efficiency of foraging at patch x, since $v_x = g_x(f/f_x)$, and there is solid evidence that g_x is proportional to the energetic efficiency of foraging at patch x (see Figure 5.10). In this case, $VR(n)$ is a measure of foraging success (accumulation of value) which combines considerations of both efficiency and rate of foraging. For example, if a colony's nectar foragers are efficient, but have a low rate of foraging trips, or if they have a high rate of foraging trips but forage inefficiently, the colony's rate of value accumulation will be low. Achieving a high rate of value accu-

3. On first inspection, the performance guarantee "no allocation is more than twice as effective" may not seem remarkable. But in comparison with other similar performance guarantees, it is actually quite impressive. For example, the best-known mathematical technique for finding an approximate solution to the traveling salesman problem (finding the shortest travel route linking several locations) invokes some of the most powerful tools of discrete mathematics and still only guarantees a route no more than 1.5 times times as long as the optimum. Simpler technique are still quite complex and can only guarantee a route that is at most twice as long as the optimum.

mulation requires that the colony's nectar foragers be allocated among patches in a way that produces at least moderate levels of both foraging efficiency and rate of foraging trips. Note that this mathematical analysis does not prove that rate of value accumulation is the appropriate currency of colonial foraging success; rather it shows how honey bee foraging behavior can lead to foraging success at the colony level.

It should also be noted that the equal value rate allocation of the foragers among nectar sources will be identical to the ideal free distribution (see Fretwell and Lucas 1970; Fretwell 1972), if we assume that foraging bees assess patch profitability in terms of rate of value accumulation, since both distributions will result in all individuals experiencing equal rates of value accumulation. This is remarkable because the behavioral mechanisms for producing these two distributions are extremely different. Whereas the equal value rate allocation arises out of the situation where each bee acts only on *local* knowledge of the available nectar sources (each bee knows about the profitability of just one flower patch and adjusts her foraging behavior accordingly), the ideal free distribution arises from the situation in which each forager possesses *global* knowledge of the available nectar sources and independently chooses among them, selecting the one where she can achieve the highest rate of value accumulation. Hence this modeling analysis reveals that the honey bees' social organization enables a colony to produce, with bees possessing only very limited information, a steady-state labor distribution among nectar sources as effective as the one that would be produced if each forager were omniscient about the nectar sources.

Summary

1. To understand how a colony responds to the information it has acquired about food sources, it is important to distinguish between employed and unemployed foragers—between foragers that are and are not engaged in exploiting a patch of flowers. Only the unemployed foragers draw on the large body of information about food sources that is displayed on the dance floor. The employed foragers ignore the information presented in dances and rely instead on information acquired while foraging at their flower patch. Thus each employed forager knows only about her particular patch of flowers,

whereas each unemployed forager has an opportunity to acquire broad knowledge of the food sources being exploited by her colony.

2. Careful observation has shown how unemployed foragers read the information on the dance floor. Dance-following bees do not conduct a thorough survey of the information available on the dance floor before leaving the hive to search for a new food source. Instead, each dance-following bee follows just one dancer closely (for several waggle runs) before leaving the hive (Figure 5.24). These observations also reveal that it is exceedingly rare for a dance-following bee to follow a bee's dance from start to finish. This finding suggests that dance followers do not acquire information about food-source profitability from dances, because, as was shown earlier (Section 5.4), information about profitability is strongly expressed only in dance duration.

The question of how unemployed foragers sample the information on the dance floor has also been addressed experimentally. The critical experiment involved presenting a colony with two sucrose solution feeders that were equidistant from the hive but different in profitability. If each unemployed forager follows multiple dances, compares them, and selectively responds to the strongest one, then dances for the more profitable feeder will be disproportionately effective per waggle run, and the proportion of recruits to the richer feeder will exceed the proportion of waggle runs for that feeder. If, however, each unemployed forager follows just one dance, chosen more or less at random, then the proportion of recruits to each feeder will match the proportion of waggle runs for that feeder. Eleven trials of this experiment were performed, and in each the proportion of waggle runs for each feeder accurately predicted the proportion of recruits arriving at the feeder (Figure 5.28). Thus it is clear that the unemployed foragers do not follow multiple dances and thus do not selectively respond to those advertising the best food source.

The results indicate instead that each bee samples just one dance, chosen basically at random, before leaving the hive to search for a new food source. This in turn implies that many bees will fail to learn about the best food source (Figure 5.29). As to why bees do not sample broadly and so increase their probability of acquiring information about the best food source, it may be better for a colony if its foragers are not broadly informed and choosy, for this situation would lead to an all-or-none response to food sources. Such a response pattern is inappropriate for a bee colony, which has many foragers and so needs to distribute its foragers among multiple food sources, rather than crowd them onto the one best source.

3. Once a bee has located a nectar source, she carefully adjusts multiple variables of her behavior in accordance with the profitability of the source. The general pattern of this behavioral modulation is as follows: as nectar-source profitability increases, the tempo of foraging increases, the volume of nectar loaded increases, the duration of dances increases, and the probability of abandoning the nectar source decreases (Figure 5.31).

4. Each forager within a honey bee colony possesses only limited knowledge about the array of food sources being exploited by her colony, yet the colony as a whole responds in a coordinated fashion to the full array of foraging opportunities. A colony's skill in allocating foragers among nectar sources arises by a process analogous to natural selection. In other words, decision-making regarding the distribution of bees among alternative nectar sources occurs not through some high-level, well-informed supervisor, but through the differential "survival" (persistence at a food source) and "reproduction" (recruitment to a food source) of the foragers, each of which knows only about her particular patch of flowers (Figure 5.32).

This idea was tested by means of a mathematical model that endows each forager with strictly limited information, and has each forager responding in a realistic fashion to this small amount of information (Fig 5.33). The model was tested by comparing its predictions to the pattern of forager allocation to two nectar sources that was observed in an experimental situation. The model has the correct qualitative behavior, tracking the better of two nectar sources, and even provides a remarkably good quantitative match between simulation and observation (Figure 5.34). Both model and experiment therefore demonstrate that it is possible for a colony to build a globally correct solution to the labor allocation problem out of the locally guided actions of its members.

5. Colonies show cross inhibition between groups of foragers gathering nectar from different patches of flowers (Figure 5.35). This inhibition enhances a colony's ability to respond differentially to nectar sources with different levels of energetic rewards. The strong exploitation of a rich new nectar source inhibits the exploitation of poorer sources in two ways. First, it *reduces recruitment* to the poorer sources by lowering the fraction of dances on the dance floor that are for the poorer sources and thus siphoning off potential recruits. Sec-

ond, it *boosts abandonment* of the poorer sources by raising the colony's nectar influx, which causes all foragers to raise their acceptance thresholds, and so increases the probability of abandonment by bees foraging at marginally attractive nectar sources (Figure 5.36).

6. The description of the mechanisms of forager allocation among nectar sources does not reveal the allocation pattern that results or the effectiveness of the nectar collection that arises from this allocation pattern. I investigated these topics by developing a system of differential equations that models the allocation process in a hypothetical colony of bees whose behavior closely approximates the observed behavior of real bees. The hypothetical colony tends toward a specific allocation of foragers among nectar sources, one in which each forager experiences the same rate of "value" accumulation regardless of which source she is visiting (the equal value rate allocation). I empirically tested the accuracy of the model by seeing whether, at a steady state, the foragers from different artificial nectar sources do indeed experience equal rates of value accumulation. The results of one test confirm this prediction of the model. Finally, I used the model to evaluate the effectiveness of a colony's allocation pattern. The result shows that no allocation brings value to the hypothetical colony's hive more than twice as quickly as does the equal value rate allocation. Evidently, the allocation pattern of a colony, even though it is generated in a distributed fashion by bees possessing only limited information, is quite good at meeting a colony's need to gather nectar as quickly and as efficiently as possible.

Coordination of Nectar Collecting and Nectar Processing

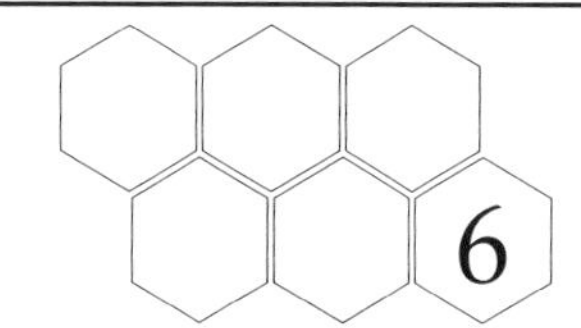

An important feature of the organization of nectar collection by honey bee colonies is the division of labor between the foragers, bees that work outside the hive to gather the nectar, and the food storers, somewhat younger bees that work inside the hive to process the nectar, either distributing it to hungry nestmates or storing it in the combs for future consumption (Figure 6.1). This specialization of bees on different parts of the overall task of nectar collection undoubtedly boosts the efficiency of a colony's energy acquisition. It means, for example, that once a forager has located a rich nectar source, she can concentrate on exploiting the source before it fades rather than dividing her efforts between collecting and processing activities. At the same time, however, this division of labor creates a problem of coordination within a colony because the rates of nectar collecting and processing must be kept in balance for the overall operation to proceed smoothly. If the collecting rate exceeds the processing rate, foragers will experience long unloading delays upon return to the hive. Reciprocally, if the processing rate—or, more precisely, the processing capacity—exceeds the collecting rate, the food storers will be underemployed.

Keeping the two rates matched is a major problem because, as shown earlier (Section 2.6), colonies experience large and unpredictable variation from day to day in the supply of nectar in the environment, a function of the plants in bloom and the weather conditions. Since a colony generally strives to acquire as much nectar as possible, it will make internal adjustments so that its nectar collection rate rises whenever the nectar supply in the external environment increases. Hence a colony's nectar collection rate can change dramatically, even from one day to the next. For instance, on a day of cool, rainy weather, the nectar intake of a colony may be zero grams, while just a few days later, on a warm, sunny day in the middle of a nectar flow, it may surge to several kilograms (Visscher and Seeley 1982; see Figure 2.15 in this book). Such strong variation in the collection rate requires, in turn, equally strong adjustments in the colony's nectar processing capacity. We shall see that honey bee colonies possess special mechanisms of feedback control which keep a colony's rates of nectar collecting and processing in balance.

How a Colony Adjusts Its Collecting Rate with Respect to the External Nectar Supply

6.1. Rapid Increase in the Number of Nectar Foragers via the Waggle Dance

A colony's rate of nectar collection (*C*) is a function of three variables:

$$C = \frac{N_c L_c}{T_c} \qquad (6.1)$$

where N_C is the number of foragers engaged in nectar collection, L_C is the average volume of a nectar load, and T_C is the average time of a collecting cycle for nectar foragers (Figure 6.1). Although detailed

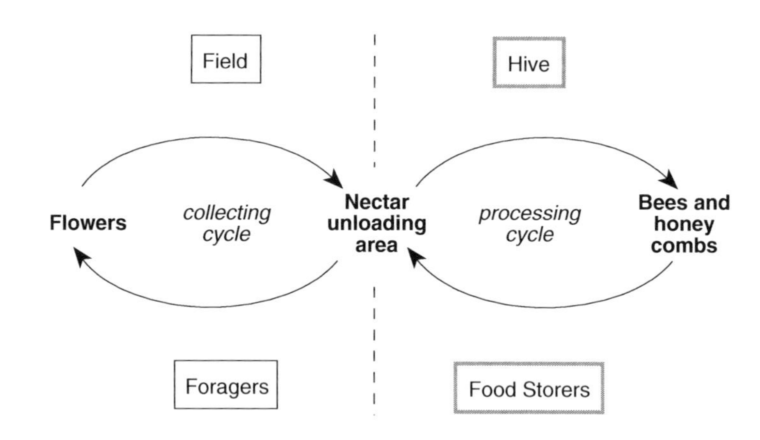

Figure 6.1 The two distinct but intersecting collecting and processing cycles that form the nectar acquisition operation. The collecting cycle occurs mainly in the field as foragers collect nectar at flowers, bring it back to their hive, and then return to the flowers to gather more nectar. The processing cycle takes places entirely within the hive as food-storer bees unload the returning nectar foragers in the unloading area (just inside the hive entrance; see Figure 5.16), transport the fresh nectar deep inside the hive to other bees for immediate use or to the honeycombs for storage, and then crawl back to the nectar unloading area to service more foragers.

data are not available on changes in T_C and L_C in relation to changes in the nectar supply in the environment, it is likely that the former often decreases and the latter often increases as nectar becomes more plentiful. Park (1929) reports, for example, that T_C for bees gathering nectar from white sweet clover *(Melilotus alba)* was 45 min under favorable conditions (study colony gained more than 2.0 kg per day), and 63 min under mediocre to poor conditions (daily weight gain less than 0.5 kg). Evidence for adjustment of L_C comes from Núñez (1966, 1970), who measured the nectar loads of bees visiting a sugar solution feeder as a function of the flow rate of the feeder. At the lowest flow rate (1.0 μL/min) the bees imbibed only about 20 μL per foraging trip on average, whereas at the highest flow rate (16.7 μL/min) they loaded nearly 50 μL on average. Indeed, beekeepers can tell when a strong nectar flow is under way by noticing when returning foragers land heavily at the hive entrance with abdomens massively swollen with nectar. Thus it seems clear that a colony's nectar influx rises under favorable conditions partly through a drop in T_C and partly through a rise in L_C.

A colony's principal means of adjusting its nectar influx in relation to nectar abundance, however, is evidently changing N_c, the number of bees actively collecting nectar. Whereas T_c probably can vary by as much as a factor of 10, and L_c by a factor of 5, N_c certainly can be adjusted by a factor of 50 or more, and in just several hours. Such a powerful adjustment within a colony is accomplished when successful nectar foragers perform waggle dances to draw previously unemployed foragers into nectar collection. One experimental result that illustrates the lability of the variable N_c is depicted in Figure 6.2. On the afternoon of 24 July 1991, at the Cranberry Lake Biological Station, 10 bees from a colony of some 4000 bees occupying a two-frame observation hive were trained to a feeder loaded with a 2.5-mol/L sucrose solution and located 350 m north of the observation hive. The time of year and location of this experiment were such that this feeder was essentially the sole source of nectar available to the study colony, as is indicated by a traffic level of less than 1 bee/min into the hive when food was not provided at the feeder. The next morning, each of the 10 bees trained to the feeder was gently captured in a plastic bag upon arrival at the feeder, to prevent the colony from exploiting the feeder. All were captured by 9:00. At this time a videocamera began recording all waggle dances performed for the feeder. Initially, of course, there were none. Then at 10:30, one of the 10 bees (the

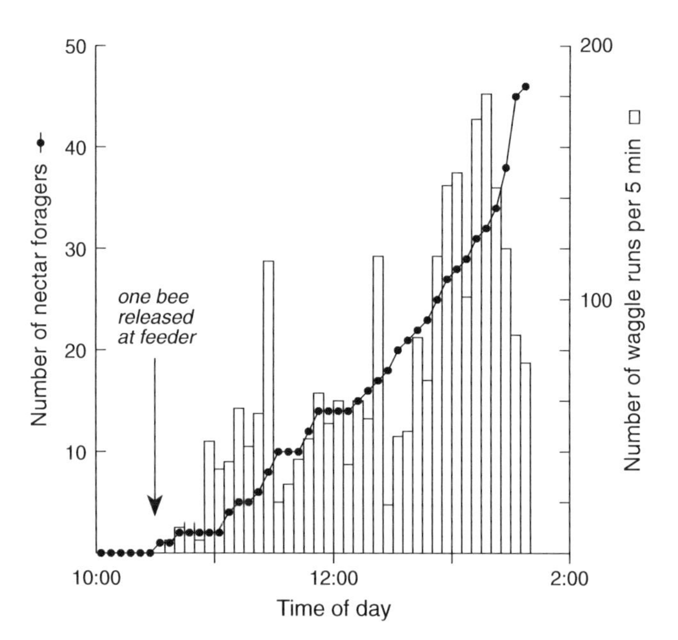

Figure 6.2 Rapid increase in the number of foragers engaged in nectar collection by means of the waggle dance. A colony of some 4000 bees was established in an observation hive at the Cranberry Lake Biological Station. On the morning of 25 July 1991, no natural nectar sources were available and no bees were collecting nectar or producing waggle dances. Then one bee was released at a feeder 350 m north of the hive that provided a 2.5-mol/L sucrose solution. Over the next 3 hr, the feeder was strongly advertised in the hive by dances containing a total of 2538 waggle runs. As a result the number of foragers exploiting the feeder skyrocketed from 1 to 46. Based on unpublished data of T. D. Seeley.

strongest dancer in the group) was released, whereupon she resumed foraging and began performing waggle dances to recruit additional foragers to the feeder. Each recruit was labeled upon her first trip to the feeder. Over the next 3 hr, the first forager conducted 28 trips to the feeder and performed a total of 871 waggle runs in the hive, which together with 1667 waggle runs performed by other bees that were recruited to the feeder, triggered an impressive explosion, from 1 to 46, in the number of bees (N_c) bringing nectar into the hive.

6.2. Increase in the Number of Bees Commited to Foraging via the Shaking Signal

Recent work is beginning to suggest that a honey bee colony can respond to a rise in the nectar supply not only by activating the existing foragers but also by increasing the number of bees in the colony commited to foraging. The mechanism of this adjustment appears to involve a striking communication behavior performed by foragers, one that von Frisch (1967) called the "jerking dance" and others have called the "vibration dance" (Schneider, Stamps, and Gary 1986a,b), "shaking dance" (Gahl 1975), "dorso-ventral abdominal vibration" (D-VAV) (Milum 1955), or simply "shaking" (Allen 1959). To produce this signal, a bee vibrates her whole body dorso-ventrally at about 16 Hz for 1–2 sec, usually while grasping another bee with her legs (Milum 1955; Gahl 1975) (Figure 6.3). After producing one such shaking signal, the bee will typically crawl a short distance across the comb and soon shake another bee. Bees show great variation in the frequency (from 1 to 20 bees shaken per min) and number (up to about 200) of signals produced consecutively (Allen 1959). The bees that receive the shaking signal simply stand motionless until released by a shaker.

Figure 6.3 One bee signaling another by means of the shaking behavior. The arrow indicates the dorso-ventral vibration of the bee's body. In transmitting this signal, a bee shakes a series of different bees, each one only briefly (1–2 sec), and at a rate of 1 to 20 bees/min. The frequency of the shaking vibration is approximately 16 Hz. Original drawing by M. C. Nelson.

Several pieces of evidence suggest that this shaking signal serves to boost the number of bees engaged in foraging at times of abundant forage or of great need for food, or both (Figure 6.4). First, when Schneider, Stamps, and Gary (1986a) compared the behavior of bees that had and had not been shaken, they found that those shaken were significantly more likely to crawl onto the dance floor during the next 30 min ($P = 0.52$) than were those that were not shaken ($P = 0.16$). This response to shaking tends to bring the shaken bees into contact with waggle dances, which should stimulate them to begin foraging. Moreover, Schneider and his coworkers found that even rather young bees, ones only 10 to 14 days old, will show the response of moving toward the dance floor. Such young bees are usually engaged in food processing rather than foraging (Section 2.2), but presumably can be induced to switch to the latter task. Thus it appears that bees other than foragers are important targets of the shaking signal. This idea derives further support from the spatial distribution of shaking signals. They are produced throughout the hive (James Nieh, personal communication), where they will be received by bees performing diverse tasks, rather than just in the dance floor, where the bees already commited to foraging are concentrated.

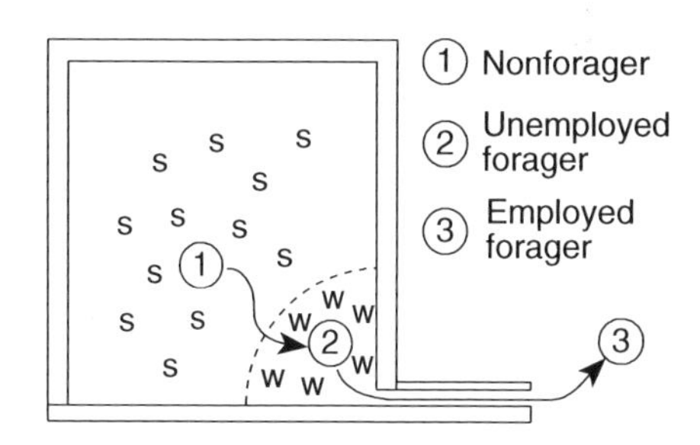

Figure 6.4 The joint effects of shaking signals and waggle dances in boosting the number of bees gathering nectar in a honey bee colony. Shaking signals (*S*) occur throughout the hive while waggle dances (*W*) occur primarily in the dance floor area near the hive entrance. Shaking signals stimulate nonforagers to move onto the dance floor, where they will encounter waggle dances, which will stimulate them to leave the hive and begin foraging.

Another piece of evidence regarding the functional significance of the shaking signal is the set of circumstances under which bees produce these signals (Schneider, Stamps, and Gary 1986a). Over the course of a day, their production in a colony peaks early in the morning, *before* foraging activity is fully under way. And over the course of a year, the morning peaks in shaking signals are seen mainly at times of abundant forage. Schneider, Stamps, and Gary (1986b) also investigated experimentally the hypothesis that shaking signal production depends on a colony's recent success in food collection. To manipulate a colony's foraging success, they attached a large flight cage (2 × 2 × 4 m) to the entrance of an observation hive, and placed

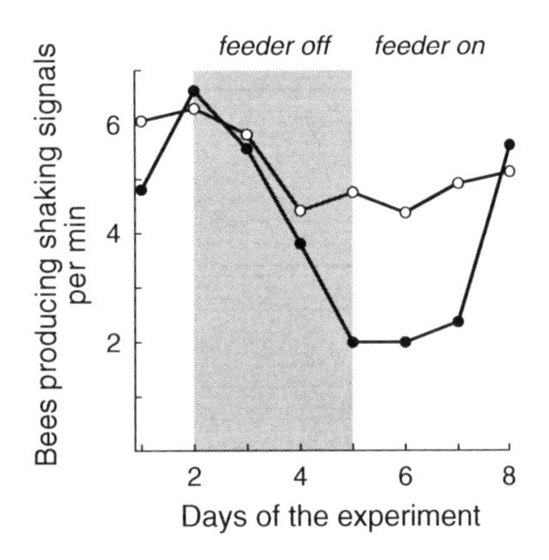

Figure 6.5 Results of one trial of an experiment testing the hypothesis that foragers are stimulated to produce shaking signals by periods of highly successful foraging. An experimental colony's foraging success was manipulated by restricting its foragers to a flight cage containing a sugar solution feeder, while a control colony's foraging was not manipulated at all. For each colony, the number of bees producing shaking signals was determined hourly between 7:00 in the morning and 5:00 in the afternoon; the values shown are the maximum values recorded each day. In the experimental colony *(filled circles)*, the production of shaking signals declined quickly when the feeder was shut off and rose again only after 2 days with the feeder back on. In the control colony *(open circles)*, the production of shaking signals changed relatively little over the 8 days of the experiment. After Schneider, Stamps, and Gary 1986a.

a sucrose solution feeder within the cage. By adjusting the amount of sugar solution presented in the feeder, they could precisely regulate the colony's daily intake of "nectar." A nonmanipulated control colony was also established in an adjacent observation hive. They measured the intensity of shaking signals for both experimental and control colonies by making hourly counts. Figure 6.5 depicts the results from one trial of the experiment. It shows that blocking the experimental colony's intake of sugar solution was followed by a steady decline in the height of the morning peak of shaking signals, and that providing a strong influx of sugar solution was followed eventually by a rise in signal production. By labeling the bees visiting the feeder in the flight cage, the experimenters also determined that these forager bees were primarily responsible for the shaking signals in the experimental colony. Taken together, these findings suggest that foragers are stimulated to produce shaking signals by several days of good foraging, apparently with the result that a colony devotes additional labor to foraging once it has experienced an extended period of highly successful foraging.

Recent observations regarding the context and pattern of shaking signal production also suggest strongly that this signal works in conjunction with the waggle dance to arouse bees to begin foraging (Seeley, Weidenmüller, and Kühnholz, unpublished results). My two colleagues and I have found that a forager bee has a high probability of producing shaking signals when she finds a rich food source after having experienced a long period of poor forage and thus little or no foraging activity. The production of shaking signals seems especially strong if the successful forager returns to a hive in which most of the foragers are still inactive, owing to the recent scarcity of forage, and so are standing essentially motionless on the combs. In nature, these circumstances will arise both when a spell of bad weather prevents a colony's foragers from venturing outside the hive and then one morning good weather again prevails, and when a nectar flow starts up following a prolonged nectar dearth. Working at the Cranberry Lake Biological Station—where food sources besides our feeders are scarce—we discovered that we can reliably induce bees to produce the shaking signal if we first train them to a sucrose solution feeder, then leave the feeder turned off for a few days so that the colony experiences a period of extremely meager nectar collection, and finally refill the feeder with a rich sucrose solution. Many of the bees trained to forage at the feeder, upon dis-

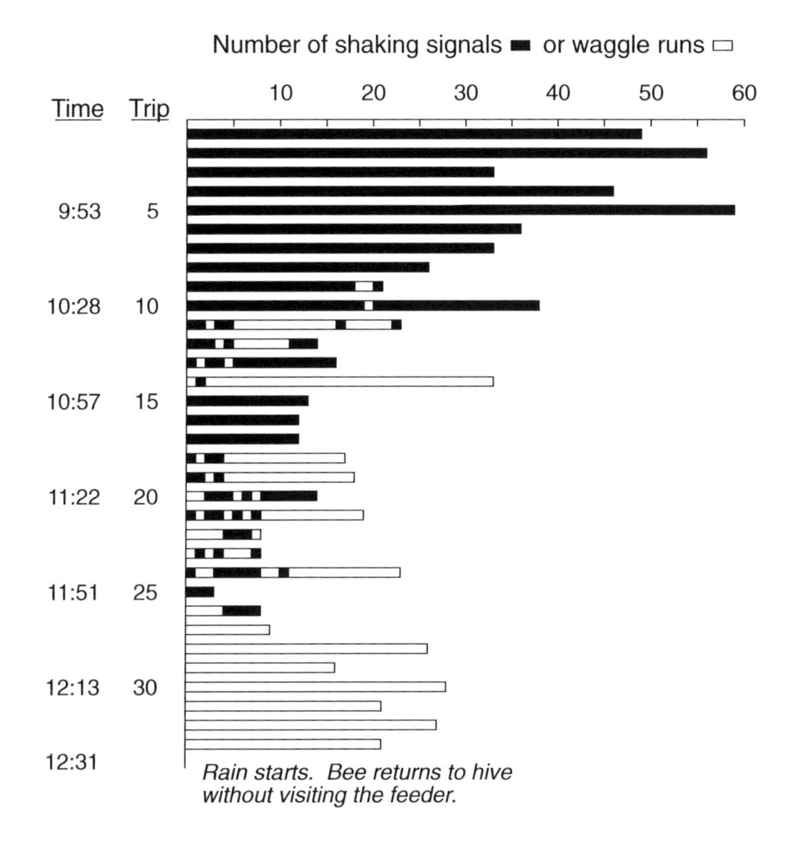

Figure 6.6 Production of shaking signals and waggle runs by one forager bee on the morning of 23 July 1994. This bee was a member of a colony established in an observation hive at the Cranberry Lake Biological Station, where forage is extremely sparse. On several days preceding the observations (17–20 July), this bee experienced successful foraging at a sucrose solution feeder located 350 m from the hive, but on the 2 days immediately preceding the observations (21–22 July) she had no success foraging because the feeder was left empty. Thus her discovery of rich forage at the feeder on the morning of 23 July followed a 2-day period of extremely low foraging activity, both by herself and by her colony as a whole. She performed mainly shaking signals on her first 10 returns to the hive from the feeder, apparently to arouse her still quiescent nestmates, but on subsequent returns she gradually switched to the production of waggle runs. Based on unpublished data of T. D. Seeley, A. Weidenmüller, and S. Kühnholz.

covering that it has been refilled, will show a behavior pattern like that depicted in Figure 6.6. Initially, when the forager bee returns to the hive she produces a lengthy string of vigorous shaking signals but no waggle runs. Eventually, she shifts to producing both shaking signals and waggle runs. And ultimately, she completes the switch between signal types and performs just waggle runs. It is a striking sight to behold a forager run excitedly into the hive and begin shaking her resting nestmates, even before she attempts to unload her sugar solution. Certainly, it appears as if she is trying hard to arouse them with her shaking signals, and there is no question that over the next hour the bees inside the hive become much more active, crawling about and following waggle dances rather than simply standing motionless on the combs.

Thus there is good, though still preliminary, evidence that the shaking signal is an important mechanism of information flow in the control of a colony's foraging operation. Certainly, it is a communication process that merits deeper investigation. Both the specific constellation of stimuli that cause a forager bee to produce the shaking signal and the full effects of this signal on individual bees and on whole colonies, remain important subjects for future studies.

How a Colony Adjusts Its Processing Rate with Respect to Its Collecting Rate

6.3. Rapid Increase in the Number of Nectar Processors via the Tremble Dance

A colony's rate of nectar processing (P), like its rate of nectar collecting, is a function of three variables:

$$P = \frac{N_P L_P}{T_P} \tag{6.2}$$

where N_P is the number of bees engaged in processing nectar (usually called receiver bees or food-storer bees), L_P is the average volume of nectar loaded by a food-storer bee, and T_P is the average time of a nectar processing cycle (see Figure 6.1). There is now no question that a colony can swiftly and dramatically raise its processing rate when its collecting rate rises, but the details of how this is accomplished are not understood fully. All three variables affecting the processing rate may be adjusted by the bees, but currently we have evidence only for the adjustment of N_P.

The first sign that a colony can speedily adjust the number of bees devoted to nectar processing came in the form of a setback to an experiment I was conducting to determine how nectar foragers acquire information about their colony's nectar influx (Seeley 1989a). As explained previously (Section 5.7.3), this experiment involved labeling all the food storers in a colony, then removing them at the end of the day, and observing the next day whether or not the nectar foragers showed a response to the increased difficulty of finding a food-storer bee (thereby distinguishing between the unloading-experience hypothesis and the nectar-odor and bee-traffic hypotheses). In particular, on 14 July 1987, I worked with a colony of some 4100 bees, and between 9:00 in the morning and 5:30 in the afternoon I labeled in the hive a total of 753 bees that were seen receiving sugar solution from foragers returning from a feeder. In the course of labeling these food-storer bees, I also gathered data on how long the returning foragers had to search inside the hive to find food-storer bees. Their searches proved quite short, only 11 sec on average, throughout the day (Figure 6.7; see also Figure 5.17). Finally, between 6:00 and 8:00 in the evening, I removed all the labeled food-storer bees from the colony by opening the hive and plucking the labeled bees off the combs with forceps and placing them in a small cage. The next morning I watched anxiously at the observation hive to see whether the foragers returning from the feeder would now have to search longer to find a food-storer bee and, if so, whether they would be less likely than on the previous day to perform waggle dances for the feeder. Initially, from 8:30 to 9:00, the bees' search times were pleasingly higher than on the previous day, 32 sec on average, but to my surprise and dismay, the average search time quickly declined, so that by 1:00 there was no significant difference between 14 and 15 July in the bees' search times (Figure 6.7). Somehow the colony had managed to replace the food storers that I had worked so hard to remove. How the colony had accomplished this I did not know, but I had noticed a striking behavior early on the morning of 15 July, when the colony was still essentially devoid of food storers: about 10% of the foragers from the feeder, after searching about inside the hive for an unloader without success, began performing an intriguing maneuver that von Frisch had called a tremble dance. The function of this dance had remained a mystery since von Frisch first described it in 1923, and I wondered if the tremble dancing was related to the increase in food storers, but did not have enough data to establish a link.

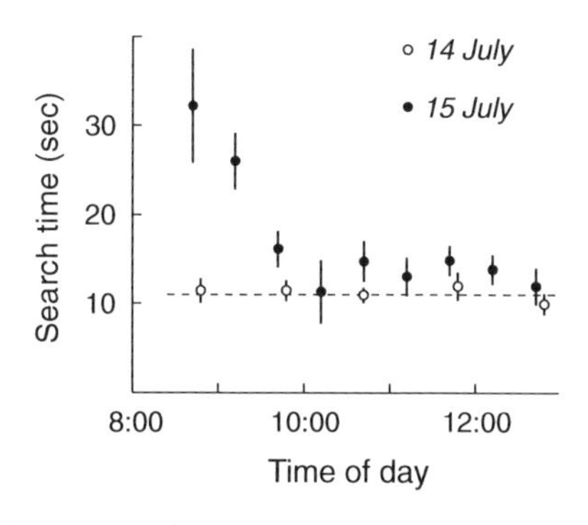

Figure 6.7 The observations that initially indicated that a honey bee colony can rapidly adjust the number of bees functioning as food storers. On 14 July 1987, the colony contained all of its food storers, and bees returning from a sugar solution feeder needed to search only 11 sec on average to find a food-storer bee to unload them. At the end of the day, most of the food storers were removed from the colony. On the following morning the foragers returning from the feeder initially experienced long search times (30+ sec), due to the colony's shortage of food-storer bees, but within a few hours the returning foragers were once again finding food storers quickly ($\bar{x} \pm$ SE). Evidently, the colony had made an internal adjustment to replace the food storers that had been removed. Based on data discussed in Seeley 1989a.

In the experiment performed on 14 and 15 July 1987, the need for additional nectar processors arose through a decrease in their *supply*, but such a decrease is unlikely to occur in nature because there is no natural analog to the experimental removal of a colony's food-storer bees. In a colony not experimentally manipulated, the need for additional nectar processors will come about through a rise in the *demand* for these bees at the start of a nectar flow. The question naturally arises whether a colony can mobilize additional bees for nectar processing in the natural context of an increased demand for these bees, as opposed to the artificial context of a decreased supply of them. Recently, with the aid of two colleagues—Susanne Kühnholz and Anja Weidenmüller—I have determined that a honey bee colony can indeed boost the number of nectar processors (N_P) to cope with a higher nectar influx. On 18 July 1994, we trained 12 bees from an observation hive at the Cranberry Lake Biological Station to forage from a sucrose solution feeder positioned 350 m south of the hive. Throughout the following day, 19 July, we allowed these 12 bees to forage from this feeder, but I captured all recruits to the feeder to keep the colony's nectar influx at a low level (only 2.7 nectar foragers per min into the hive). Meanwhile, Susanne and Anja sat beside the hive and applied paint marks to the food-storer bees seen unloading the sugar solution from these 12 foragers. They also scanned the hive every 15 min, counting any bee performing the tremble dance. After 12 hr, virtually

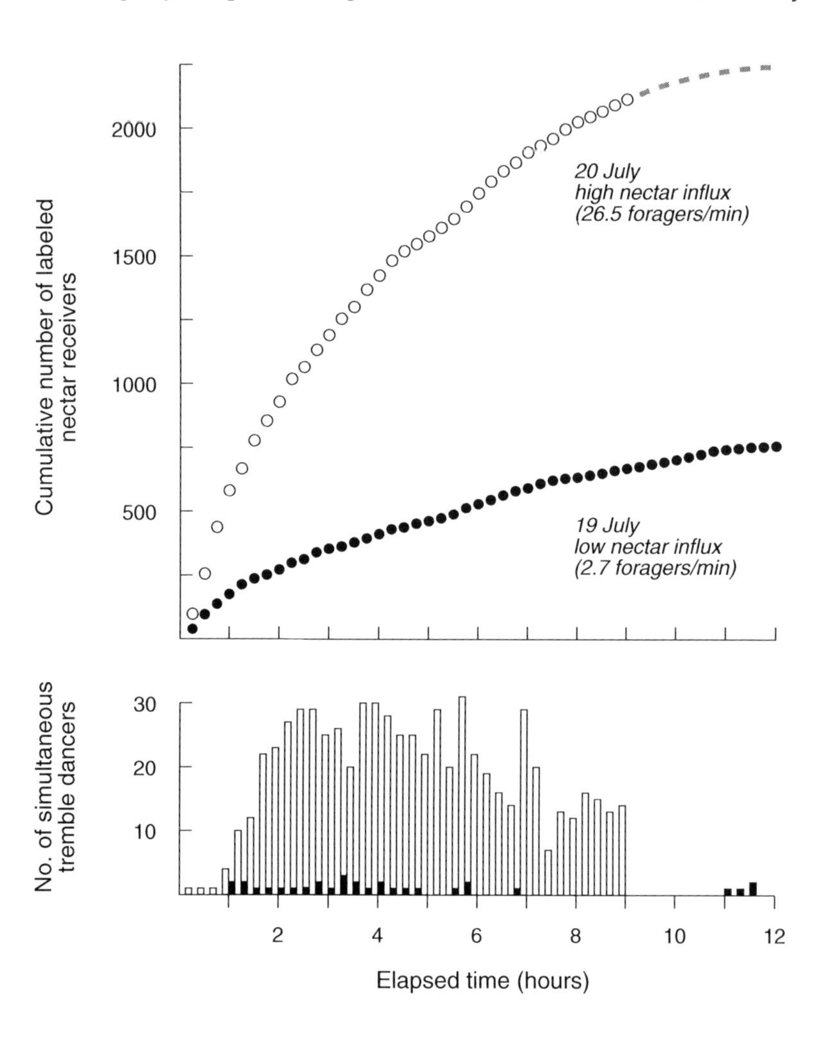

Figure 6.8 Results of one trial of an experiment testing the hypothesis that a colony increases its nectar processing rate during a nectar flow by increasing the number of food-storer bees. The colony was established in an observation hive, and the number of bees allowed to gather sucrose solution from a feeder was regulated to adjust the colony's nectar influx. All the bees in the hive that were observed receiving nectar from the foragers visiting the feeder were labeled with a paint mark on the thorax, and the total number of receiver bees so labeled was recorded every 15 min. Also, the number of tremble dancers in the observation hive was counted every 15 min. (*filled bars:* 19 July counts; *open bars:* 20 July counts). On 19 July 1994, only 12 bees were allowed to gather food from the feeder; hence the nectar influx was low and the level of tremble dancing was low. The total number of nectar receivers in the colony was approximately 770. On the following day, however, more than 100 bees were allowed to bring back food from the feeder; hence the nectar influx was 10 times higher and many of the foragers produced tremble dances. Most important, the number of nectar receivers rose dramatically, to approximately 2250 bees, or about triple the count of the previous day. Based on unpublished data of T. D. Seeley, S. Kühnholz, and A. Weidenmüller.

all the food storers in the colony had been labeled (Figure 6.8). The total was some 770 bees, representing 17% of the colony, whose population was determined to be approximately 4450 bees. Also, Susanne and Anja had observed that there was very little tremble dancing in the hive that day. Finally, on the third day, 20 July, I allowed recruitment to the feeder to proceed without interference, with the result that the colony's influx of sugar solution from the feeder rose quickly to a level 10 times higher than that on the previous day (26.5 incoming nectar foragers per min). This tremendous surge in the food influx triggered a dramatic increase in the number of bees involved in nectar reception: some 2250 bees—approximately 50% of the colony's population! It also elicited a breathtaking display of tremble dancing inside the hive, with 10–30 of the foragers performing tremble dances at all times. A second trial a few days later yielded essentially identical results. It therefore is clear that a colony can boost the number of bees functioning as food storers when the demand for these bees rises. It also appears that the mechanism which increases the number of food storers is likely to involve the tremble dance.

6.3.1. DESCRIPTION OF THE TREMBLE DANCE

In the early 1920s von Frisch wrote a lovely description of the curious tremble dance: "At times one sees a strange behavior by bees who have returned home from a sugar water feeder or other goal. It is as if they had suddenly acquired the disease St. Vitus's dance [chorea]. While they run about the combs in an irregular manner and with a slow tempo, their bodies, as a result of quivering movements of the legs, constantly make trembling movements forward and backward, and right and left. During this process they move about on four legs, with the forelegs, themselves trembling and shaking, held aloft approximately in the position in which a begging dog holds its forepaws. If they have brought in sugar water . . . often [they] will retain it until they have quieted down. The duration of this `tremble dance' is quite variable. I have seen instances where the phenomenon has died away after three to four minutes, then the bee appeared normal again and flew out of the hive. Usually, however, this dance lasts much longer and three times I have observed a bee tremble on the combs without interruption for three quarters of an hour" (von Frisch 1923, p. 90; my translation).

The complex behavior pattern that von Frisch described is depicted graphically in Figure 6.9. Here we see that the movements of a bee

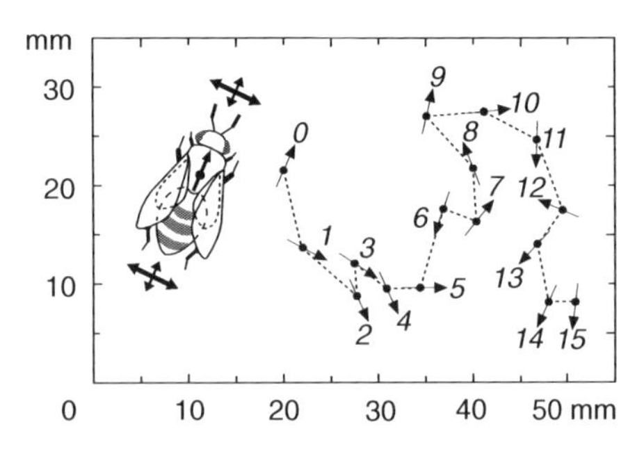

Figure 6.9 A bee's behavior while performing a tremble dance. The bee on the left illustrates the strong side-to-side, vibrational movement of the body, while the diagram on the right shows the rotational and translational movements of the body. The *numbered arrows* on the right denote, at 1-sec intervals, the bee's position on the comb and the angle of her body with respect to vertical. During this 15-sec segment of a dance, the bee walked continuously, with a mean rotational velocity of 58°/sec and a mean translational velocity of 5.7 mm/sec. After Seeley 1992.

performing the tremble dance can be dissected into three distinct components: (1) *vibrational*—the strong side-to-side, and sometimes front-to-back, shaking of the body; (2) *rotational*—the constant changing of direction of the body axis; and (3) *translational*—the slow walking forward across the comb (described in detail in Seeley 1992). The vibrational movement appears to have approximately the same frequency as that of the waggle dance, which is 10-15 Hz, but a markedly different form. Whereas in the waggle dance the bee's abdomen swings back and forth as the bee pivots around a point approximately at the front of her head, in the tremble dance the bee's whole body shakes to and fro: there is no pivot point. This side-to-side shaking is punctuated—every second or so—by momentary pauses; hence the dance has a rather jerky, nonrhythmic appearance. During each pause, the bee rotates her body to face a different direction. These rotational movements are large and frequent. A series of measurements of the inter-fix angle of one typical dancer, with fixes taken off a video-recording at 1-sec intervals, shows an average inter-fix angle of 48 ± 37° (without regard to the direction of the turn, clockwise or counterclockwise). These frequent turns produce a random orientation of the bee's body with respect to gravity. While performing the lateral shaking and rotational movements, the bee also moves across the comb. This translational component of the tremble dance has a low velocity, only 6.0 ± 2.3 mm/sec, but the movement is continuous, so that a tremble dancer is constantly on the move. The low translational velocity, combined with the high rotational velocity, means that each dancer traces out a highly convoluted, frequently criss-crossed travel path, one which nevertheless can easily cover an area of more than 100 cm^2 in a 2-min period (see fig. 3 in Seeley 1992).

Recent work by Nieh (1993) has revealed an acoustic component to the tremble dance. Approximately four times a minute, a tremble dancer will lunge forward and butt her head against another bee. At the same time, she uses her flight muscles to generate a sound, first described by Esch (1964), which has a fundamental frequency of 320 Hz and lasts up to 100 msec. It is not yet clear whether tremble dancers produce this acoustic signal throughout the hive, for Nieh's observations were limited to bees on the dance floor. But here he found that the most common recipients of the signal were waggle dancers (44%) and other tremble dancers (25%). When he then compared which bees received this signal with which bees were present on the dance floor, he found that tremble dancers seem to selectively direct their signal *to-*

ward waggle dancers, tremble dancers, and bees receiving nectar from another bee (probably food storers), and *away from* dance followers and bees simply standing around on the dance floor (Nieh 1993).

Typically, a forager begins her tremble dance in the vicinity of the hive entrance, where she has tried without success to find a food-storer bee. Once she has started her dance, she will continue to dance for a long time, some 27 min on average (observed range: 2 to 82 min). During this time the bee will travel deep inside the hive, far deeper than she would if she had instead performed a waggle dance (Figure 6.10). As a result, tremble dances are distributed throughout the broodnest portion of the hive, in contrast to waggle dances, which are concentrated near the hive entrance. When I first observed the bees performing tremble dances, I thought that they were doing this behavior to attract a food storer. But it quickly became clear that this is not the case, for as they travel slowly about the hive, shaking their bodies, they rarely stop to unload nectar. Indeed, even though a dancer's abdomen may be swollen with nectar, generally she will not even attempt to unload it until the end of her tremble dance. At this point, she will regurgitate her nectar to another bee, groom herself, beg a bit of food, and then fly out of the hive to resume foraging.

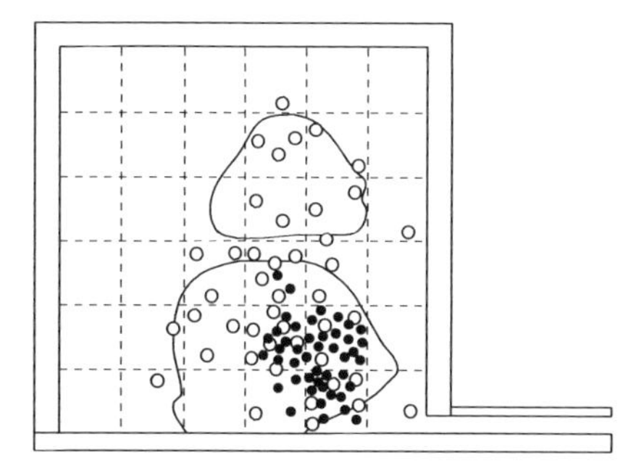

Figure 6.10 Spatial distributions in a two-frame observation hive of 44 waggle dances and 44 tremble dances, obtained by plotting the positions of dancers observed in scan samples made at 2-min intervals over 60 min. Outlined areas in the hive denote regions containing brood. After Seeley 1992.

6.3.2. THE CAUSE OF TREMBLE DANCES

The significance of the tremble dance was a mystery to von Frisch, for although it—like the waggle dance—seemed clearly to be a communication signal, he could neither identify its cause nor detect any effect on other bees in the hive. This situation led him in 1923 to the tentative conclusion that the tremble dance gives the other bees no information, a view which, for lack of better information, he maintained in his masterwork on the bees' dances (von Frisch 1967). But my chance observations in July 1987 provided me with an important hint of the true significance of the tremble dance as a communication signal. Having witnessed a curious correlation between the removal of a colony's food storers, the performance of tremble dances by the nectar foragers of the colony, and eventually the replacement of the missing food storers, I was led to suspect that the cause of tremble dances is long delays experienced by nectar foragers in finding food storers, and that the effect of tremble dances is to recruit additional bees to the task of storing nectar (increase N_p), or to stimulate the existing food-storer bees to work harder (increase L_p or decrease T_p), or both. In short, I had the hunch that the tremble dance helps remove a bottleneck in the nectar acquisition process by signaling the need in the colony for additional labor devoted to nectar processing.

Four years later, in the summer of 1991, I undertook experiments designed specifically to test these ideas. First came the test of my hypothesis about the cause of the tremble dance. This involved experimentally raising a colony's nectar influx (to simulate the onset of a nectar flow) and observing whether tremble dancing was triggered whenever the colony's rate of nectar intake was so high that returning foragers experienced long searches to find food storers. In performing this test, I made sure that the quality of the foragers' food source did not deteriorate when their colony's nectar influx rose, so that I could distinguish between my hypothesis and one proposed many years earlier by Lindauer (1948; see also Schick 1953), which is that the stimulus that causes foragers to start tremble dancing is a marked deterioration of their food source.

Figure 6.11 illustrates the experimental design and results of this test. In late June 1991, a colony of bees in an observation hive was taken to the Cranberry Lake Biological Station, and foragers from

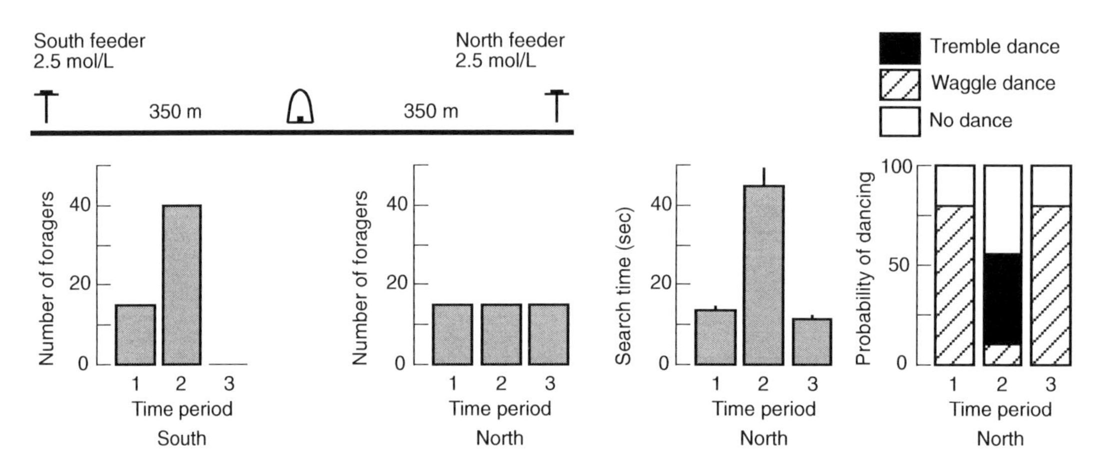

Figure 6.11 Experimental array and results of one trial of the test that a lengthy search in the hive to find a food-storer bee is a critical stimulus causing a forager to perform tremble dances. The colony's nectar influx was varied by adjusting the number of foragers (0–40) visiting the south feeder, and data were taken on the search times and dance behaviors of the foragers (always 15) visiting the north feeder. Whenever the nectar influx from the south feeder was kept low (time periods 1 and 3), the north-feeder bees experienced short search times and they performed waggle dances. In contrast, when the nectar influx from the south feeder was raised to a high level (time period 2), so that the north-feeder bees had to perform lengthy searches in the hive to find food-storer bees, they performed mainly tremble dances. Time periods: 1 = 11:00–12:00, 2 = 1:30–2:15, 3= 2:30–3:00. Both the average search time and the probability of tremble dancing were significantly ($P < 0.001$ for both) different between time periods 1 and 3 and time period 2. After Seeley 1992.

this colony were trained to two feeders approximately north and south of the hive (layout as shown in Figure 5.26). The first trial of the experiment began on the morning of 10 July, when both feeders were loaded with a highly concentrated (2.5-mol/L) sucrose solution, and I began gathering data on the behavior of the foragers from the north feeder as they returned to the observation hive. The two most important variables of each bee's behavior were (1) how much time she spent searching for a food-storer bee and (2) whether or not she performed a dance, either waggle or tremble. From 11:00 to 12:00, only 15 bees were allowed to forage from each feeder (all recruits were captured upon arrival at the feeders), and thus the colony's nectar influx was kept low. The bees from the north feeder behaved as expected: they experienced search times averaging only about 15 sec and they performed only waggle dances. Then from 12:00 to 1:30, the assistant at the south feeder stopped capturing the recruits there so that the number of bees bringing in food from this feeder would increase. By 1:30 there were 40 bees busily bringing home nectar from the south feeder. At this point, the colony's nectar influx had become moderately high and the behavior of the foragers from the north feeder had changed dramatically. Observations from 1:30 and 2:15 revealed that these bees were now searching much longer than before to find an unloader, about 45 sec on average, and that they were now performing mainly tremble dances! To fully appreciate this result, it is important to note that the only stimulus changes experienced by the north-feeder bees between the first period (11:00–12:00) and the second period (1:30–2:15) were those sensed inside the hive as part of the unloading experience (increased difficulty of finding a food-storer bee, and so on). Finally, between 2:15 and 2:30 the colony's nectar influx was lowered by shutting off the south feeder, whereupon the north-feeder bees again began to experience short search times and perform only waggle dances. A second trial of the experiment was performed the following day with essentially identical results.

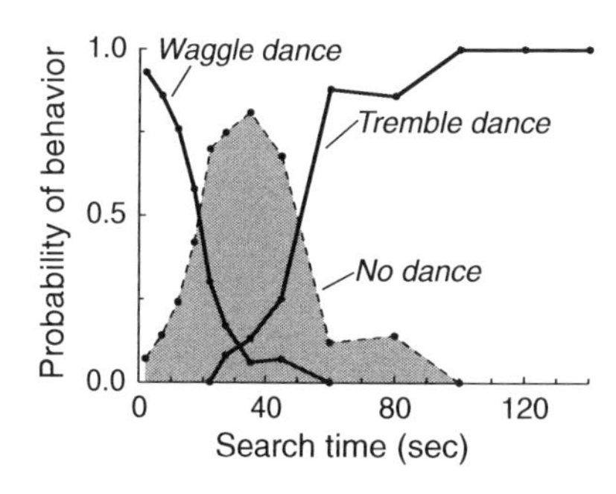

Figure 6.12 Dance behavior as a function of in-hive search time for foragers visiting a rich food source. Data were gathered from 15 bees visiting a feeder providing a 2.5-mol/L sucrose solution and located 350 m north of the hive. As is shown in Figure 6.11, search times were varied for these bees by altering the number of bees bringing in nectar from a second feeder, south of the hive. Hence the only changes underlying the switch from waggle dancing to tremble dancing were those experienced inside the hive: search time and other variables of the unloading experience. After Seeley 1992.

Thus it became clear that if a forager returns to the hive from a rich nectar source and finds that she must search extensively to find a food storer, she will perform tremble dances. Figure 6.12 shows the general relationship between duration of search time and the probability of tremble (and waggle) dancing for bees visiting a highly profitable feeder. Whereas the majority of foragers experiencing a search time of 20 sec or less performed a waggle dance, the majority of foragers experiencing a search time of 50 sec or more performed a tremble dance. Interestingly, there was a rather broad range of intermediate search times (20–50 sec) in which the foragers tended not to perform either dance, indicating that it is unlikely that a forager will be motivated to perform both dances simultaneously.

A recent study by Kirchner and Lindauer (1994) confirms the importance of long search times in triggering tremble dances. This study also points out that evidently what is most important to a nectar forager is not how long she must search to find the food-storer bee that first receives her nectar *(initial search time),* but the total amount of time spent searching to find food storers in the course of getting rid of a load of nectar *(total search time).* (Sometimes a forager must locate several food storers in a series—whenever each one takes only a portion of the forager's nectar load.) The strongest indication that total search time is more important than initial search time comes from an experiment in which Kirchner and Lindauer provided a solution of 2-mol/L sucrose plus 1-mol/L salt (NaCl) to bees visiting a feeder. These bees, upon return to their hive, had no trouble finding food storers (mean initial search time: 10 sec), but the food storers refused to accept much of the strange food, so that each forager had to find, on average, some 21 different food storers to complete her unloading. The total search time was therefore extremely long—146 sec on average—and more than 75% of the bees performed tremble dances, despite experiencing quite short initial search times. These results indicate that returning nectar foragers make note of the overall difficulty of unloading, not just the delay in starting the unloading.

6.3.3. THE EFFECTS OF TREMBLE DANCES

Recall my hypothesis for the effect of the tremble dance: that it would trigger an increase in the colony's capacity for processing nectar, either by recruiting additional bees to function as food storers or by stimulating the existing food storers to work harder (take larger loads or process them faster), or both. One way to test this hypothesis experimentally was to manipulate a colony's foragers so that the colony would experience a sudden, stressful boost in its nectar influx, and then observe whether the tremble dancing that resulted was followed consistently by a rapid decay in the average search time of the returning foragers, indicative of an increase in the colony's capacity for processing nectar.

The first of two trials of this "stress test" was performed on the morning of 19 July 1991. The layout was identical to that shown in Figure 6.11: bees from an observation hive were trained to forage from two feeders located 350 m north and south of the hive and each was loaded with a rich, 2.5-mol/L sucrose solution. Initially, from 8:15 to 9:45, only 15 bees were allowed to forage from each feeder; the two assistants at the feeders gently captured in plastic bags all additional foragers arriving at the feeders. Thus at first the colony had a low rate of forager arrival at the hive, and as a result these foragers experienced short searches (average of 9 sec) and performed only waggle dances (see Figure 6.13). Then at 9:45, the assistants quickly released all the captured foragers at the two feeders, a total of 88 bees. Despite being confined for an hour or more, the vast majority of the released foragers flew immediately to their feeder to resume loading, and within another minute or two they were winging their way back to the hive, thereby producing an almost instantaneous quadrupling of the forager arrival rate at the hive and a corresponding multiplication of the colony's nectar influx. Fifteen minutes later, I resumed my data collection at the hive. These observations revealed that the release of the foragers and the resultant rise in the colony's nectar influx were sufficient to create long in-hive search times (average of 30 sec) and to trigger tremble dancing. Most important, however, I observed over the next 2 hr that despite a forager arrival rate that continued to rise and eventually reached a level five times that recorded at the beginning of the experiment, the average search time experienced by the foragers dropped to 10 sec, *exactly what was seen at the start of the experiment!* This finding implies that the colony's nectar processing capacity had been raised to match the new, higher level of nectar influx. Also, the level of tremble dancing was high when the search time was being lowered—when the nectar processing capacity was being raised—and then fell to essentially zero once the search time stabilized at its normal, low level. Hence there was a strong correlation between the occurrence of tremble dances and the rise in the colony's capacity for processing nectar.

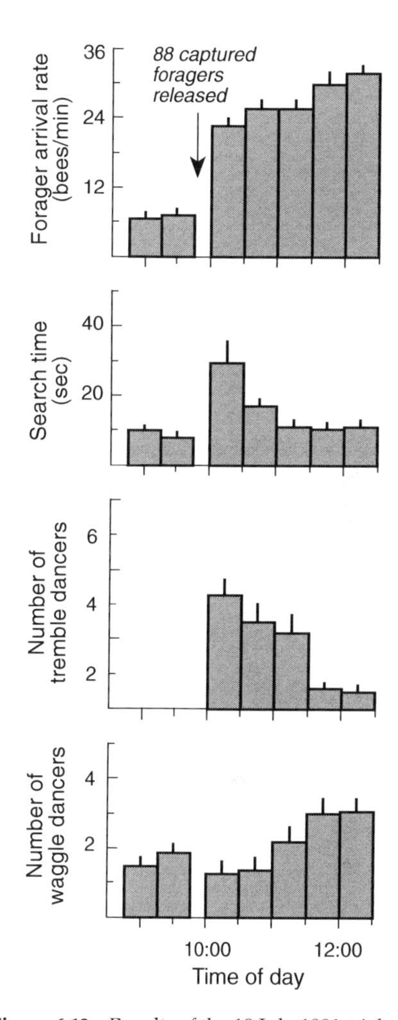

Figure 6.13 Results of the 19 July 1991 trial of the "stress test" of the hypothesis that the effect of tremble dances is an increase in a colony's nectar processing capacity. Data were gathered from bees visiting two feeders, each providing a 2.50-mol/L sucrose solution, as depicted in Figure 6.11. Initially, only 15 bees were allowed to forage from each feeder and all additional foragers were captured at the feeders. When the captured foragers were released, the colony was suddenly stressed with a much higher nectar influx. All measurements of search time were based on the original 30 bees ($x \pm$ SE). After Seeley 1992.

This correlation suggests strongly that there is a cause-effect relationship between tremble dances and a rise in a colony's ability to process nectar, but it is not definitive proof that the former is causing the latter, for some other cue or signal, itself tightly correlated with the tremble dance, might be the critical stimulus for greater nectar processing. To test more stringently for an effect of the tremble dance on the colony's nectar processing capacity, one needs to perform an experiment that involves manipulation of the tremble dance independently of other variables. One possibility would be to repeat the "stress test" described above, but with the modification of removing the tremble dancers as they appear. Then, if the colony's nectar processing rate rises after the colony is stressed with a higher nectar collection rate, despite the absence of any tremble dances in the colony, it will be clear that the tremble dance is not a necessary stimulus for increased nectar processing. But, if the colony's nectar processing rate does not rise, this finding will provide further support for the idea that the tremble dance is necessary in signaling the need for increased nectar processing.

A second effect of the tremble dance is indicated by the recent experimental findings of Nieh (1993) and Kirchner (1993). When they played back the sound produced by tremble dancers, either by loading the sound directly onto bees (Nieh) or onto the combs of a hive (Kirchner), they found that bees performing waggle dances for nectar sources tended to stop dancing and leave the hive. This tends to shut down a colony's recruitment of additional bees to nectar sources. As a control, they played back signals composed of white noise or 100 Hz sine waves and found no response. Thus it appears that the tremble dance helps a colony achieve a match between nectar collecting and nectar processing not only by stimulating a rise in the processing rate but also by inhibiting any further rise in the collecting rate.

The recent discoveries about the causes and effects of the tremble dance have revealed that we can think of waggle dances and tremble dances as playing complementary roles in keeping a colony's rates of nectar collecting and nectar processing well matched (Figure 6.14). Both are performed by foragers returning from rich nectar sources worthy of greater exploitation, but whereas waggle dances appear

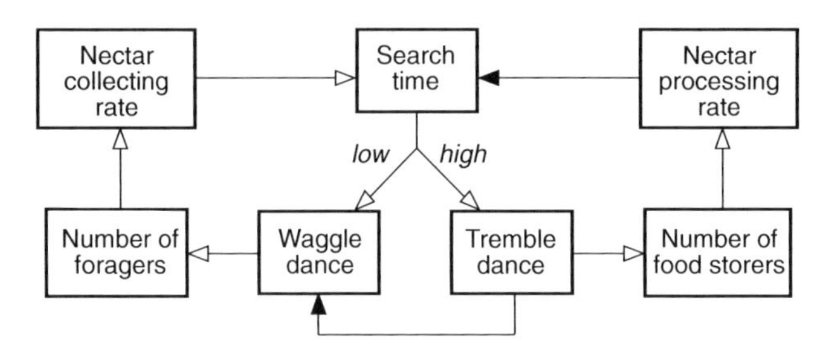

Figure 6.14 The complementary effects of waggle dances and tremble dances whereby a colony maintains a match between its rates of nectar collecting and nectar processing. Foragers returning from rich nectar sources choose between the two types of dances depending on the time spent searching to find a food-storer bee, a reliable indicator of the relative rates of collecting and processing. Each dance type has an excitatory effect on the number of bees performing either foraging or food storing, and the tremble dance also has a supplementary inhibitory effect on waggle dancing.

when the average search time is low (hence collecting rate < processing rate) and raise the collecting rate, tremble dances appear when the average search time is high (hence collecting rate > processing rate) and raise the processing rate.

6.4. Which Bees Become Additional Food Storers?

Given that colonies allocate additional bees to the task of food storing at the start of a nectar flow (see Figure 6.8), the question arises: Which bees become additional food storers? One possibility is that each colony maintains a pool of *inactive* workers, or reserves, upon which it can draw when conditions change and additional labor is suddenly needed for a particular task such as nectar processing. Another possibility, not exclusive of the first, is that colonies cope with changes in their labor needs by shifting *active* workers from one sector of a colony's economy to another. For example, the additional food storers might come from the set of bees engaged in brood care. One indication that this second idea is correct comes from additional results of the experiment performed on 18–20 July 1994 and described above (Seeley, Kühnholz, and Weidenmüller, in preparation). Approximately 10% of the members of the colony involved in this study were bees that had been labeled and then introduced into the observation hive when they were 0 days old, hence were of known age (see Section 4.7). Thus when the food-storer bees in the colony were censused in the colony on 19 and 20 July, it was possible to obtain an age distribution for these bees. The results, shown in Figure 6.15, reveal a distinct rise in the proportion of young bees when the nectar influx increased and the colony raised the number of food-storer bees. On 19 July, only 5% of the food storers were less than 18 days old, but one day later, when the nectar influx was suddenly much higher, this proportion rose to 28%. Indeed, a sizable fraction of the bees receiving nectar at the time of high influx were only 10 or fewer days old. Overall, there was a significant downward shift in the mean age, from 24.9 to 19.6 days ($P < 0.03$). It therefore seems that the tremble dances had stimulated many relatively young bees, perhaps previously active in brood care, to switch to the task of nectar reception. This result, confirmed in a repetition of the experiment, is perhaps not so surprising, given the fact that the tremble dance is primarily performed within the broodnest region of the hive (Figure 6.10), and hence the target of its message seems to be mainly the colony's nurse bees.

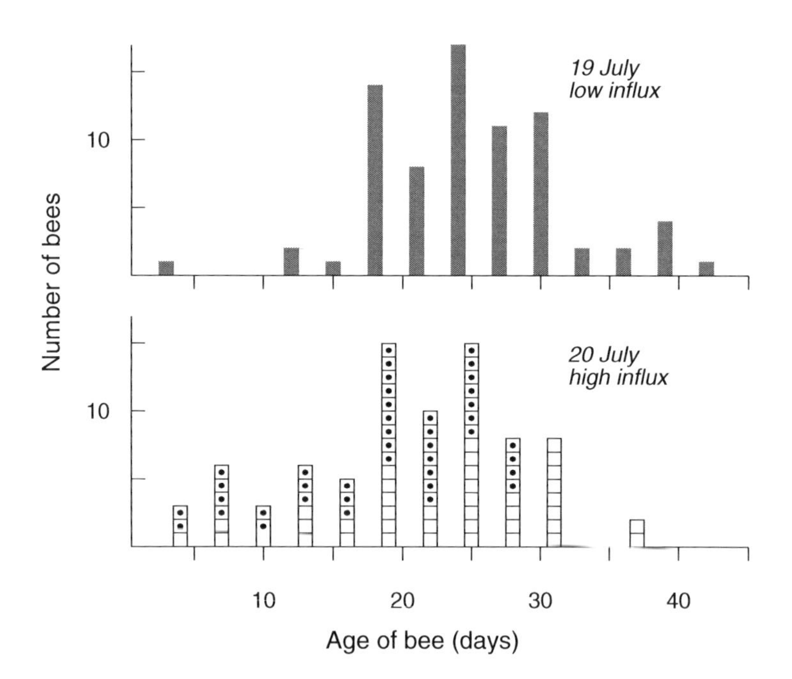

Figure 6.15 Age distributions of the bees observed receiving nectar from foragers on 2 consecutive days, one where the colony's nectar influx was low and the other where it was high (see Figure 6.8). When the nectar influx was adjusted from low to high, the age distribution of the receiver bees shifted downward. Even young bees, those 10 or fewer days old, which are normally engaged in brood care, were observed functioning as food storers. Based on unpublished data of T. D. Seeley, S. Kühnholz, and A. Weidenmüller.

This finding raises the puzzle of how a colony solves the problem of *correctly allocating labor among multiple, competing needs,* such as nectar processing and brood care. Here we gaze forward over a region of scientific terrain which remains as yet uncharted (but see Section 10.1 for one hypothesis on how task allocation *may* occur in colonies).

Summary

1. Honey bee colonies possess a division of labor in nectar acquisition between the foragers, which work in the field collecting the nectar, and the food storers, which work in the hive processing the newly gathered nectar (Figure 6.1). This organization boosts the efficiency of a colony's energy acquisition, but requires coordination of the two labor groups to keep the rates of the collecting and processing in balance. This problem of coordination is a major one because a colony experiences large and unpredictable variation from day to day in its rate of nectar collection, a result of variation in the nectar supply outside the hive due to changes in weather conditions and floral resources.

2. A colony adjusts its collecting rate with respect to the external nectar supply in part through changes in each forager's rate of foraging trips and size of nectar loads, but mainly through changes in the number of active, employed foragers. When the nectar supply increases, a colony boosts the number of bees gathering nectar. This involves not only rapidly activating unemployed foragers, by means of the waggle dance signal (Figure 6.2), but apparently also stimulating nonforagers to begin foraging, by means of the shaking signal (Figure 6.3). Foragers perform the shaking signal throughout the hive (Figure 6.4), evidently in response either to a prolonged period of successful foraging (Figure 6.5) or to a return of rich forage following a dearth (Figure 6.6), or both. When the nectar supply decreases, a colony lowers the number of nectar collectors—probably simply by having foragers shut off their recruitment signals (waggle dances and shaking behaviors) as they experience poorer foraging, with the result that the dropout rate exceeds the recruitment rate among employed foragers.

3. A colony raises its nectar processing rate when its collecting rate increases by boosting the number of bees engaged in the task of processing nectar (Figure 6.8). Additional bees are stimulated to function as nectar processors by the tremble dance, a behavior performed inside the hive by foragers in which the bee walks slowly about the nest (Figure 6.10), constantly making trembling movements forward and backward, and right and left, and constantly rotating her body to face different directions (Fig. 6.9). When a colony's collecting rate increases markedly, so that its nectar foragers must conduct long searches to find food storers upon return to the hive, the foragers are stimulated to perform tremble dances (Figure 6.11). In general, bees returning from highly profitable nectar sources will perform tremble dances if their search times average more than 50 sec, whereas they will perform waggle dances if their search times average less than 20 sec (Figure 6.12). The performance of tremble dances has both an excitatory effect, so that a colony's nectar processing capacity is increased, and an inhibitory effect on waggle dancers, so that its nectar collecting rate is stabilized (Figure 6.13).

4. Waggle dances and tremble dances play complementary roles in keeping a colony's rates of nectar collecting and nectar processing well matched (Figure 6.14), for the former enables a colony to boost its collecting rate while the latter enables it to boost its processing rate.

5. Although many of the bees stimulated by tremble dances to function as food storers are probably bees that were previously inactive in the hive, a significant fraction of the new food storers are relatively young bees who otherwise probably would be engaged in brood care (Figure 6.15). This capacity to quickly switch bees among different tasks raises the important, and still unsolved, puzzle of how a colony can *correctly* allocate its labor among competing needs.

Regulation of Comb Construction

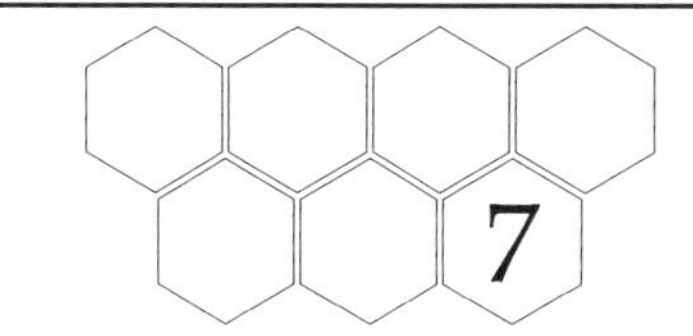

An essential component of a colony's overall foraging process is the building of the beeswax combs in which the colony's reserve supplies of pollen and honey are stored. These combs are energetically costly—the synthesis of 1 g of beeswax consumes at least 6 g of honey (Hepburn 1986)—hence a colony does not construct a full set of combs when it first occupies a homesite. Rather, a colony of bees limits its investment in comb building to what is absolutely necessary; it initially builds only a small set of combs and subsequently enlarges these combs only when it critically needs additional storage space (see Section 3.6). Thus comb construction in honey bee colonies presents us with puzzles about the controls of the *timing* and the *amount* of a production process. In addition, one can raise the question of control of the *type* of production, for a honey bee colony constructs two distinct types of comb—worker and drone (Figure 7.1)—and it rather precisely limits the drone comb to some 15% of the total comb within its hive. One can also pose the questions of control of the *location* and the *form* of production, that is, how a colony manages to construct a set of parallel, uniformly spaced combs with each comb a regular array of hexagonal cells. The questions of type, location, and form of comb building, however, bear on issues separate from the social organization of a colony's foraging, and so will not be discussed in this book.

7.1. Which Bees Build Comb?

The logical first step toward unraveling the mechanisms controlling the timing and amount of comb construction is to identify which

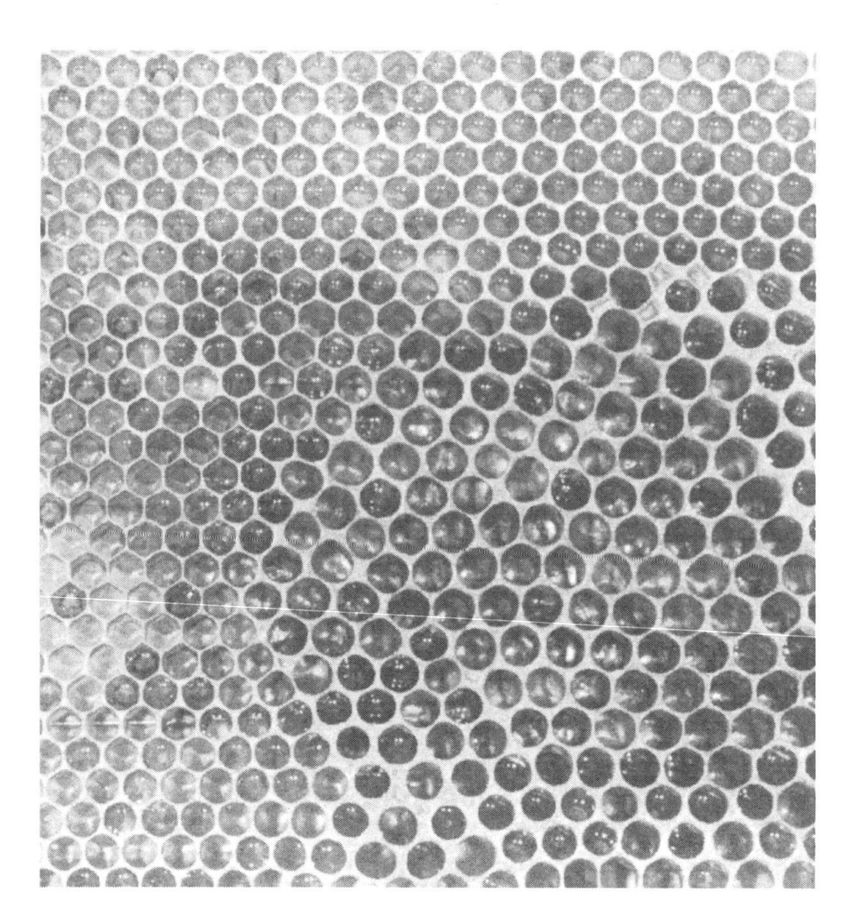

Figure 7.1 Beeswax comb showing the two sizes of cells found in the combs of a honey bee colony: the smaller worker cells *(top and left)* and the larger drone cells *(lower right)*. This difference in cell size reflects the difference in body size between workers and drones, which are reared in these cells. Both types of cell are used for food storage as well as brood production. Photograph by T. D. Seeley.

members of a colony are actually responsible for building the comb. These bees must lie at the heart of the control mechanisms. To date, the best information on which bees build comb comes from a study published nearly 70 years ago by G. A. Rösch, a student of Karl von Frisch. In essence, Rösch (1927) identified the age group to which the comb builders belong. To accomplish this, he worked with a small hive containing 12 frames of comb and a colony of some 10,000 bees, to which he added 100 0-day-old worker bees each day, starting on 9 May 1926. Each bee added to the colony received a paint mark coding her date of eclosion (emergence from her brood cell), so that her age could be precisely determined during later observations. After adding bees for 54 days, on 2 July Rösch removed one frame with comb from the hive and replaced it with another frame without comb, thereby providing the bees with an open space in the hive in which to build comb. After 12 hr, he gently lifted the frame from the hive so as to minimally disrupt the mass of bees building a new comb within the frame (Figure 7.2), anesthetized the cluster of comb-building bees, and removed the labeled bees from the cluster. Rösch assumed that all these bees were participating in the task of comb building, but of course it is possible that some were actually performing other tasks or simply resting in the building cluster and were incidentally removed along with the comb builders.

Figure 7.3 shows the age distribution of the 202 labeled bees that were recovered from the building cluster. Although their age range is quite broad, from 2 to 52 days, the majority (67%) of the bees came from a relatively narrow subrange, from 10 to 20 days. Two replications of this experiment yielded essentially the same results. Thus the evidence at hand indicates that the comb-building bees come primarily from the ranks of the middle-aged bees in a colony. As already shown (Section 2.2), these are bees which are largely finished with the tasks associated with the central broodnest region of the hive—such as feeding brood, capping brood, and tending the queen—but often have not yet begun to work outside the hive as foragers. Instead, they are performing tasks associated with the peripheral, food-storage region of the hive—such as receiving nectar, storing nectar, packing pollen, and, evidently, building comb.

Rösch produced a second piece of evidence that the comb builders are the middle-aged bees in a colony when he examined the histological status of the wax gland epithelium for each labeled bee collected in the building cluster. (He reasonably assumed that the height of a bee's wax gland epithelium is a good indicator of her capacity for wax production, and this has been abundantly confirmed by subsequent studies, which consistently report a tight correlation between wax secretion and epithelium height (reviewed in Hepburn 1986)). As is shown in Figure 7.3, Rösch found that when his bees were young (less than 10 days old) their wax glands were developing, that when they were middle aged (about 10 to 18 days) their wax glands were of maximum size, and that when they were old (more than 18 days old) their wax glands were either rapidly degenerating or already of minimal size. Although this pattern is not universal—for example, in colonies that have recently swarmed the age range of bees with large wax glands can extend to 30 or more days (Turrell 1972)—subsequent studies report similar schedules of wax gland rise and fall (Hepburn

Figure 7.2 Comb-building bees. The comb construction starts when individuals with well-developed wax scales deposit their wax on the construction front after chewing each wax scale to mix it with a salivary gland secretion which renders the wax more plastic. Photograph by T. D. Seeley.

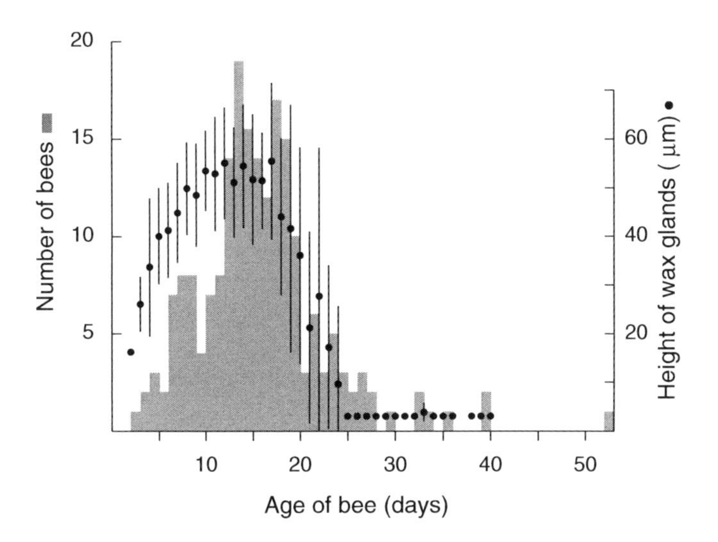

Figure 7.3 The age distribution of the comb-building bees in a typical colony in mid-summer. *Shaded histogram bars:* frequency distribution for 202 known-age bees recovered from a cluster of bees building a new comb. Presumably most, though not necessarily all, of these bees were engaged in comb construction. *Data points:* height of the wax gland epithelium in relation to age, based on histological examination of 736 known-age bees recovered from comb-building clusters in three separate experiments ($x \pm$ SD). The age range of peak wax gland development, 10–18 days, matches that of the bees most commonly found in a building cluster. Based on data published in Rösch 1927.

1986; Muller and Hepburn 1992), indicating that the pattern shown in Figure 7.3 is typical for established colonies during the summer. This finding implies, in turn, that generally it is the middle-aged bees in a colony which function as the wax producers and comb builders.

7.2. How Comb Builders Know When to Build Comb

To explain how bees know when to build comb, it is fruitful first to identify the specific conditions under which comb building occurs within a beehive, and then to examine how the comb builders sense whether or not these conditions are fulfilled. Thus in this section I will proceed from a look at colony-level patterns to an analysis of individual-level processes.

7.2.1. THE CONDITIONS UNDER WHICH COMB IS BUILT

In Chapter 3 (Section 3.6), I reviewed the results of one study by Kelley (1991) which suggests that bees in an established colony will begin building additional comb only when two conditions are met: (1) the colony is gathering nectar at a high rate and (2) the colony's combs for honey storage are nearly full. In general terms, then, the *timing* of comb building appears to be determined by two variables, one that is external to the colony (nectar availability) and one that is internal (comb fullness). It was also argued in Section 3.6 that having the control of comb building tied to both these variables makes good func-

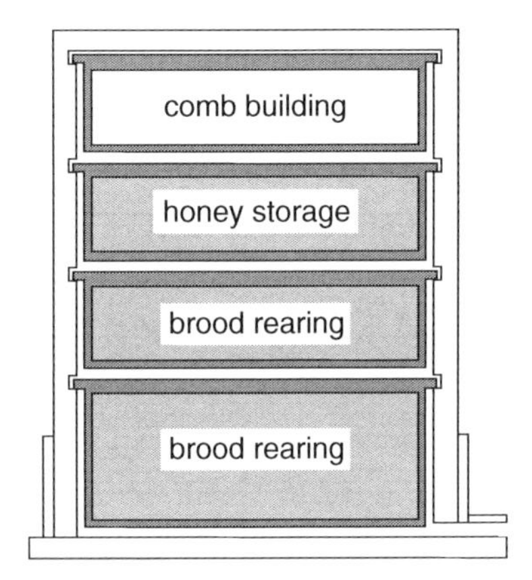

Figure 7.4 Design of the observation hive used to investigate the conditions under which a colony initiates and maintains comb building. The two lower frames contained fully built combs filled with brood plus some pollen and honey. The honey storage frame also contained a fully built comb. Depending on the needs of the experiment, this comb was either kept essentially empty (by periodically inserting a new frame of empty comb) or left unmanipulated, so that the bees could gradually fill it with honey. The top, comb building, frame always started out completely devoid of comb, hence it provided the bees with open space in which to build additional comb.

tional sense, because whereas neither variable by itself is a good indicator of the need for more comb, the two variables considered jointly provide a reliable indicator of the need for additional comb. Only when a colony experiences both abundant nectar outside the hive and little empty storage space inside the hive does it clearly need additional comb.

The findings of Kelley regarding the conditions that trigger comb building have been corroborated recently by an unpublished study conducted by Stephen Pratt during the summer of 1993. Pratt used basically the same experimental set-up as Kelley: a colony of some 5000 bees occupying a four-frame observation hive in which the bottom two frames contained fully built combs filled with brood, the third frame provided a fully built comb for honey storage, and the fourth frame, on top, was empty, to provide a space where the bees could build new comb (Figure 7.4). Also, Pratt's experimental design was similar to Kelley's in that he regulated his colony's "nectar" influx by controlling the amount of sucrose solution (2.5 mol/L) available from a feeder, and he varied the empty storage space in his colony's hive by replacing its storage comb with one either empty or partially filled with honey, whichever was required for each phase of his experiment. Finally, by monitoring the colony's comb building under various combinations of experimental treatments, he could determine the set of conditions that triggers comb building. The principal improvement of Pratt's experimental procedure over Kelley's was that Pratt performed his study at the Cranberry Lake Biological Station, where natural sources of nectar are generally sparse (see Section 5.7.2), and hence his bees were largely limited to gathering the sugar solution he provided; as a result he worked with far tighter control of his colony's rate of nectar intake.

Figure 7.5 depicts the results of Pratt's experiment. Throughout the 11-day-long experiment, the colony experienced a high influx of "nectar," but during the first 3 days (19–21 July) the colony's storage comb was kept nearly empty and the colony constructed no comb. Over the next 4 days (22–25 July) the colony was allowed to fill its storage comb, and when some 60% of the storage comb's cells contained honey, the colony began to build additional storage comb. Thus Pratt's experiment, like Kelley's, demonstrates that the conditions that induce comb building are a high influx of nectar combined with storage combs that are almost full.

Pratt's experiment also yielded evidence on how the amount of comb building is controlled. Pratt made the surprising observation that his study colony continued building comb until the very end of his experiment, even though during the last 4 days (26–29 July) the colony's storage comb was manipulated so that it remained largely devoid of honey (Figure 7.5). This continuation of comb building is intriguing because it contrasts with what Kelley had observed in his experiment: comb building stopped almost immediately (within 24 hr) when he shut off his colony's nectar influx (see phase C in Figure 7.6). Clearly, a colony can promptly shut down comb building when it wants to. These two sets of observations suggest that the conditions required to *maintain* comb building are somewhat different from those required to *initiate* comb building. Evidently, a colony needs only to continue experiencing a high nectar influx for it to continue building comb. Why should this be? In nature, a colony will probably never experience an abrupt emptying of its storage combs, especially when a nectar flow is under way. Thus under normal conditions it seems that the comb-building bees can afford to ignore the state of their honey stores once comb building has begun. In contrast, a colony living in nature has a rather high probability of experiencing an abrupt drop in its nectar influx, as would occur if bad weather arises, and in this situation a colony probably should shut down comb construction for it will no longer need the additional storage space. Thus it seems that it will pay the comb-building bees to remain keenly sensitive to the state of their colony's nectar influx once comb building has started. All things considered, it seems to make good sense that a colony's continued comb building depends only on the continuation of a strong flow of nectar into the hive. If this is correct, the amount of comb built by a colony is controlled mainly by the duration, and perhaps also the intensity, of the nectar flow that stimulates the colony to build comb.

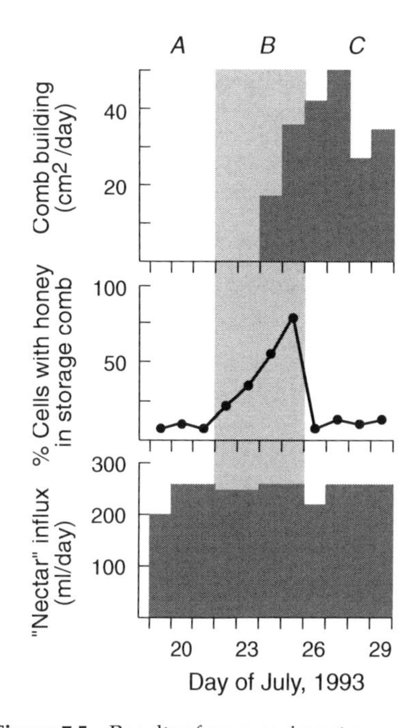

Figure 7.5 Results of an experiment conducted in July 1993 to identify the conditions under which a colony of bees will start and then continue building new comb. Two variables of a colony's foraging situation were controlled: the daily influx of "nectar" (2.5-mol/L sugar solution) and the fullness of the honey storage comb. The colony built no comb during phase A, when the colony's nectar influx was high but its storage comb was kept nearly empty. The colony began building comb during phase B, when the nectar influx remained high and the colony was allowed to fill its honey storage comb. Comb building began when some 60% of this comb's cells contained honey. Finally, the colony continued building comb during phase C, when the nectar influx was maintained at a high level but the storage comb was again rendered virtually empty. Based on unpublished data of S. Pratt.

Before concluding this section, I should point out that nectar influx and comb fullness are not the only variables of a colony's condition that determine whether or not it begins to build comb. Pratt (unpublished) has found that in autumn colonies will build little new comb despite experiencing full storage combs and high success in nectar collection. Rather than begin to build more comb, colonies seem instead to begin to store honey in the central, broodnest region of the hive, where vacant cells are becoming available for food storage as the queen shuts down her egg laying in preparation for winter. Also, as is shown in Figure 7.7, Kelley (1991) observed no comb building in

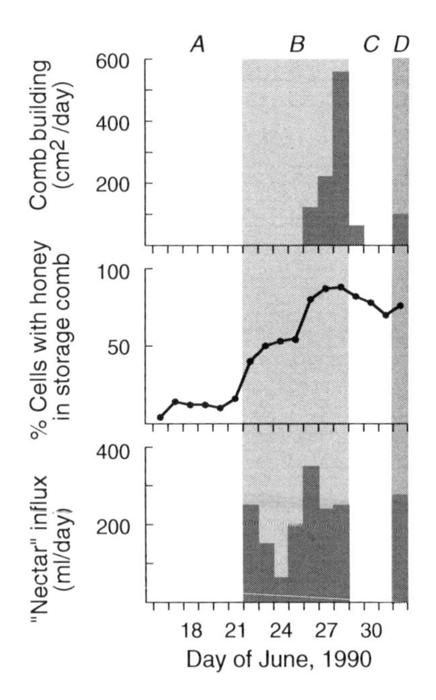

Figure 7.6 Results of an experiment conducted in June 1990 to investigate the conditions required for comb building. This experiment differs from that shown in Figure 7.5 in that this time the manipulated variable was the colony's "nectar" (2.5-mol/L sugar solution) influx rather than the fullness of its honey storage comb. The colony built no comb in phase A, when the nectar influx was low and the storage comb remained nearly empty. The colony began building comb late in phase B, when the colony's nectar influx was high and its storage comb had become nearly full (some 80% of the cells contained honey). During phase C, the colony's nectar collection was lowered, and its comb building shut down almost immediately. The building activity promptly resumed, however, in phase D, when the nectar influx was raised and the colony's storage comb still contained much honey. After Kelley 1991.

one experiment in which he fed a colony 400 mL of sucrose solution daily for 7 days and by the third day more than 80% of the cells in the storage comb were brimming with honey. The reason for this puzzling pattern became clear on the seventh day, when Kelley noticed five sealed queen cells in his hive, indicating that the colony was preparing to swarm. Evidently, when bees are in the swarming mode, they refrain from building comb at their old nest site so as to save their resources for the large task of constructing a fresh set of combs at their new nest site. Thus the conditions of high nectar influx and nearly full honeycombs will trigger comb building only if other conditions are also fulfilled: the colony is neither preparing to swarm nor preparing to overwinter. In the following section, assume that we are dealing with a nonswarming colony during the spring or summer.

7.2.2. HOW COMB BUILDERS SENSE THE CONDITIONS UNDER WHICH COMB IS BUILT

It is highly doubtful that the bees responsible for comb building directly monitor their colony's rate of nectar collection and the fullness of their colony's storage combs in order to know when they should build comb. Presumably this would require powers of information collection far greater than those actually possessed by worker bees. After all, a colony's success in nectar collection reflects the activities of thousands of bees, and the fullness of its honeycombs reflects the status of thousands of cells spread over several combs. It seems far more likely, therefore, that the comb builders sense one or more cues that provide an indirect, yet reliable, indication of their colony's level of nectar influx and remaining space for honey storage.

Two hypotheses have been proposed for how this simpler mechanism of information acquisition might work. The first, which can be called "the honey stomach distension hypothesis," was originally formulated by Ribbands (1952). In a discussion of how food transmission helps inform individuals of their colony's requirements, he wrote: "Park (1923) demonstrated that the water reserves of a colony can be stored in the honey-sacs of house bees, and it could be supposed that when comb space is insufficient the ripening nectar has to be similarly stored; bees would be diverted to this task and their stored loads assimilated and converted into wax, which would then be used to remedy the lack of comb space." Since we now know that it is the middle-aged, or food-storer, bees that are the principal comb builders, we can state this hypothesis somewhat more precisely as follows: when there is insufficient space for the food-storer bees to store the incoming nectar, these bees will be forced to retain nectar in their honey stomachs, and the prolonged distension of the honey stomach triggers the secretion of wax and the construction of comb. The second hypothesis can be called the "storing difficulty hypothesis" and was first stated by Kelley (1991). He suggested that the critical stimulus for comb building by the food-storer bees is not a prolonged distension of the honey stomach, but is instead a difficulty in finding cells in which to deposit nectar. According to this hypothesis, the food storers do not wait to start building comb until their existing storage combs are full and they are thus forced to hold large volumes of nectar in their bodies; rather, they begin building comb when the storing of nectar reaches a certain level of difficulty. This mechanism would enable the bees to begin to solve the storage space problem before it becomes acute.

Which, if either, hypothesis is correct? To check the honey stomach distension hypothesis, Kelley (1991) looked to see if food-storer bees do indeed become increasingly full of nectar prior to comb building. He sampled bees 12–18 days old from his study colony on 20 June 1990, when the colony's storage comb was largely empty, and again on 25 June, when the colony's storage comb had filled considerably and the colony was about to start building comb (see Figure 7.6). Each bee that he collected was anesthetized by cooling, and then its honey stomach was dissected from its body and weighed. Kelley found that the mean stomach size of the food storers did significantly increase from the time when the hive was relatively empty to the time when it was relatively full (5.9 ± 6.3 mg versus 15.9 ± 12.3 mg [x ± SD], $P < 0.001$). This finding provides, however, only weak support for the honey stomach distension hypothesis, for the observed difference in stomach size could simply reflect the fact that the study colony was being fed on 25 June but not on 20 June (see Figure 7.6). In other words, the increase in mean honey stomach size that Kelley recorded could have arisen because the food-storer bees were processing more nectar on 25 June than on 20 June, not because their storage combs were more full on 25 June than on 20 June. Moreover, the fact that the mean honey stomach weight on 25 June was only 15.9 mg, whereas a bee's honey stomach can weigh 60 or more mg when bulging with nectar, indicates that on average the colony's food storers were not experiencing greatly distended stomachs on the day before they began to build comb. It is easy to understand why they were not more

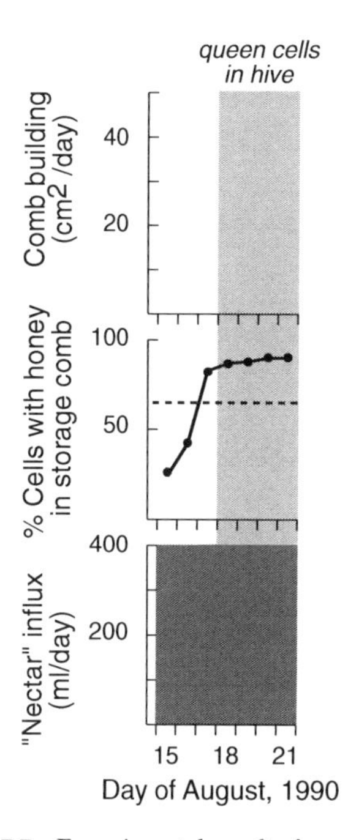

Figure 7.7 Experimental results from August 1990 which reveal that bees in a colony preparing to swarm will avoid building comb under conditions that would normally trigger massive comb construction. The colony was given a high nectar influx for a week, and consequently it filled its storage comb with honey well above the level normally associated with the start of comb building *(dashed horizontal line, middle panel);* but surprisingly the colony built no comb at all. Sealed queen cells were noticed in the hive on the final day of the experiment, which indicates that queen cells must have been present in the hive during the previous several days. Evidently, the colony was preparing to swarm throughout most, if not all, of this experiment. After Kelley 1991.

full, for as is shown in Figure 7.6, even when the colony began to build comb, some 20% of the cells in the storage comb remained vacant. Obviously, the colony had not yet reached the point where its food-storer bees were forced to use their own bodies as storage containers. The same holds true for the colony studied by Pratt, in which comb building began when some 40% of the cells in the storage comb remained empty. Thus the evidence at hand, although preliminary, suggests that the honey stomach distension hypothesis is wrong.

Turning to the storing difficulty hypothesis, unfortunately we find that no explicit test of this idea has been performed. What little evidence there is regarding this hypothesis comes from observations that I have made of the behavior of food-storer bees (Seeley 1989a). Working with a 4000-bee colony living in a two-frame observation hive, I followed 10 food-storer bees, each one for up to 60 min, under two sets of conditions: first, when the hive was largely empty of honey (27–29 May 1987; only 13–32% of the cells in the storage comb contained honey), and second, a few days later, when the hive was essentially full of honey (1–2 June 1987; 97–99% of the cells contained honey). Between these two observation periods, the black locust trees *(Robinia pseudoacacia)* had come into bloom and the resultant nectar flow enabled the bees to fill the observation hive with honey. I observed that when the hive was almost empty, the food-storer bees could quickly locate a cell in which to store a nectar load, and so had a mean processing cycle time (Figure 6.1) of only 10.2 min (SD = 4.0 min). In contrast, when the hive was packed nearly full of honey, the food-storer bees had great difficulty finding a place to deposit nectar. Indeed, none did so, and all instead spent time either standing still concentrating the nectar or walking about dispensing the nectar to nestmates, or both. In this situation, the mean duration of a processing cycle was far higher than before, 28.3 min (SD = 19.5 min). Clearly, when the hive was almost full of honey the food storers experienced much difficulty finding a place to put nectar. It should be noted, however, that these observations apply to bees living in a hive already packed nearly full of honey, hence one in which comb building usually would already have begun (in fact, my study colony had started building burr comb on the glass walls of my observation hive).

But what we really need to know to test the storing difficulty hypothesis is whether or not food-storer bees experience increasing difficulty *before* their hive is crammed full of honey, for as just noted (Figures 7.5 and 7.6) the onset of comb building normally precedes the complete filling of a hive. Ideally, we would have a clear picture of how the difficulty of storing nectar increases as a function of the fullness of a colony's combs, and we would have experimental evidence indicating whether or not the food-storer bees actually respond to the difficulty of storing nectar. I look forward eagerly to a rigorous examination of this attractive hypothesis for how an individual bee acquires information about her colony's need to build comb.

7.3. How the Quantity of Empty Comb Affects Nectar Foraging

We now know that a colony's rate of nectar collection strongly influences its comb-building activity, with high nectar influx an essential stimulus for comb construction. Recent studies indicate that the cause-effect relations between combs and nectar foraging can operate in the other direction as well, such that the amount of empty comb in a colony's hive influences its nectar foraging activity. More specifically, there is now good evidence that a colony with much empty comb will gather nectar at a higher rate, compared with a colony with little empty comb, all else being equal. This influence of empty comb on the colony's rate of nectar collection may be adaptive, for it may help colonies avoid starvation whenever their honey reserves fall to dangerously low levels. In nature, if a colony has much empty comb in its hive, probably it is approaching starvation, and in this situation its odds of survival will be boosted if its nectar foragers work more vigorously, for this will help a colony amass a honey reserve as quickly as possible. But why shouldn't nectar foragers always work at their highest possible rate? Perhaps because doing so would actually lower the lifetime contributions of forager bees to their colony's economy, since achieving a high foraging rate may entail a reduction in foraging efficiency (Houston, Schmid-Hempel, and Kacelnik 1988). In short, natural selection may have shaped bees to respond to extreme shortages of food, indicated by large quantities of empty comb inside their hives, by temporarily adjusting the foraging strategy to maximize the rate, not the efficiency, of nectar collection.

7.3.1. THE COLONY-LEVEL PATTERN

The first indication that the amount of empty comb in a colony's hive influences the colony's rate of nectar collection comes from an experiment performed by Rinderer and Baxter (1978). During the spring nectar flow in Louisiana, they assembled 20 colonies of bees in one

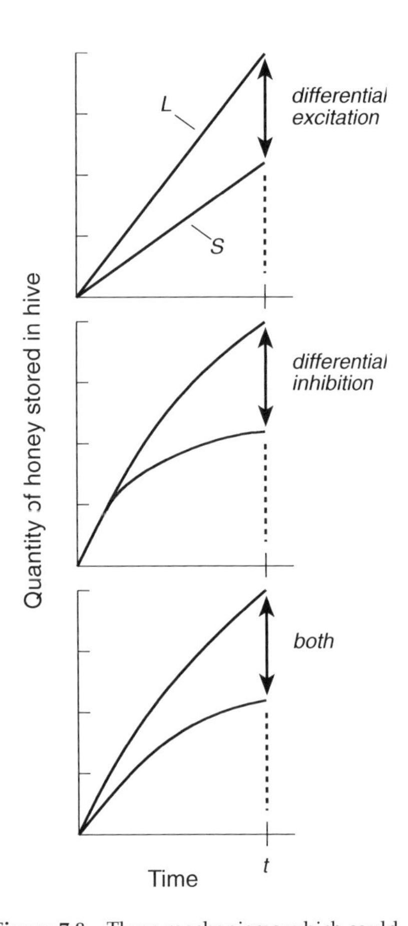

Figure 7.8 Three mechanisms which could produce the observed pattern of colonies with a large amount of empty comb (L) storing more honey in their hives within a given time t than colonies with a small amount of empty comb (S). *Top:* differential excitation of nectar foraging results in consistent differences between colonies in the rate of nectar collection. *Middle:* differential inhibition of nectar foraging, a result of colonies with little empty comb having less honey storage capacity and so experiencing feedback inhibition sooner than colonies with much empty comb. *Bottom:* differential excitation and differential inhibition occurring simultaneously.

apiary and randomly assigned each colony to one of two treatment groups. In one group, each colony's hive received 36 frames of empty comb (total empty comb area = 1.88 m^2), while in the other group each colony's hive received 72 frames of empty comb (total comb area of 4.06 m^2). (Some of the frames used in the second treatment group were slightly larger than those used in the first treatment group.) Each colony's hive was weighed at the start and at the end of the 15-day experiment to measure how much honey each colony had gathered. Then the experiment was repeated by removing the added combs from each colony's hive, moving all the colonies to a new location, and finally repeating the manipulations as before, but reversing the treatments for the two groups of colonies. In both trials of the experiment, the difference in amount of empty comb resulted in a significant ($P < 0.008$) difference in mean colony weight gain: trial 1, 36 versus 51 kg; trial 2, 47 versus 58 kg (36-frame versus 72-frame treatments, respectively). Rinderer (1982a) also repeated this experiment several years later and obtained essentially identical results. Since honey is basically concentrated nectar, a colony's weight gain is an indication of the level of its activity in nectar collection. Thus the finding of greater weight gain by colonies with more empty comb implies that having lots of empty comb somehow causes a colony to collect more nectar than it would otherwise.

7.3.2. THE INDIVIDUAL-LEVEL PROCESSES

What mechanisms might underlie the colony-level pattern of differential nectar collection in relation to different amounts of empty comb? Rinderer and Baxter (1978) suggest that empty comb stimulates nectar foraging. On this view, the difference in mean colony weight gain that they reported reflects differential *excitation* of the foragers in the two sets of colonies (Figure 7.8). Alternatively, however, this difference in colony weight gain might reflect differential *inhibition* of the foragers. In other words, it is possible that the foragers in both treatments started out with the same rate of nectar collection, but that this declined more quickly in the colonies with smaller hives because their hives more quickly reached the level of fullness at which a colony is forced to lower its rate of nectar collection. One explanation for this forced drop in nectar influx could be that increasingly full combs make it harder for the food-storer bees to process nectar, and this, in turn, makes it harder for the foragers to collect nectar. We know, of course, that a nectar forager only completes a collecting trip once she has succeeded in passing her nectar load to a food-storer bee.

The information supplied by Rinderer and Baxter (1978) indicates that differential excitation was mainly responsible for the differential nectar collection in their experiment. They point out that differential inhibition of the nectar foragers was unlikely because "at all times during the experiment, every colony had empty storage space available." Since the storage capacity of one frame of empty comb is 2.2 kg of honey (Rinderer 1982a), the total storage capacity of the 36 frames of empty comb given to each small-hive colony was about 79 kg; and on average the small-hive colonies stockpiled 36 kg of honey in the first trial and 47 kg in the second. Thus it is clear that the small-hive colonies never filled their combs with honey. But even though the combs of the small-hive colonies were never filled with honey, it is possible that differential foraging inhibition contributed to the outcome of this experiment because it is likely that a colony's capacity for nectar processing declines, and therefore its rate of nectar collection is forced to drop, long before its combs are completely filled. One indication of this comes from the investigations of the control of comb building discussed above. Here I noted that comb building begins when a colony's combs are only 60–80% full (Figures 7.5 and 7.6), and that comb building apparently is performed by the same bees that perform nectar processing (Figure 7.3). Thus the process of building comb is likely to reduce the number of bees processing nectar, and this could hinder the activity of the foragers by forcing each one to search longer for a food-storer bee every time she returns to the hive.

That comb building does in fact lead to inhibition of nectar foraging is demonstrated by one result recorded by Pratt in his study of the control of comb building. As is shown in Figure 7.9, he observed a marked surge of the in-hive search times of returning nectar foragers precisely when his study colony began to build comb. As shown previously (Section 5.7.3), a greater mean in-hive search time not only forces nectar foragers to complete fewer foraging trips per day, but also induces them to raise their acceptance threshold for nectar sources. Both effects lower a colony's rate of nectar collection. Given this link between comb building and diminished nectar foraging, it would be useful to know whether or not Rinderer and Baxter observed either comb building in only their small-hive colonies or comb building starting sooner in the small-hive colonies than in the large-

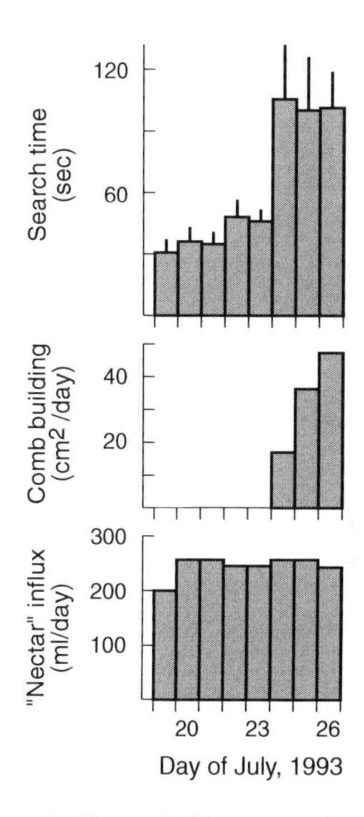

Figure 7.9 The marked increase in the search time of nectar foragers when their colony began to build comb. Note that this surge in the search time did not reflect an increase in the colony's nectar collection *(bottom panel)*, and thus it was almost certainly caused by a decrease in the colony's capacity for nectar processing. These results support the idea that comb building will produce a temporary inhibition of nectar foraging. Based on unpublished data of S. Pratt.

hive ones. Either phenomenon would have provided indirect evidence of differential inhibition of nectar foraging between their small-hive and large-hive colonies. Unfortunately, they give no information about comb building by their study colonies.

Additional evidence that empty combs do indeed stimulate bees to collect nectar more rapidly comes from a series of laboratory experiments in which groups of 30–50 bees were placed in "hoarding cages" where they were given fixed amounts of empty comb together with a feeder providing sucrose solution ad libitum. After 3 to 7 days the amount of sugar solution that each group of bees had transferred from its feeder to the combs was measured (Rinderer and Baxter 1978, 1979, 1980). For example, in one experiment the two treatments consisted of giving each group of 50 bees either one piece of comb (47 cm^2, with approximately 180 cells) or three such pieces of comb. After 3 days, the mean volume of nectar transferred by groups of bees with three combs was 30% higher than by groups with one comb: 7.25 versus 5.58 mL ($P < 0.01$). It is exceedingly doubtful that differential inhibition of nectar collection influenced the outcome of this experiment since for both treatments, on average, more than 93% of the cells remained empty at the end of the experiment.

One might still argue that such studies involve highly artificial conditions and that hence the stimulatory effect of the empty comb that was observed is merely a consequence of more combs *generally* stimulating bees to work harder by giving them a more natural environment, rather than of more combs *specifically* stimulating bees to gather more nectar. The findings of a follow-up experiment performed by Rinderer (1982b) suggest, however, that the first of these two interpretations is incorrect. In this experiment, he again worked with the experimental system of hoarding cages, but instead of comparing the effects of different amounts of empty comb in a cage, he compared the effects of different types of comb odor pumped into each cage. Rinderer found not only that groups of bees receiving volatile odors from *empty* combs stored 32% more sugar solution than did control groups receiving no volatile odors from empty combs, but also that groups of bees receiving volatile odors from *honey-filled* combs stored no more sugar solution than did the control groups. These results demonstrate that the stimulation of hoarding by empty comb that was observed in the preceding hoarding-cage experiments was not simply a broad response to combs in general, but instead was a specific response to empty combs in particular.

Summary

1. A honey bee colony needs beeswax combs to hold its reserve supplies of honey and pollen, but because these combs are energetically expensive to construct, a colony builds additional combs only when absolutely necessary. Thus comb building presents us with puzzles about the controls of the timing and the amount of a production process.

2. Two pieces of evidence indicate that the comb builders come mainly from the ranks of the middle-aged bees in a colony. First, analysis of the age distribution of bees in building clusters (Figure 7.2) reveals that the large majority are 10–20 days old (Figure 7.3). Second, analysis of wax gland activity as a function of age shows that these glands are of maximum size in bees 10–18 days old (Figure 7.3). Because the middle-aged bees in a colony are known to perform other tasks in the peripheral, food-storage region of the hive, it appears that comb building is done by the bees that also handle nectar processing, the food-storer bees.

3. Comb-builder/food-storer bees will begin building comb only when two conditions are fulfilled: (1) the colony has a high influx of nectar and (2) its storage combs are getting nearly full (Figures 7.5 and 7.6). Thus the *timing* of comb building reflects both an external variable (nectar availability) and an internal variable (comb fullness). Curiously, once a colony has started building comb it apparently requires only a continued strong nectar influx for it to continue its building (Figure 7.5), possibly because in nature a colony will always have well-filled combs once it has commenced comb building. This suggests that the *amount* of comb built during any particular building period reflects mainly an external variable: the availability of nectar.

4. It is doubtful that the comb builders have the powers of information collection needed to directly monitor their colony's rate of nectar collection and the fullness of its storage combs. It seems more likely that instead they sense some incidental cue which reliably indicates that the time is right for comb building. One possibility is prolonged distension of the honey stomach. Arguing against this hypothesis, however, is the observation that bees begin building comb when 20% or more of the cells in the storage comb remain vacant, hence long before bees are compelled to use their own bodies as storage containers. A second possibility for the critical cue is difficulty in finding cells in which to deposit nectar. Although it is clear that

bees experience difficulty finding a place to put nectar once their hive is packed full of honey, it remains unknown whether they do so *before* the hive is completely filled, the time when comb building normally begins. Thus this second hypothesis remains to be tested.

5. A colony with much empty comb in its hive will gather nectar more rapidly than one with little empty comb, all else being equal. This colony-level pattern could reflect a situation in which different amounts of empty comb lead to differential excitation of foragers or differential inhibition of foragers, or both (Figure 7.8). Probably both are involved. It seems likely that differential inhibition must play some role, for when a colony's combs become nearly full, its capacity for nectar processing shrinks and consequently its rate of nectar collection must drop (Figure 7.9). However, tests of the hoarding performance of bees in cages where the combs never approach fullness reveal a correlation between the amount of empty comb and the quantity of sugar solution stored, suggesting that extensive empty comb in a colony's hive does excite the colony's foragers to work more vigorously. If so, this response may be an adaptation to help colonies avoid starvation.

Regulation of Pollen Collection

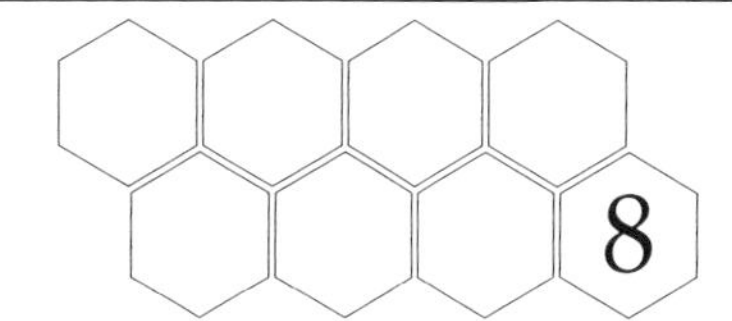

One key to understanding the control system regulating a honey bee colony's intake of pollen is recognizing that a colony's *supply* of pollen undergoes far greater fluctuations than does its *demand* for pollen (Figure 8.1). A colony's supply of pollen—the amount available outside the hive—can vary dramatically from day to day depending upon the weather conditions and the flowers in bloom. Consider, for example, what happens when the weather shifts from a sunny, warm day to a rainy, cool day: the colony experiences an overnight collapse in its external supply of pollen. In contrast, the colony's internal demand for pollen—the amount consumed by the adult and immature bees inside the hive—does not vary dramatically from day to day. Rather, the daily consumption of pollen changes only gradually as the colony's population of brood and adults slowly rises and falls over the course of a summer. A colony copes with the phenomenon of larger dynamics in supply than in demand by building a modest stockpile of pollen, approximately one kg, inside the hive (Figure 2.7). This reserve supply is sufficient to buffer the colony against a failure in the external pollen supply lasting a week or so, and therefore it is large enough to keep a colony properly nourished throughout a long spell of bad weather. Presumably, the reason that a colony does not amass a still larger reserve of pollen—several kilograms instead of just one—is that the benefits of a larger reserve (stronger buffering against disruptions in the pollen supply) do not outweigh the costs (fewer cells available for brood rearing and honey storage, for a given investment in comb).

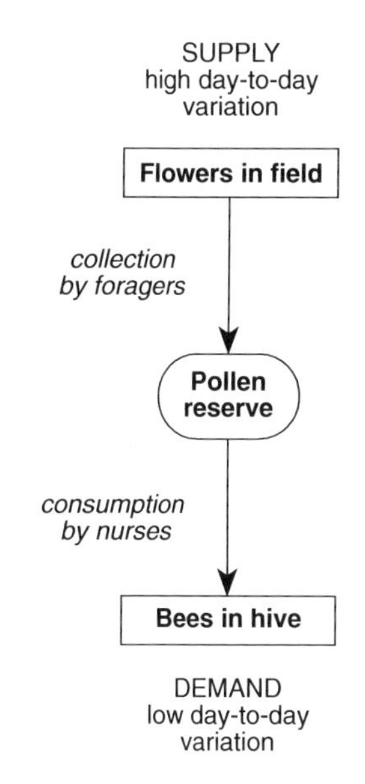

Figure 8.1 The main features of the pollen sector of a colony's economy. The external supply of pollen varies greatly from day to day, because of changes in the weather conditions, whereas the internal demand for pollen varies little from day to day, because the number of bees (brood and adults) needing protein from the pollen-consuming nurse bees changes only gradually. A colony buffers itself against dips in the pollen supply by building up a pollen reserve inside the hive.

8.1. The Inverse Relationship between Pollen Collection and the Pollen Reserve

A pollen reserve stored safely inside the hive solves the colony's problem of discrepancy between external supply and internal demand, but it creates a new problem for the colony: namely adjusting the rate of pollen collection in accordance with the amount of pollen stored in the hive. This regulation is essential if the colony is to maintain a sufficient, but also not excessive, reserve of pollen. If this reserve is too small, a result of consumption having exceeded collection, the collection rate needs to be raised. Conversely, if the pollen reserve is too large, a consequence of collection having exceeded consumption, the collection rate should be lowered.

One experiment that demonstrates the inverse relationship between the size of a colony's pollen stores and its rate of further pollen intake has already been described in Chapter 3 (Section 3.7). There we saw that the pollen foragers in a colony will respond to a marked rise in their colony's pollen reserve by slowing, if not stopping, their collection of pollen (Figure 3.12). We saw too that this response is rather rapid, such that a change in the pollen reserve at the end of one day can trigger a change in the foragers' behavior the following day.

Another recent study, this one by Fewell and Winston (1992), documents the same pattern of an inverse relationship between pollen reserve and pollen collection, but in still finer detail. They began their experiment in May 1989 by measuring the pollen stores in six full-size colonies, each one containing approximately 35,000 bees and inhabiting a standard beehive. On average, these hives each contained a total of 2240 ± 360 (x ± SE) cm^2 of cells holding pollen. Fewell and Winston manipulated the amount of pollen in these hives, boosting it to a high level (4455 ± 53 cm^2) in three and dropping it to a low level (240 ± 48 cm^2) in the other three. At the same time, they equalized the amount of honey, unsealed brood (eggs and larvae), sealed brood (pupae), and empty comb in the six hives, so that the only marked difference in internal conditions between the two sets of colonies was the size of their pollen stores. Over the next 16 days they monitored each colony's rate of pollen collection (by measuring the arrival rate of pollen foragers at each hive and the size of their pollen loads), and periodically they surveyed the contents of each colony's hive to assess changes in the levels of pollen, honey, and brood. This work revealed several striking differences in the patterns of pollen collection between the two sets of colonies. First, the average rates of pollen collection for colonies with low and high pollen stores were 22.3 and 14.5 g/hr, thus showing an inverse relation to the colony's pollen stores. Moreover, these two pollen collection rates were evidently well above and well below, respectively, the mean pollen consumption rates for the two types of colonies since, as is shown in Figure 8.2, the pollen stores of the two sets of colonies steadily rose or fell during the 16-day period of monitoring the colonies. Indeed, by the end of the observations, the pollen reserves in colonies of both treatments had returned to a level not significantly different from what was observed prior to the manipulations. Thus it is clear that each of the six colonies had regulated its rate of pollen collection in relation to its reserve supply of pollen.

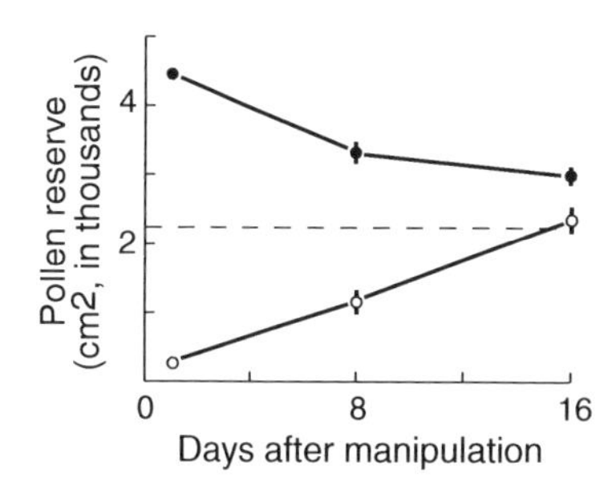

Figure 8.2 Homeostasis in the size of a colony's pollen reserve. Six equal-sized colonies were divided into two groups of three. Each colony in one group had its pollen reserve raised to 4460 cm^2, while each colony in the second group had its pollen reserve lowered to 240 cm^2. The colonies then lowered or raised their rates of pollen collection in accordance with the size of their pollen reserves so that by the end of 16 days their pollen reserves had all converged to a level close to the mean pre-manipulation size of 2240 cm^2 (indicated by the dashed line). After Fewell and Winston 1992.

8.2. How Pollen Foragers Adjust Their Colony's Rate of Pollen Collection

A colony's rate of pollen collection (C) is a function of three variables:

$$C = \frac{N\,L}{T} \tag{8.1}$$

where N is the number of foragers engaged in pollen collection, L is the mean pollen load gathered on a foraging trip, and T is the average foraging trip time for a pollen forager. Evidently, pollen foragers can adjust all three variables in accordance with their colony's pollen reserve. To investigate changes in the number of a colony's pollen foragers (N) in relation to its pollen stores, Camazine (unpublished) censused the number of pollen foragers in a colony at times of large and small pollen reserves. For this experiment he used a three-frame observation hive in which the upper two frames were completely filled with brood and the bottom frame was reserved for manipulations of the colony's pollen stores. To establish and maintain a small pollen reserve in the hive, he removed the bottom frame at the end of each day and replaced it with a new, empty frame. Because the upper two frames of the observation hive contained almost no empty cells, the returning pollen foragers were forced to deposit their pollen in the lower frame, and thus replacing this frame removed essentially all the colony's stored pollen. To establish and maintain a large pollen reserve, Camazine again removed the bottom frame at the end of each day, but replaced it with a frame packed full of pollen. He censused

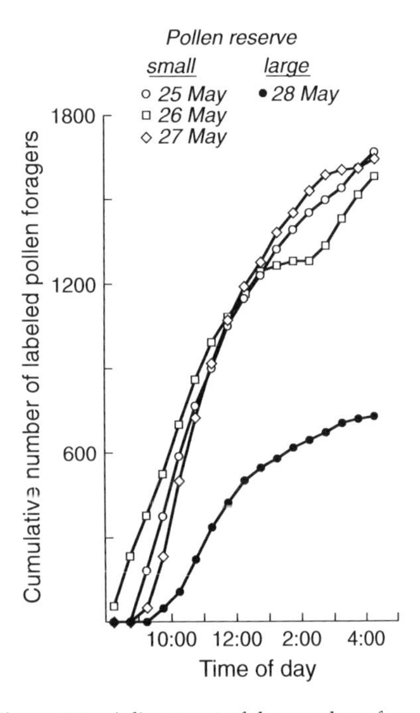

Figure 8.3 Adjustment of the number of pollen foragers in relation to the size of the colony's pollen reserve. For the first 3 days, 25–27 May 1990, the colony was deprived of pollen by removing daily the combs containing pollen, and each day's census of the pollen foragers revealed a total of some 1700 bees. Then on the evening of 27 May the colony was given a frame packed full of pollen. The census for 28 May revealed only 737 pollen foragers. Since the weather was hot and sunny on all 4 days, the drop in the number of pollen foragers was evidently caused not by some external change in the colony's environment, but by the internal change in the colony's pollen reserve. Based on unpublished data of S. Camazine.

the colony's pollen foragers daily by daubing paint on the thorax or abdomen of each pollen-bearing bee that entered the hive between 8:00 in the morning and 4:15 in the afternoon and recording the number of different bees so labeled. The first trial of this experiment was performed on 25–28 May 1990. The colony was deprived of pollen on 25, 26, and 27 May and on these 3 days the census counts totaled 1684, 1597, and 1662 pollen foragers (Figure 8.3). Then in the evening, at 6:00, on 27 May the colony was given a bottom frame packed full of pollen, and the following day a census was performed. This revealed only 737 pollen foragers, indicating a 56% decline in the number of bees collecting pollen. Repeating the experiment on 19–24 July 1990 revealed a 26% decline, from 1455 to 1074, in the mean number of pollen foragers between days of low and high pollen stores. Clearly, a large store of pollen in a colony's hive somehow causes many of the pollen foragers to stop collecting pollen, thereby reducing the number of pollen foragers (N).

Presumably the opposite situation, a small pollen store in the hive, causes the pollen foragers to recruit nestmates to the task of pollen collection, thereby increasing the number of pollen foragers (N). One indication that a small pollen store does indeed stimulate recruitment by pollen foragers comes from an experiment by Camazine (unpublished) in which he measured the proportion of pollen foragers performing waggle dances under the conditions of large and small pollen stores. When he manipulated a colony's pollen stores as described above, and followed pollen foragers one by one upon their return to his observation hive, he found that only 14 out of 174 (8%) pollen foragers performed dances on days when the colony's pollen reserve was large, but that 29 out of 162 (18%) did so on days when the pollen reserve was small.

A study by Fewell and Winston (1992) provides clear evidence that pollen foragers can also adjust trip time (T) and load size (L) in relation to their colony's pollen reserve. They labeled 50 pollen foragers for individual identification from four of their study colonies (two in each treatment group, Figure 8.2) and monitored these bees' departures and arrivals at their hives. They also collected and weighed the pollen loads brought back by unlabeled pollen foragers from each of the four colonies. These measurements revealed a markedly lower mean trip time and a slightly higher mean load size for foragers from colonies with small pollen reserves relative to those for foragers from colonies with large pollen reserves: 53.6 versus 79.9 min, and 16.7 versus 14.0 mg, respectively ($P < 0.05$ for both differences). Most of the difference in trip time was a consequence of the fact that pollen foragers from colonies with much pollen spent far more time inside the hive between trips compared to those from colonies with little pollen (30.0 versus 15.6 min). Camazine (unpublished) also observed in his studies of the in-hive behavior of pollen foragers that if a colony has a large pollen reserve its pollen foragers appear halfhearted in their foraging (Figure 8.4). Compared with pollen foragers in a colony whose pollen reserve was small, they entered the hive less quickly (walking rather than running through the entrance tunnel), spent more time grooming before searching for a cell in which to deposit their pollen, and remained longer in the hive after depositing their pollen.

The empirical findings just described indicate that a colony alters its pollen collection by adjusting both the number of pollen foragers (N) and the per capita collection rate of these pollen foragers (L/T). Fewell and Winston (1992) provide information on the relative importance of these two distinct mechanisms for modulating a colony's pollen collection, at least for one particular experimental situation. They found that the total rate of pollen collection (C, in g/hr) was 54% higher for their colonies receiving the low pollen treatment than for those receiving the high pollen treatment. They also found that the mean rate of pollen collection by individual foragers (L/T, in g/hr/bee) was approximately 43% higher in the low pollen relative to the high pollen colonies. Thus in this study the total boost in a colony's rate of pollen collection came largely from an increase in the per capita collection rate of pollen foragers (L/T), and only slightly from an increase in the number of pollen foragers (N), since the latter evidently increased only 8% ($1.54 = 1.43 \times 1.08$). This is an interesting result, for it shows that even a large adjustment in a colony's rate of pollen collection can be accomplished with a relatively small change in the number of pollen foragers. Other studies, however, such as the census experiments conducted by Camazine (Figure 8.3) and various experiments performed by Free (1967), show that the number of pollen foragers can decline by 50% or more overnight when a colony's need for pollen is suddenly lowered through an experimental manipulation of the colony's pollen reserve. Thus in these studies it appears that the change in a colony's total rate of pollen collection came primarily from a change in the number of pollen foragers.

Whether a colony relies mainly on changing the number of pollen foragers (N) or on changing the per capita collection rate of pollen foragers (L/T) to adjust its rate of pollen collection may depend strongly on whether the colony is raising or lowering its pollen intake, and on whether the colony is making a large or small adjustment in pollen intake. This will be true if the two variables (N and L/T) differ in their modulation characteristics. For example, it seems possible that the number of pollen foragers is quickly decreased (through abandonment) but only relatively slowly increased (through recruitment), whereas the per capita collection rate is both rapidly decreased and rapidly increased. If so, then a rapid increase in a colony's total rate of pollen collection might be generated mainly by raising the per capita rate of pollen collection, whereas a rapid decrease in a colony's pollen collection might involve lowering both the number of pollen foragers and the per capita rate of pollen collection. These two variables may also differ in the amplitudes of their modulation, such that the number of pollen foragers can be varied more broadly than can the per capita rate of pollen collection. If this is so, the larger the adjustment in a colony's rate of pollen collection, the more important the change in the number of pollen foragers. Clearly, these ideas are mostly speculative. The full story of how pollen foragers change their behavior to adjust their colony's rate of pollen collection remains untold, and the subject deserves further investigation.

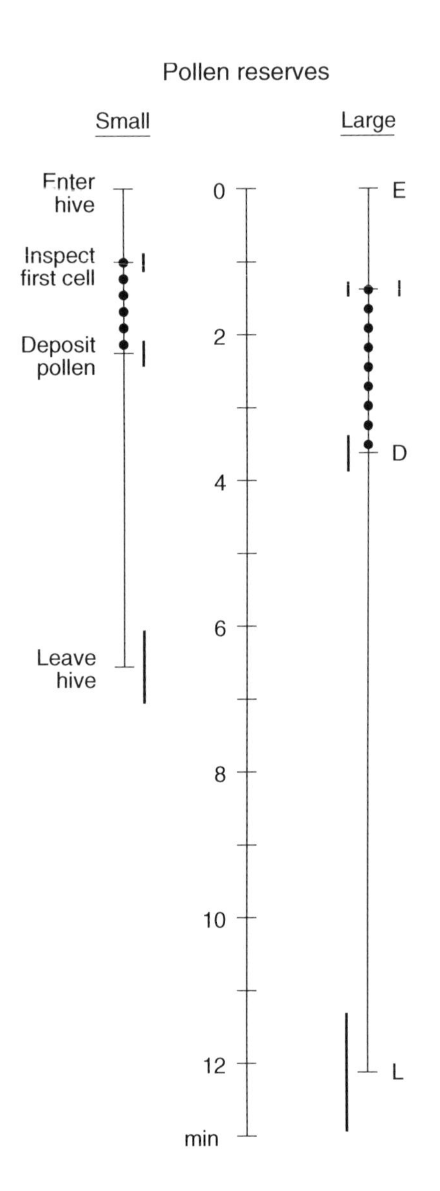

8.3. How Pollen Foragers Receive Feedback from the Pollen Reserves

The fact that a colony modulates its pollen collection in relation to its pollen reserve indicates that there must be an information link between a colony's pollen reserve and its pollen foragers, one which

Figure 8.4 Reduction in the tempo of foraging as the colony's pollen reserve increases. The two time lines depict the mean time budgets for the in-hive behaviors of pollen foragers from one colony at times of small (< 50 cells) and large (> 1000 cells) pollen reserves. Thick vertical lines beside the points *I*, *D*, and *L* depict the standard errors of the times to inspect the first cell, deposit pollen, and leave the hive, respectively. Notice that each phase of the in-hive behavior—entering the hive, searching for a cell in which to deposit pollen, and preparing to leave the hive after depositing the pollen—is performed more slowly and less enthusiastically when the colony's pollen reserve is large. Notice too that the mean number of cells that a forager inspects before depositing her pollen loads is significantly smaller when the pollen reserve is small than when it is large (5.6 versus 9.3 cells inspected, $P < 0.001$). In all, 84 and 92 bees were followed for the conditions of small and large pollen reserves, respectively. Based on unpublished data of S. Camazine.

provides either excitatory feedback when the reserve is small, or inhibitory feedback when the reserve is large, or both. A useful first step in analyzing this process of feedback control is to determine whether the link between pollen stores and pollen foragers is direct or indirect, that is, whether or not the pollen foragers directly sense the amount of pollen stored in their hive in order to tune their collecting behavior in accordance with their colony's needs.

Older studies of pollen collection by honey bee colonies, although not designed to address the question at hand, report findings suggesting that the pathway of feedback between pollen reserve and pollen foragers is not direct. For example, several studies (Free 1967; Todd and Reed 1970; Al-Tikrity et al. 1972; Hellmich and Rothenbuhler 1986) have shown a strong stimulatory effect of brood, especially larvae, on pollen foraging. This finding raises the possibility that pollen foragers respond not to the size of the pollen reserve per se, but to some indicator of their colony's overall *need* for pollen, which might combine information about both the pollen reserve and the pollen demand inside the hive. It is possible though, that the strong effect of brood on pollen foraging arises simply because the brood causes heavy consumption of pollen, which in turn causes a colony's pollen reserve to drop, and it is the depression of the pollen reserve that is actually sensed by the pollen foragers. A second piece of evidence militating against the hypothesis of direct feedback from pollen reserve to pollen foragers is the observation that feeding a colony pollen by placing it in glass petri dishes, not in the beeswax combs, causes a decrease in a colony's pollen collection (Free 1967; Free and Williams 1971). Here, though, one could argue that even though the pollen was not present in the hive in its normal way (packed into cells), the pollen foragers were nevertheless directly sensing the presence of pollen, perhaps through its odor.

Camazine (1993) recently performed an ingenious pair of experiments designed to answer conclusively the question of direct versus indirect feedback from pollen stores to pollen foragers. His experimental procedures are presented diagramatically in Figures 8.5 and 8.6. First he set up a pair of three-frame observation hives, with the two hives mounted side by side, hence in a common environment. Both hives were mounted in a laboratory room, but their entrances were connected to tubes leading outdoors so that the bees could forage naturally. Both hives were stocked with a colony of about 8500 bees, two frames containing brood and honey, and a third frame

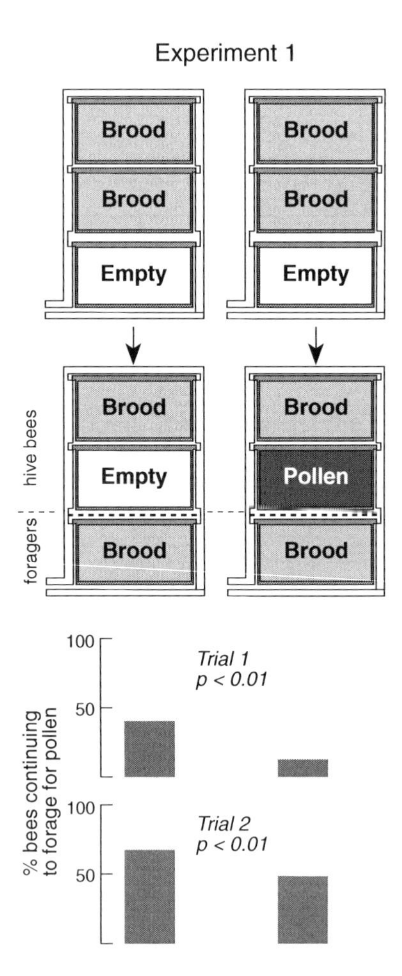

Figure 8.5 Design and results of the experiment performed to determine whether pollen foragers need contact with their colony's pollen reserve to obtain feedback regarding its size. Two colonies were established in identical observation hives and deprived of pollen for 5 days *(top)*. Then the pollen foragers of each colony were segregated below a single screen, and a frame of pollen was placed above the screen in one hive *(bottom)*. The following day, in both trials, a significantly smaller percentage of the pollen foragers continued collecting pollen in the colony that received the pollen, which indicates that the foragers were able to receive feedback through the screen. After Camazine 1993.

whose contents were varied during the experiment. The first experiment began by depriving both colonies of pollen for 5 days using the technique already described—removing most of a colony's pollen at the end of each day by replacing its bottom frame with a new, empty frame. During the fifth day of this pollen deprivation period, Camazine and his assistants daubed paint on every pollen-bearing bee entering either hive, and so labeled all the pollen foragers in each colony. Then at the end of the fifth day, Camazine segregated the pollen foragers from the hive bees in each colony by plugging the entrance of each hive, removing its glass walls, and then gently plucking from its combs every labeled pollen forager. When released outside the laboratory building, these bees immediately flew to their hive's entrance, but because it was plugged they were unable to enter their hive. At this point Camazine inserted either an empty or a pollen-filled frame into the middle position of each each hive, moved the second brood frame to the bottom position, then gently smoked all the bees inside each hive onto its upper two frames and inserted a wire screen of 4 mm mesh which blocked the passage of these bees (but not the flow of food) back down to the bottom frame. Last, he replaced the glass walls on each hive and removed the barrier at each hive's entrance to allow the pollen foragers to enter their hives. Thus at the end of the fifth day, all the pollen foragers in each hive were on the bottom frame, separated by a screen from the hive bees above. Also, in one hive there was a large store of pollen (but isolated from the pollen foragers by a screen), while in the other there was little or no pollen.

Throughout the sixth day all the bees returning to each hive that had been marked the previous day as pollen foragers were again labeled with paint. These bees were marked with a different color depending upon whether they returned to their hive with or without pollen. Thus at the end of the day it was possible to calculate what fraction of the previous day's pollen foragers had continued pollen foraging, had switched to nectar foraging, or had not left the hive, hence had not foraged at all. In both trials of this experiment the pattern was clear (Figure 8.5). In the colony given pollen relative to the colony not given pollen, a significantly smaller percentage of the foragers continued to forage for pollen on the sixth day: 12% versus 40%, and 48% versus 67%. Thus it is clear that pollen foragers do not need direct contact with the pollen reserve to respond to changes in the size of this reserve.

The results of this first experiment, however, leave open the possibility that pollen foragers adjust their behavior by sensing the level of pollen odor in the hive. To test this hypothesis, Camazine performed a second experiment, essentially identical to the first except that a double screen with a 2-cm space between the two layers was used instead of a single screen (Figure 8.6). This double screen prevented contacts and food exchange between the pollen foragers and hive bees, but it did not block the passage of odors to the pollen foragers. Thus if pollen foragers respond to pollen odor, the proportion of pollen foragers continuing to forage for pollen on the sixth day should have been smaller in the colony that received pollen. Both trials of this experiment, however, yielded the opposite result. In the colony given pollen relative to the colony not given pollen, the percentage of the foragers continuing to forage for pollen was either significantly *larger* or no different: 55% versus 39%, and 30% versus 30%. These results demonstrate that pollen foragers are not using the level of pollen odor in the hive to adjust their behavior in relation to their colony's pollen reserve.

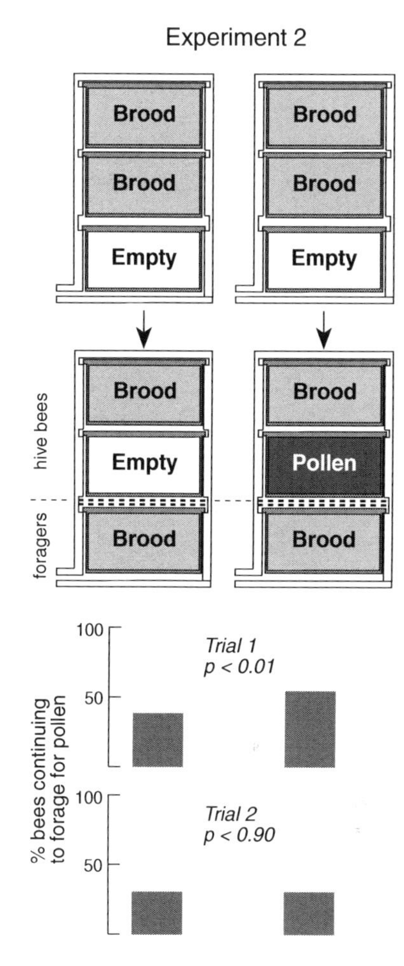

Figure 8.6 Follow-up to the experiment depicted in Figure 8.5, conducted to determine whether pollen foragers respond to the size of their colony's pollen reserve by sensing the level of pollen odor in their hive. The experimental plan matches that shown in Figure 8.5 except that a double screen was used. In neither trial was the percentage of pollen foragers that continued foraging any lower in the hive that received a pollen frame relative to the hive that received an empty frame. Hence foragers do not use the level of pollen odor to adjust their behavior in relation to their colony's pollen reserve. After Camazine 1993.

In summary, these two experiments show that pollen foragers do not need direct contact with their colony's pollen reserve to respond to a rise in that reserve. Hence they show that this response is not mediated by the foragers' senses of touch or taste. These experiments also show that the foragers' response is not mediated by the sense of smell. Also, because pollen neither emits light nor produces sound, it seems certain that pollen foragers do not measure the pollen reserve inside the dark hive through vision or hearing. Thus all the evidence at hand indicates that foragers acquire information about the size of their colony's pollen reserve not through direct sensory perception of the stored pollen, but through some indirect mechanism of information flow. What might this be?

8.4. The Mechanism of Indirect Feedback

In Camazine's first experiment (Figure 8.5), the bees above the screen in the hive that received pollen were somehow able to convey to the foragers below the screen the information that their colony's pollen reserve had grown large. How did they do this? Nearly 30 years ago Free (1967) hypothesized that when the nurse bees in a colony experience a shortage of pollen in the hive, they might inform the pollen foragers of the shortage by preparing more cells for pollen storage,

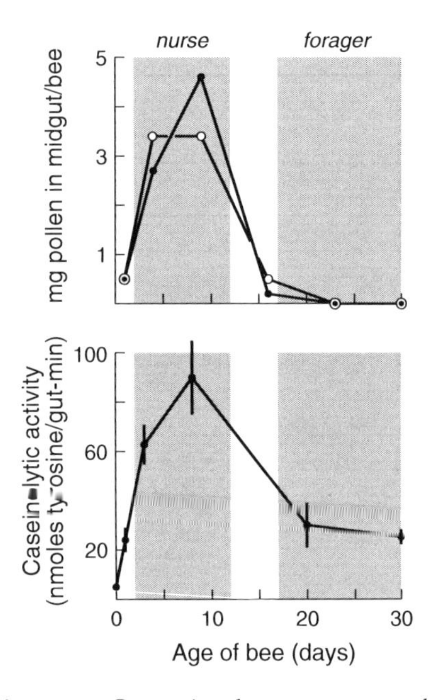

Figure 8.7 Comparison between nurses and foragers in terms of the pollen content *(top)* and the caseinolytic activity *(bottom)* of a bee's midgut, the site of pollen digestion. The much higher values for nurses indicate that they are the bees in a colony mainly responsible for digesting pollen. The two curves in the top figure represent data from two distinct colonies. After Crailsheim et al. 1992 *(top)* and Moritz and Crailsheim 1987 *(bottom)*.

thus making it easier for the returning pollen foragers to unload their pollen. Ease in locating a suitable storage cell would signify to a pollen forager that her colony's pollen reserve has been depleted and that she should take action to boost the colony's pollen collection. In essence, this hypothesis expresses the idea that the nurse bees produce an *excitatory* cue at times of pollen shortage in the hive. Although Free's hypothesis is eminently plausible, and indeed we now know that pollen foragers do find it easier to locate suitable storage cells when their hive contains little pollen (Figure 8.4), it cannot explain the difference between colonies in pollen forager behavior that was found in Camazine's first experiment. The screen in each hive prevented the nurse bees from reaching the cells contacted by the pollen foragers; hence if Free's hypothesis were the full explanation for the feedback control, there should not have been any difference between the two colonies in pollen forager behavior. We must conclude, therefore, that although the feedback mechanism hypothesized by Free may well play a role in regulating a colony's pollen collection (and definitely deserves a rigorous examination), it is certainly not the full explanation.

A second hypothesis for how hive bees convey information about the pollen reserves to pollen foragers was suggested by Camazine (1993). He proposed that when a colony has an ample supply of stored pollen, the nurse bees distribute a large amount of protein-rich hypopharyngeal gland secretion to the pollen foragers. The receipt of plentiful proteinaceous food would signify to a pollen forager that her colony has accumulated a large pollen reserve and that she should reduce her pollen collection. In essence, this hypothesis expresses the idea that nurse bees produce an *inhibitory* cue at times of pollen abundance in the hive. This hypothesis was inspired by the recent discoveries by Crailsheim and his colleagues about the physiology and sociology of pollen consumption by honey bees. They have found that the nurse bees in a colony are its principal pollen-digesting units. Specifically, the alimentary tracts of nurse bees have the highest pollen content and their midguts (where pollen is digested) have by far the highest proteolytic activity (Moritz and Crailsheim 1987; Crailsheim et al. 1992) (Figure 8.7). The nurse bees use the protein extracted from pollen to produce a protein-rich secretion from their hypopharyngeal glands. When Crailsheim (1991) injected 8-day-old (nurse) bees with ^{14}C-phenylalanine and traced the dispersal of their ^{14}C-labeled hypopharyngeal gland secretions, he found that approximately 75% was transferred to the brood while the other 25% went to the adult workers of the colony, including the foragers. The foragers have little ability to digest pollen and indeed consume little pollen, yet they have a high rate of protein turnover associated with their high flight activity; hence it appears that foragers rely on processed proteinaceous food from nurse bees to maintain a protein balance (Crailsheim 1986; Crailsheim 1990). *If* the amount of hypopharyngeal gland secretion received by foragers is correlated with the size of their colony's pollen reserve, and *if* individual foragers can sense the state of their protein nutrition and will adjust their pollen foraging accordingly, the flow of protein from the pollen reserves through the nurses and to the foragers could serve as a mechanism of negative feedback (literally!) for regulating a colony's pollen collection.

Two lines of evidence support the hypothesis that nurse bees produce an inhibitory cue at times of pollen abundance in the hive. The first is an experiment performed by Camazine (1993), in which he used basically the same procedures as those described above (Figures 8.5 and 8.6), but with the modification that one of the hives received a single screen, allowing the passage of the feedback from hive bees to foragers, while the other hive received a double screen, blocking the feedback (Figure 8.8). Also, neither hive was given pollen. This experimental design can reveal whether the feedback that passes through the screen is inhibitory or excitatory. Evidently, it is inhibitory. This is most easily understood by assuming, for the moment, that the feedback is excitatory. If this were the case, then the pollen foragers in the colony with the double screen would get no excitatory signal, and hence relatively few of them should continue to forage for pollen. But as is shown in Figure 8.8, just the opposite pattern was found. The proportion of pollen foragers continuing to forage for pollen was lower in the single screen hive than in the double screen hive: 60% versus 80% in the first trial and 25% versus 65% in the second. This result is consistent with the hypothesis that proteinaceous food from nurses inhibits pollen collection by foragers. Evidently, the pollen foragers in the single screen hive received some proteinaceous food from the nurses, even though there was little or no pollen in their hive.

The second piece of evidence supporting the hypothesis that proteinaceous food from nurse bees inhibits pollen collection by foragers comes from an unpublished study by Crailsheim, Camazine, and

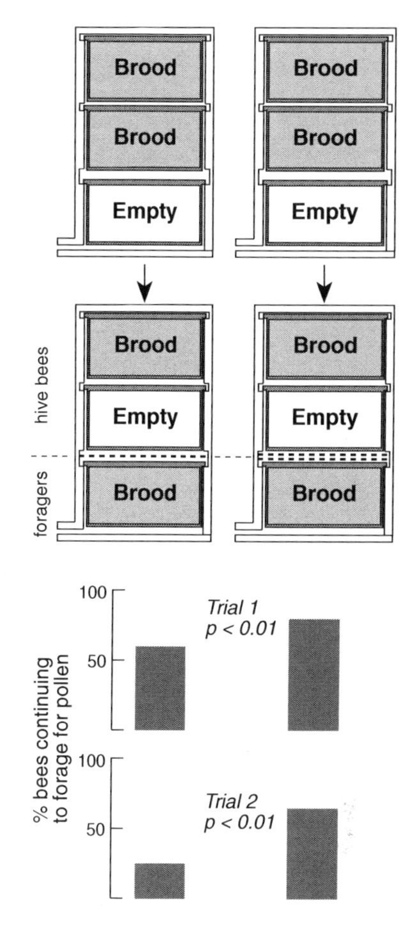

Figure 8.8 An experiment designed to reveal whether the feedback from the pollen reserve that comes to pollen foragers via the hive bees is inhibitory or excitatory. The experimental plan follows that shown in Figure 8.5, except that one hive receives a single screen (which permits feedback to the foragers) while the other receives a double screen (which blocks feedback to the foragers). The percentage of bees continuing to forage for pollen was lower in the single screen hive, indicating an inhibitory effect of the feedback. After Camazine 1993.

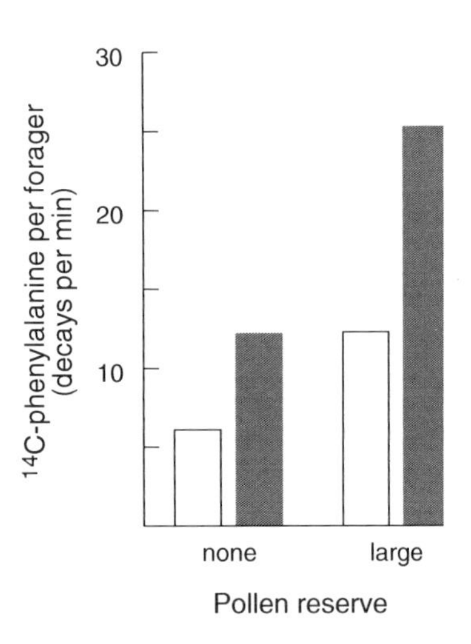

Figure 8.9 Comparison between pollen foragers that continued collecting pollen *(open bars)* and those that ceased collecting pollen *(filled bars),* in terms of the amount of ^{14}C-phenylalanine received overnight from the nurse bees. This comparison was made twice with two colonies. One colony was given a large pollen reserve in the course of the experiment, and one was kept deprived of pollen throughout the experiment (these two treatments are depicted in Figure 8.5). Based on unpublished data of S. Camazine, K. Crailsheim, and G. Robinson.

Robinson. This involved two colonies in three-frame observation hives that were manipulated according to the procedure shown in Figure 8.5. The main difference between this experiment and the one represented in Figure 8.5 is that on day 5 the experimenters not only labeled the pollen foragers in each colony, but also fed ^{14}C-phenylalanine to 8-day-old nurse bees in each colony. Also, on day 6, they not only recorded which of the previous day's pollen foragers continued or abandoned pollen collection, but also removed these two groups of bees from the hive at the end of the day and measured the radioactivity level of each bee. To date they have performed just one trial of this experiment, but it provides noteworthy correlational evidence supporting Camazine's hypothesis. As is shown in Figure 8.9, in both colonies the mean radioactivity counts were markedly higher for the foragers abandoning pollen collection than for those continuing pollen collection. This result demonstrates that the bees that abandoned pollen foraging did indeed receive more protein from the nurse bees than did the bees that continued pollen foraging. Moreover, it is clear that there was more transfer of the radioactive label (via proteinaceous food) from nurses to foragers in the colony that was given a large pollen reserve.

The ultimate test of Camazine's hypothesis, however, remains to be performed. It will consist of thorough experimental investigations of the two key assumptions of the hypothesis: (1) the flow rate of hypopharyngeal gland secretions from nurses to foragers varies with the size of a colony's pollen reserve, and (2) the amount of hypopharyngeal gland secretion that a pollen forager receives causes her to adjust her collecting activity.

8.5. Why the Feedback Flows Indirectly

If we knew nothing about the controls underlying the homeostatic pattern in a colony's pollen reserve (Figure 8.2), we would probably hypothesize a simple system of direct feedback control like that shown in the upper part of Figure 8.10, whereby pollen foragers directly perceive the size of their colony's pollen reserve and adjust the strength of their collecting behavior accordingly. Feedback in this manner would be straightforward and would easily account for the colony-level pattern of regulated pollen collection that is observed. As we have seen, however, the control system evidently utilizes a more complex, indirect pathway of feedback, one that involves pollen digestion and protein distribution by nurse bees, as shown in the lower part of Figure 8.10. The lower diagram in Figure 8.10 specifies no direct influence of the pollen reserve on the pollen foragers. However, this remains to be demonstrated conclusively. We know from experiment 1 of Camazine (1993), shown in Figure 8.5, that pollen foragers tend to ignore the *absence* of pollen in the hive as long as they receive food from hive bees, and this suggests that they pay little or no attention to the amount of pollen stored in the hive. One could explicitly test that bees do not respond directly to the *presence* of pollen by performing an experiment of the design shown in Figure 8.6 (double screen in both hives), but with the pollen-laden and empty frames positioned below the double screen. In both hives the pollen foragers will be deprived of proteinaceous food from the hive bees; so if pollen foragers do truly ignore any stored pollen in their hive, both hives should show the same high value for the percentage of bees continuing to forage for pollen.

Given that a simple system of feedback control would apparently meet a colony's needs for regulating its pollen collection, why does a honey bee colony possess such a complex system of feedback control?

One possibility is that the bees' indirect mechanism for feedback leads to a steadier, more reliable reserve supply of pollen in the hive than would be achieved by the simpler system of direct feedback. The important idea here is that the bees' indirect feedback mechanism may enable a colony to adjust its pollen collection in response to a change in its pollen consumption even before this change has caused a significant rise or fall in its pollen reserve. For instance, when the number of larvae increases in a colony, the nurse bees may feed more

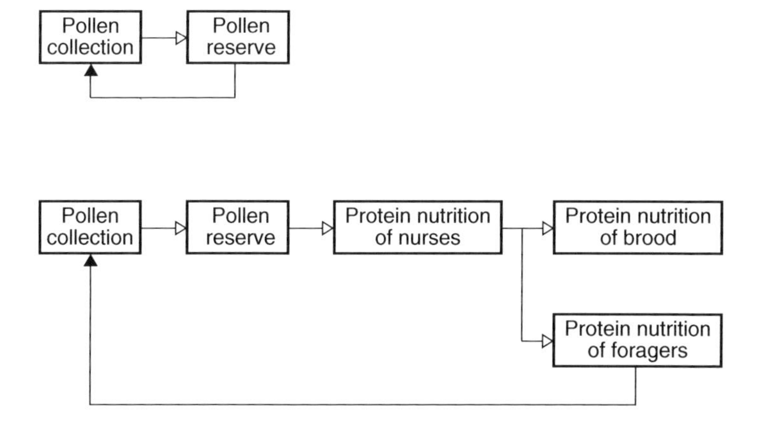

Figure 8.10 Two designs for the negative feedback control underlying the homeostasis in the pollen reserve of a honey bee colony. White and black arrows denote pathways of excitation and inhibition, respectively. *Top:* A simple control system in which the size of the pollen reserve is directly assessed by the pollen foragers and they adjust their pollen collection accordingly. The larger the reserve, the more the pollen foragers are inhibited from further collection. *Bottom:* The more complex control system that is actually found in a honey bee colony. It involves indirect negative feedback from the pollen reserve to the foragers by means of the nurse bees, which consume the pollen and feed proteinaceous food to the foragers. Foragers apparently sense the state of their protein nutrition and adjust their pollen collection accordingly. The larger the pollen reserve, the greater the pollen consumption by the nurses, the better the protein nutrition of the foragers, and the more the foragers are inhibited from further pollen collection.

of their hypopharyngeal gland secretions to the brood and less to the adults, and this change in behavior could stimulate pollen collection even before the colony's pollen reserve has begun to shrink. If so, then one beneficial effect of this feedback mechanism, relative to what would arise from the simpler mechanism of direct feedback, is a lower variance over time in a colony's pollen reserve. This, in turn, could mean that a colony has a higher probability of possessing a full reserve supply of pollen when it really needs it, such as at the start of several days of rainy weather. Thus the bees' indirect pathway of feedback may be an adaptation that helps a colony achieve stability in its pollen reserve despite fluctuations in its pollen demand, and this stability in its pollen reserve should strengthen a colony's ability to cope with the large, unpredictable fluctuations in the pollen supply that are an inevitable part of its foraging ecology. This hypothesis can be tested experimentally by seeing whether a honey bee colony, when forced to suddenly increase its pollen consumption (for example, when the amount of larval brood in its hive has been artificially increased), is indeed able to boost its pollen collection *before* there has been a drop in its pollen reserve.

A second hypothesis for why evolution has produced an indirect feedback mechanism in the control of pollen collection focuses on the issue of ease of information collection by the pollen foragers. If these bees acquired information about their colony's pollen reserve directly, presumably each forager would have to survey all the combs in the hive, somehow estimating the amount of stored pollen in each comb and ultimately summing her estimates for all the combs. Such a feat of information collection seems virtually impossible for a forager bee, especially since a typical colony's hive contains some 100,000 cells arranged in combs whose total surface area is about 2.5 m^2 (Figure 2.7). Within this expanse of comb the number of cells containing pollen can range widely, from zero to several thousands (Figure 3.11). There can be little doubt that direct feedback to foragers from a colony's pollen reserve would require the evolution of sophisticated—or at least extremely time-consuming—techniques of information acquisition. In contrast, the mechanism of indirect feedback apparently uses an extremely straightforward means of information acquisition by a forager: she simply senses her own hunger for protein. Indeed, this sensory ability is likely to have existed in the presocial ancestors of honey bees, because solitary insects must possess internal sensory processes for maintaining a protein balance in their bodies (Waldbauer and Friedman 1991). Thus it is possible that indirect feedback arose without the evolution of any novel mechanisms of information acquisition by the pollen foragers.

All in all, it seems reasonable to hypothesize that honey bee colonies possess a convoluted, indirect mechanism of feedback from pollen reserve to pollen foragers because this mechanism—relative to a mechanism of direct feedback—entails far simpler processes of information collection by the pollen foragers and hence it was far easier to evolve. The first step toward testing this hypothesis will be to elucidate the precise mechanisms by which pollen foragers receive feedback regarding their colony's pollen reserve, for doing so will reveal whether the means of information collection underlying the mechanism of indirect feedback are truly as simple as we now believe.

8.6. How a Colony's Foragers Are Allocated between Pollen and Nectar Collection

To efficiently utilize the labor it has available for food collection, a colony must be able to allocate its foragers between the tasks of pollen and nectar collection in accordance with its nutritional needs. We have seen already one indication that colonies do indeed possess this ability. When colonies with a small pollen reserve are suddenly supplied with a large pollen reserve, the number of bees engaged in pollen collection plummets (Figure 8.3). Presumably, this response enables a colony to devote more labor to the task of nectar collection. Conversely, when colonies are forced to deplete their pollen reserves during a period of inclement weather, the number of bees engaged in pollen collection rises (Section 8.2). It seems likely that swelling the ranks of the pollen foragers entails some thinning of the ranks of the nectar collectors, but this is not certain because the additional pollen foragers could be recruited from the pool of unemployed foragers or even from the large pool of nonforagers in the colony. The latter possibility is supported by studies which suggest that honey bee colonies possess a special communication process—the shaking signal (Section 6.2)—whereby foragers can stimulate nonforagers to become foragers. Unfortunately, we still lack detailed information on either the patterns or the processes of forager allocation between nectar and pollen collection, even though these arrangements are surely an important part of the overall organization of foraging by honey bee

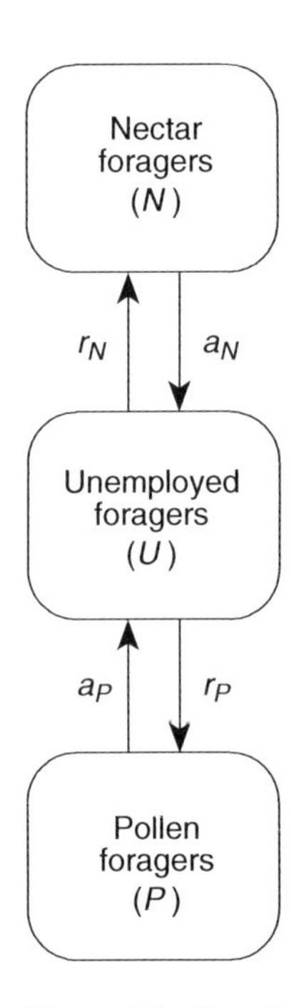

Figure 8.11 The model of how honey bee colonies allocate their foragers between nectar collection and pollen collection. At any given moment, each forager is in one of the three compartments shown. The model predicts a steady-state distribution of the foragers among the three compartments over time as a function of the set of rate constants (r_p, a_p, and so on) associated with the four transitions shown. In this situation, a colony as a whole can achieve a wise labor allocation if each of its foragers simply modifies her behavior slightly, in ways that tune the rate constants in accordance with the colony's nutritional needs.

colonies. Thus at present we are limited to examining the matter theoretically, but this is worthwhile for it provides a conceptual framework for future empirical investigations.

If we assume—for simplicity—that the number of foragers in a colony is fixed, the basic allocation phenomenon can be represented graphically as shown in Figure 8.11. At any given moment, each of a colony's foragers is in one of the three compartments shown: nectar forager (*N*), pollen forager (*P*), or unemployed forager (*U*). Foragers can shift themselves among these three compartments, but must pass through *U* in order to shift from *N* to *P*, or vice versa. The biological basis of this requirement is simply the observation that when foragers shift between forage sites, the vast majority do so by following dances inside the hive for information about a new site (Section 5.1). Each of the transitions depicted in Figure 8.11 is characterized by a rate constant (r_n and r_p for recruitment to *N* and *P*, and a_n and a_p for abandonment of *N* and P). This model of the allocation process is expressed mathematically by the following three differential equations, which express the rate of change in the number of bees in each compartment:

$$dN/dt = r_N U - a_N N \tag{8.2}$$

$$dP/dt = r_P U - a_P P \tag{8.3}$$

$$dU/dt = a_N N + a_P P - (r_N + r_P)\,U \tag{8.4}$$

The model predicts a steady-state distribution of the foragers between nectar collection, pollen collection, and unemployed status as a function of the set of four rate constants associated with the transitions.

What this abstract view of the allocation phenomenon makes clear is that the investigation of how a whole colony manages to wisely allocate its foragers between pollen and nectar collection—when the total number of foragers remains constant—centers on understanding how the individual forager bees manage to tune the rate constants in relation to their colony's nutritional needs. This intricate subject remains largely unexplored, but we can nevertheless enumerate the *possible* ways in which the foragers could change their behavior to adaptively adjust the rate constants. Consider, for example, a colony that has depleted its pollen reserve and so needs to allocate more bees to pollen foraging. The colony will accomplish this if its foragers raise

r_p and a_n, and lower r_n and a_p, or at least bring about some subset of these four adjustments. Is it reasonable to propose that the foragers can perform the various behavioral adjustments needed to produce all four of these changes? I think the answer is yes. The evidence presented above (Section 8.4) indicates that foragers almost certainly sense their colony's pollen need by sensing their personal protein hunger. This insight implies that probably all of the foragers within a colony, not merely the pollen foragers, possess information about the colony's need for pollen. Therefore, we can reasonably hypothesize that the bees in all three compartments of the model contribute to the adaptive tuning of the rate constants. Raising r_p and lowering r_n, for example, might involve the pollen foragers producing more and longer recruitment dances, together with the nectar foragers producing fewer and shorter recruitment dances, plus the unemployed foragers selectively following dancers bearing pollen. Lowering a_p, to cite another example, would presumably involve the pollen foragers lowering their acceptance threshold for pollen sources, thereby reducing their tendency to quit pollen foraging. Clearly, the mystery of how the foragers alter their behavior to tune the rate constants offers a rich subject for investigation. We know nothing at all, for example, about the possibility that unemployed foragers preferentially follow the dances of bees bearing pollen or nectar as a function of their colony's nutritional needs. I hope we will not have to wait long to learn what behavioral adjustments the foragers do actually make so that their colony maintains a proper labor allocation between nectar and pollen collection.

Summary

1. A honey bee colony's external supply of pollen undergoes far greater day-to-day variation than does its internal demand for pollen (Figure 8.1). To buffer itself against this fluctuation in the supply outside the hive, a colony stores a modest reserve supply of pollen, about a kilogram, inside the hive. Maintaining this pollen reserve at a proper size—neither too small nor too large—requires that a colony adjust its rate of pollen collection in accordance with its pollen reserve. Colonies do indeed show an inverse relationship between pollen reserve and pollen collection (Figures 3.12 and 8.2).

2. A colony adjusts its collecting rate with respect to the pollen re-

serve in part through changes in the total number of pollen foragers (Figure 8.3) and in part through changes in the per capita collecting rate of pollen foragers (Figure 8.4). In some experiments it is clear that changes in the per capita work rate account for most of the total change in a colony's rate of pollen collection, while in others it appears that changing the number of pollen foragers is the more important adjustment mechanism. Perhaps these two mechanisms provide complementary means of adjusting a colony's pollen intake, one (per capita collecting rate) which has a high adjustment speed but only a low range of settings, and the other (number of collectors) which has a low adjustment speed but a high range of settings.

3. Pollen foragers do not receive feedback about the size of their colony's pollen reserve through direct sensory perception of the pollen. This fact is demonstrated by experiments which show that foragers do not need to touch or taste the pollen to respond to a rise in their colony's pollen reserve (Figure 8.5), and that this response is not mediated by the foragers' sense of smell (Figure 8.6). Also, because pollen does not emit light and does not produce sound, it is clear that pollen foragers do not sense the amount of pollen inside the dark hive through vision or hearing.

4. The mechanism whereby pollen foragers acquire feedback from their colony's pollen reserve evidently employs an indirect pathway of information flow that involves the nonforager bees in the colony. Probably the critical bees in this feedback mechanism are the nurse bees, for they are the principal pollen-digesting units within a colony (Figure 8.7). The nurses may provide *excitatory* feedback when there is little pollen in the hive, perhaps by preparing more cells for pollen storage, with the result that returning pollen foragers find it easier to unload their pollen. At present, however, there is evidence only of *inhibitory* feedback from the nurses, when there is abundant pollen in the hive (Figure 8.8). Preliminary evidence suggests that the inhibitory cue is the proteinaceous hypopharyngeal gland secretion of nurses, some of which is fed to the foragers (Figure 8.9). Future studies are needed to test the two key assumptions underlying this hypothesis: (1) the amount of the secretion received by the foragers varies in relation to the size of a colony's pollen reserve, and (2) the amount received by a pollen forager influences her collecting activity.

5. Several explanations may account for why the mechanism of feedback from pollen reserves to pollen foragers is so convoluted

(Figure 8.10). One possibility is that the bees' complex feedback mechanism helps stabilize a colony's pollen reserve by making it possible for a colony to match its pollen collection to its pollen consumption without any change in the pollen reserve. Reduced variance in the pollen reserve should increase the probability that a colony has a full reserve when it is needed. Another possibility is that the complexity in the feedback mechanism reflects the fact that a simple, direct mechanism would require extremely difficult, if not impossible, feats of information collection by the pollen foragers, whereas the indirect mechanism apparently requires only modest, easily achieved information collection by the foragers. These hypotheses are not mutually exclusive, and each can be tested through further investigation of the functional design of this control system.

6. To forage efficiently, a colony must allocate its foragers between pollen and nectar collection in accordance with its nutritional needs. One can conceptualize this allocation problem in terms of a three-compartment model (nectar foragers, pollen foragers, and unemployed foragers) in which the steady-state distribution of the foragers is determined by the rate constants for the transitions between the compartments (Figure 8.11). This abstraction of the allocation phenomenon pinpoints a rich topic for future studies: how the foragers modify their behavior to adaptively tune the rate constants in relation to their colony's nutritional status.

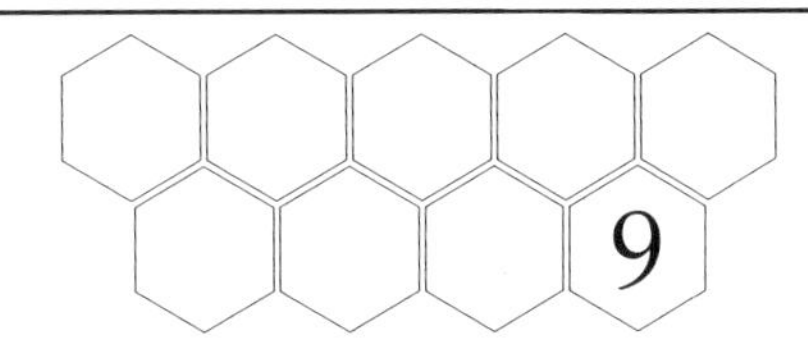

9 Regulation of Water Collection

On hot summer afternoons, one often sees worker bees drinking water at the edge of a pond, stream, or other damp spot, and then flying directly back to their hive. Also, on cool mornings, following several days of rainy weather that have kept the bees at home, one commonly sees bees sipping water from the dewy grass in front of their hives. Such acts of water collection are crucial to two parts of a colony's physiology: thermoregulation of the broodnest and nutrition for the immature bees. Consider first the need for water for temperature control. From late winter to early autumn, the annual period of brood rearing by honey bee colonies, the temperature in the broodnest region of each colony's hive is precisely regulated between 33° and 36°C, averaging about 34.5° and varying by less than 1°C across a day. This impressive temperature stability is accomplished through a set of mechanisms whereby colonies either heat or cool the broodnest, depending on the ambient temperature (reviewed in Seeley 1985 and Heinrich 1993).

When overheating threatens, the bees move farther apart on the combs and start to fan their wings, thereby cooling the hive interior through forced convection. If these measures prove inadequate, then they will also spread water, especially within the broodnest, for evaporative cooling. Water is spread as small puddles in depressions on the capped cells containing pupae, as thin layers over the roofs of open cells containing eggs and larvae, or as hanging droplets in these cells (Figure 9.1). Water may also be rapidly evaporated through "tongue-lashing," whereby bees hang over the brood cells and steadily extend their tongues back and forth. Each time a bee does this it expresses a drop of water from its mouth and pulls the droplet between mandibles and tongue into a film that has a large surface for evaporation. These various ways of using water for nest cooling can be referred to collectively as "water spreading."

The second general need for water, in preparing food for the brood, arises because the food fed by the nurse bees to larvae has a high water content—for example, the food given to the very young, larvae under 4 days old, is 70–80% water—whereas the honey that the nurse bees usually feed upon generally contains less than 20% water (von Rhein 1951). Clearly, the nurse bees have a great need for water. Meeting this need is accomplished in large measure by the collection of nectar, which is 30–90% water (see Figure 2.12), but sometimes it also requires the collection of water, especially after a string of cool or rainy days during which the bees have been prevented from gathering any nectar (see Figure 3.14 and Kiechle 1961).

Figure 9.1 The spreading of water droplets by nurse bees when a colony's broodnest is threatened by overheating. Spreading water, combined with fanning the wings to expel hot air from the hive, causes evaporative cooling of the brood combs. After Park 1925.

9.1. The Importance of Variable Demand

Water is collected not by a colony's nurse bees and food-storer bees, which ultimately make use of it inside the hive, but by the forager bees, which fly out to whatever puddle, brook, or other water supply is near their hive, fill their honey stomachs with water, and return home. Thus for water, as for nectar, there exists a division of labor between the bees that work outside the hive collecting a material and the bees that work inside the hive processing and consuming this material. This division of labor implies that in gathering water, as in gathering nectar, a colony must solve the problem of keeping a collection process and a consumption process in balance. Indeed, for water, a prolonged imbalance between collection and consumption can be disastrous. If consumption exceeds collection on a very hot day, probably the colony will overheat, causing abnormal development of the brood if not a complete meltdown of the combs.

Thus the water and nectar sectors of a colony's economy are basically similar in having a division of labor between collectors and consumers, but at the same time they are fundamentally different in having their dynamics driven from opposite sides of the supply-demand relationship (Figure 9.2). This reflects the fact that these two commodities have complementary patterns of variation in supply and demand. With nectar, unpredictable variation arises mainly on the collection, or supply, side of the operation, because the demand

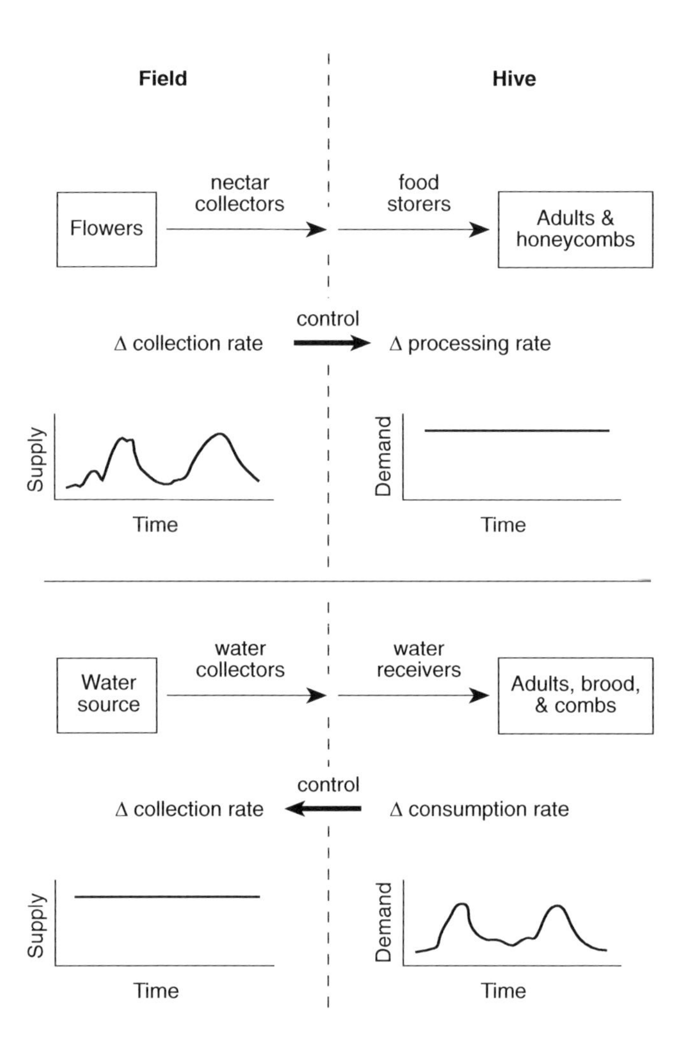

Figure 9.2 Comparison of the nectar and water sectors of a honey bee colony's economy. They are similar in that both involve a division of labor between bees working outside the hive collecting the material and bees working inside the hive processing or consuming the material. But the two sectors are different in that the dynamics in the nectar sector are driven by variation in supply, whereas the dynamics in the water sector are driven by variation in demand. Accordingly, the controls which keep a colony's collection and consumption rates in balance work in opposite directions for nectar and water.

for nectar remains high until the hive is packed with honey-filled combs, whereas the supply of nectar varies daily if not hourly as the foraging conditions change. But with water, unpredictable variation arises on the consumption, or demand, side of the operation, because most natural water sources provide an essentially infinite supply, whereas the demand can vary hourly as the ambient temperature changes. In short, the nectar sector's activity is *supply driven* while the water sector's activity is *demand driven.*

This contrast surely underlies many of the organizational differences between these two parts of a colony's economy. For example, a colony maintains a store of nectar, but not of water, inside its hive.

The functional significance of this difference seems clear: a colony needs an internal, reserve supply of honey to buffer itself against wide swings in nectar availability outside the hive, but it does not need an internal, reserve supply of water because water is always plentiful outside the hive. A second example of an organizational difference that traces to differential variation in supply and demand concerns the controls coordinating collection and consumption. These work in opposite directions for nectar and water (Figure 9.2). For nectar, as shown earlier, a colony possesses devices, such as the tremble dance, which enable it to modulate its processing rate in response to changes in its collection rate (Section 6.3). For water, as we shall see, a colony possesses several elegant mechanisms for adjusting its collection rate in accordance with changes in its consumption rate.

9.2. Patterns of Water and Nectar Collection during Hive Overheating

In Chapter 3 (Section 3.8), I reviewed the results of an experiment by Lindauer (1952) in which he heated the combs of a colony within an observation hive and observed a dramatic rise in the colony's traffic at a water feeder, starting just half an hour after the onset of heating (Figure 3.13). This experiment demonstrates that a honey bee colony can quickly and powerfully boost its water collection when overheating threatens. But because this experiment was done under highly artificial conditions—in a greenhouse in midwinter, and with probably only water (no nectar) available to the colony's foragers—we need to examine additional examples of a colony's foraging response to overheating, ones involving colonies living under more natural conditions. In particular, we need to know whether overheating of the broodnest affects not only a colony's water collection, but also its nectar collection. This point is important because it has generally been believed that the mechanisms regulating water collection and nectar collection are coupled in such a way that when a colony increases its intake of water it also decreases its intake of nectar, especially nectar with a high sugar concentration (see Section 9.6). It may be, however, that the mechanisms regulating water collection have been slightly misunderstood. A good starting point for taking a fresh look at these controls is to consider the overall patterns of water and nectar collection by a colony as it copes with a heat stress.

In the summer of 1993, Susanne Kühnholz (unpublished) repeated

Lindauer's experiment in which he heated the combs of an observation hive colony, but she used a colony whose foragers were free to fly outdoors to natural sources of water and nectar. To monitor the colony's collection of water and nectar, she and an assistant captured bees one at a time as they were about to enter the hive, squeezed each bee's abdomen to induce it to regurgitate the contents of its honey stomach, and then measured the percent of sugar in the fluid with a refractometer. The colony under study occupied a two-frame observation hive and, like Lindauer, Kühnholz overheated the colony's combs by directing the radiation from a 100-W bulb against the glass on each side of the hive. Thermistor probes embedded in the combs provided information on the combs' temperature. The results of one trial of this experiment, performed on 13 September 1993, are depicted in Figure 9.3. During the initial control period, the temperature of the broodnest was 33–34°C and there was no collection of water, only collection of nectar, whose sugar concentration was generally in the range of 30–60%. During two subsequent experimental periods, when the temperature of the brood combs had risen to 37–38°C, there was strong water collection by some 10–14% of the returning foragers, but there was no noticeable change in the range of sugar concentrations of the nectar. It remained steady at 30–60%. The experiment ended with a second control period, and it is interesting to note that even though the lamps were shut off at 4:02 in the afternoon, the water collection continued for at least another hour, even when the broodnest temperatures were no longer elevated. Evidently, it can take a colony considerable time to eliminate its water deficit after a severe overheating. The most important finding of this experiment, though, is actually something that was not found: any sign that excitation of water collection is coupled with an inhibition of the collection of highly concentrated nectar.

The same conclusion applies to a study conducted by Lensky (1964) in which he placed a hive of bees in a greenhouse in Israel on a sunny summer day, and monitored the traffic of the colony's foragers at a water feeder and a sugar solution (30%) feeder, both located inside the greenhouse. Lensky also recorded the temperature inside and outside the hive. As shown in Figure 9.4, over the course of the day the temperature outside the hive reached nearly 48°C, yet the bees were able to limit the rise of the broodnest temperature to below 38°C, no doubt by means of evaporative cooling. At the same time, the bees generated an interesting pattern of collecting water and sugar solu-

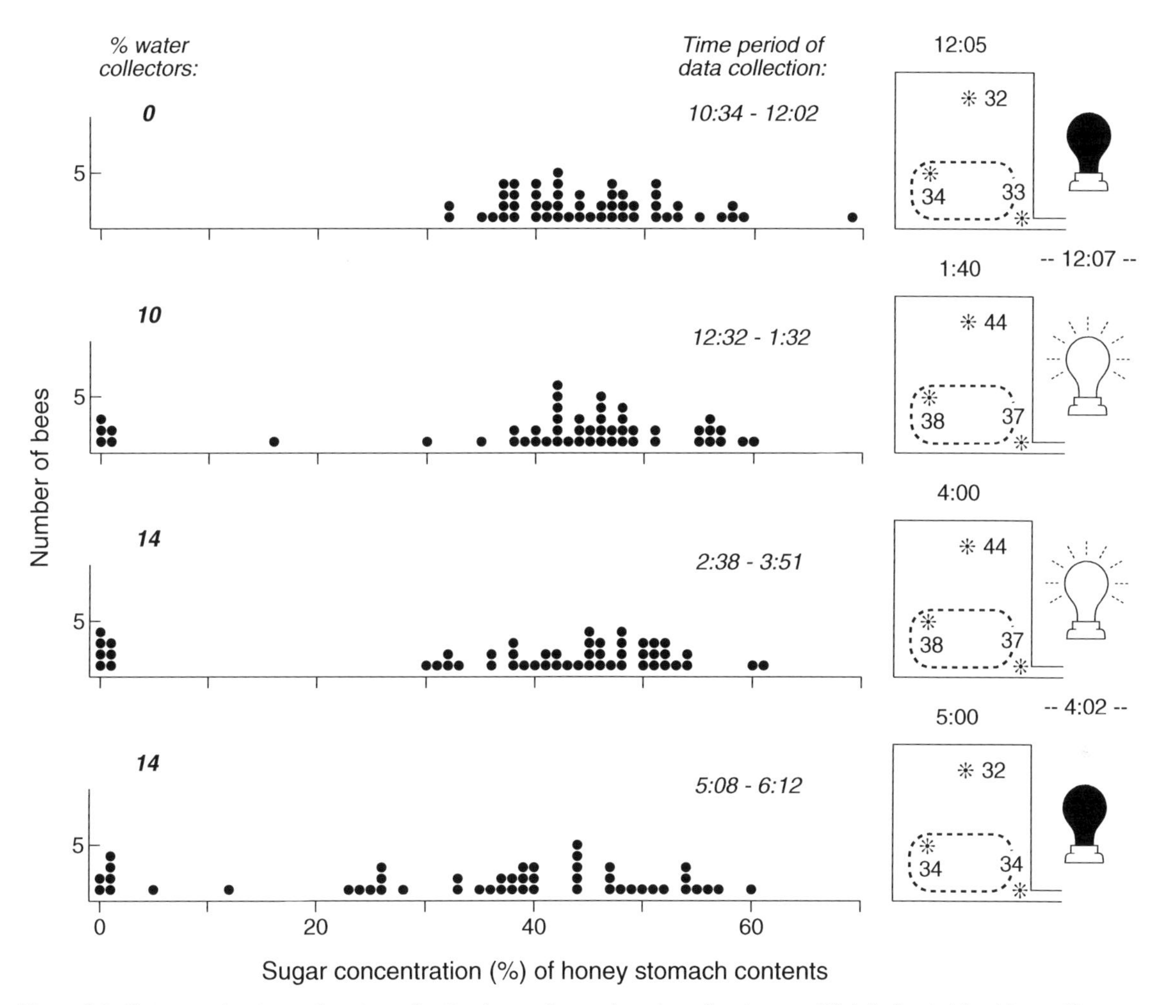

Figure 9.3 Patterns of water and nectar collection by a colony whose broodnest was artificially heated for 4 hr, on 13 September 1993. The colony occupied an observation hive; radiation from a 100-W incandescent lamp on each side provided the heat stress. Temperatures inside the hive, both in the upper honey storage area and in the lower broodnest region (indicated by a dashed line), were monitored with thermistor probes. Returning foragers were captured and the contents of their honey stomachs assayed for sugar content with a refractometer. When the lamps were turned on, the temperature in the broodnest rose, but the colony limited its rise to 38°C, in large measure by establishing a strong water flow into the hive for evaporative cooling. Note that the colony increased its water collection without stopping its collection of highly concentrated nectar. Based on unpublished data of S. Kühnholz and T. D. Seeley.

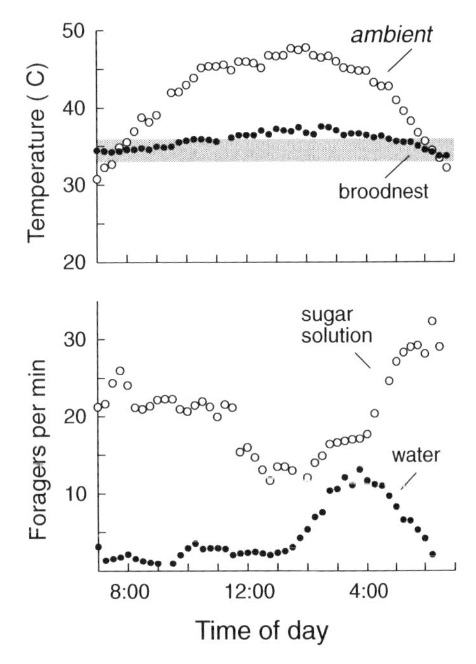

Figure 9.4 Patterns of temperature and foraging for a colony installed in a greenhouse throughout a sunny day. The temperature outside the hive rose quickly to more than 45°C, but the bees managed to keep the broodnest temperature to within 2°C of its normal range (33–36°C, indicated by shading). No doubt they accomplished this at least partly through evaporative cooling, using the water and sugar solution (30% sucrose) brought into the hive. After Lensky 1964.

tion. Water was collected throughout the day, but began to be collected strongly only at about 1:00, once the collection of sugar solution had ebbed. (The marked drop in the collection of sugar solution between 11:00 and 2:00 may have been caused by the inability of the nectar foragers to achieve sufficient evaporative cooling of their bodies while in flight at temperatures over 45°C [Heinrich 1980]. The water foragers probably had an advantage under these circumstances, for water provides stronger evaporative cooling than does a 30% sugar solution.) What happened next is most important: shortly after the collection rate for water began to rise, the collection rate for sugar solution also began to rise, and from about 2:00 to 4:00 *the colony's intake of both fluids rose.* Thus there is no indication that the two collection processes were regulated jointly in such a way that when the colony's water collectors received encouragements to forage more, its nectar collectors simultaneously received discouragements to forage less. As we shall see below, this small fact may be an important signpost on the way to understanding the mechanisms regulating a colony's water collection.

One finding common to these experiments by Lindauer, Kühnholz, and Lensky is that a honey bee colony can appropriately modulate its rate of water collection, increasing it when the broodnest begins to overheat and decreasing it when the danger of overheating has passed. The next thing to consider is which bees within a colony perform the task of water collection and how these bees know when they should and should not collect water.

9.3. Which Bees Collect Water?

Only a very small fraction of the bees in a colony ever gather any water. This was demonstrated by Lindauer (1952, 1954) when in early April 1951 he moved a full-size colony to a location outside Munich where there were no nearby sources of water, and provided the colony with a watering place just 4 m from the hive. The following day the bees discovered Lindauer's water source and for the next 5 months they evidently gathered most of their water there. Periodic checks of the nearest two alternative watering places, some 200 and 450 m from the hive, revealed no bees. Each day, from 6 April to 18 September, Lindauer labeled the bees visiting his watering place with paint marks for individual identification, and made a morning and afternoon record of the individuals collecting water. Throughout this

5-month period only 507 different bees were seen at the feeder, even though over the same time period the colony must have fielded at least 50,000 forager bees (Figure 9.5). Evidently less than 1% of this colony's bees served as water collectors.

It is clear that the water collectors are a subset of the foragers and hence are among the older bees in a colony (see Figure 2.5). Kiechle (1961) observed labeled individuals switch back and forth between water collection and nectar collection when he changed the contents (either water or a 1-mol/L sucrose solution) of a feeder exploited by a colony housed in a flight room. Moreover, when studying colonies living outdoors, Kiechle observed individuals that served as water collectors at a time of high water need and then became pollen or nectar collectors when the water need subsided. This finding raises the question of what determines which foragers function as water collectors and which ones serve as nectar or pollen collectors. One hypothesis is that which material a particular forager gathers is determined simply by the recruitment dance she happens to follow in locating her current forage site. In other words, water collectors may be simply those foragers that, totally by chance, have followed recruitment dances of bees visiting water sources. It may be, however, that the full story of how foragers become allocated among the different sectors—nectar, pollen, and water—of a colony's foraging operation is not this simple. For example, it is possible that the unemployed foragers preferentially follow the dances of bees bearing a certain material depending on their colony's needs (this idea was also raised in the context of pollen foraging, Section 8.6). Moreover, it is possible that certain foragers have a genetic predisposition to collect water, just as some foragers of certain patrilinies favor pollen or nectar collection (Robinson and Page 1989). Presumably, such effects on forage type must mean that unemployed foragers innately prefer to follow dancers bearing pollen or nectar. Whether some unemployed foragers preferentially follow dancers bearing water and, if so, whether this preference reflects external factors (the colony's needs) or internal factors (such as the bee's genes or her memories of previous foraging experiences), or both, remain unsolved mysteries.

Whatever the mechanisms that lead certain foragers to gather water, it is clear that once a bee is engaged in this task, she can become highly specialized in it, gathering water for many days in a row and performing dozens of collecting trips each day (Figure 9.5). For ex-

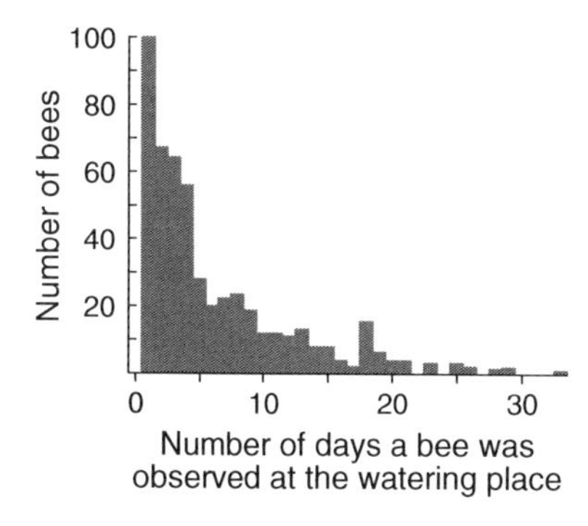

Figure 9.5 Duration of service as a water collector. A full-size colony was moved to a location without water sources nearby, an artificial watering place was established, the bees visiting it were labeled for individual identification, and records of the water collectors were made each morning and afternoon. Between 6 April and 18 September 1951, 507 bees were observed visiting the water source. The durations of their water-collecting activity ranged widely, as shown, but the majority (54%) functioned as water collectors for 4 days or more. After Lindauer 1952.

ample, when Park (1929) caught 7 randomly chosen water collectors from a colony on the afternoon of 17 August 1921, labeled them for individual identification, and then followed their behavior throughout 18 and 19 August, he found that these bees were amazingly single-minded in their work. On both days, all but one of the bees spent the entire day gathering water, and altogether the bees averaged more than 46 collecting trips per bee per day, at a per capita rate of just under 5 trips per hour. Robinson, Underwood, and Henderson (1984) report similar statistics for a bee that evidently specialized exclusively on water collection throughout her 14-day foraging career.

9.4. What Stimulates Bees to Begin Collecting Water?

Some bees, perhaps most, are stimulated to start fetching water by the recruitment dances of hivemates that have already begun to collect water. But what stimulates the very first water collectors and hence starts the entire water-collection process? In the case of water collection for cooling purposes, one might suppose that it is the sensation of high temperatures inside the hive which initially tells bees that their colony needs water. Two pieces of evidence suggest, however, that this hypothesis is wrong. One is an anecdotal report (1954) by Lindauer that he could trigger water collection in a colony occupying an observation hive even if he heated just a restricted part of the hive, one far removed from the entrance and where none of the colony's labeled water collectors were located. The second clue is Lindauer's thoroughly documented finding (1954) that a colony's foragers become strongly motivated to gather water when confined inside the hive by cool, rainy weather (see Figure 3.14). Obviously, under these conditions the water collectors must be responding to some cue other than high temperatures inside the hive. For reasons of parsimony, it seems reasonable to suppose that water collectors respond to the same nontemperature cue when they start gathering water on hot days. What might this be?

Lindauer (1954) proposed that what stimulates the first water collectors to become active is the presence of highly concentrated sugar solution in their honey stomachs. A bee might sense this either directly, as the fluid passes over her taste organs during food exchanges, or indirectly, as a feeling of thirstiness. Lindauer further suggested that water collectors will have this experience whenever their colony suffers a negative water budget—when the colony's water collection has been kept unusually low (when cool weather prevents foraging) or when its water consumption has suddenly risen (when hot weather triggers evaporative cooling). This hypothesis was tested by Kiechle (1961), who checked for a correlation between the foraging conditions of a colony, the mean sugar concentration in the bees in the colony, and the bees' motivation to collect water. To analyze the honey stomach contents of the colony's members, he sampled 10–15 bees daily around 9:00 in the morning from both the broodnest (nurse bees) and the hive entrance (mostly unemployed foragers). To assay the foragers' motivation for water collection, he placed a small, water-soaked cloth at the hive entrance and recorded what fraction of the bees encountering this water source also drank from it for at least 10 sec. As is shown in Figure 9.6, he found that during periods of cool, rainy weather (22–25 July and 7–8 August), the sugar concentration of the honey stomach contents rose dramatically, especially in the bees caught at the hive entrance, and that such rises were matched by rises in the percentage of bees drinking water at the hive entrance.

Kiechle also performed one experiment with his study colony. On the morning of 11 August, after taking his 9:00 samples of bees for honey stomach analyses and assaying the bees' desire to collect water, he fed the colony 100 ml of a dilute (15%) sucrose solution, and then in the afternoon he repeated his measurements at 3:00. He found that between 9:00 and 3:00 the mean sugar concentration of the nurse bees fell from 37% to 25%, and that the percentage of bees drinking at the hive entrance fell from 50% to 8%. Thus Kiechle found precisely the correlations between weather conditions, mean sugar concentration of honey stomach contents, and motivation to collect water that are predicted by Lindauer's hypothesis. In the future, it may be possible to devise a more incisive test of the hypothesis that foragers respond to high sugar concentration in the honey stomach as an indicator of the need to start gathering water.

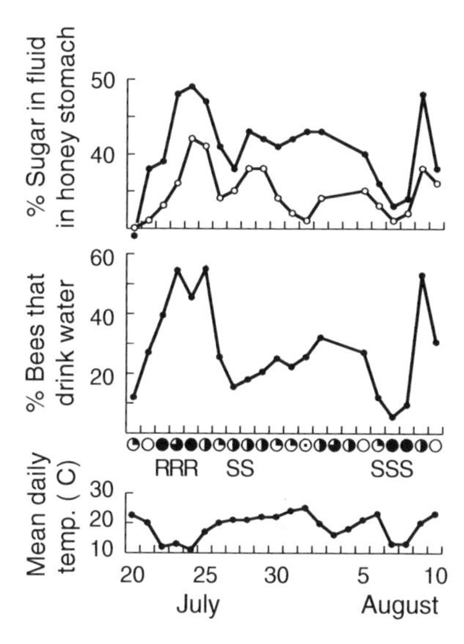

Figure 9.6 The sugar concentration in the honey stomachs of broodnest bees *(open circles)* and hive entrance bees *(filled circles)* varies markedly in relation to the weather. *Symbols: R* = heavy rain, *S* = showers, and circles indicate the degree of cloud cover. The sugar concentration remains low while the bees are able to forage, but after a couple of days of rainy or cool, flightless weather (honey bees rarely fly from the hive at temperatures below 10°C), the sugar concentration becomes quite high. Correlated with this condition is a rise in the fraction of the bees that drink upon encountering a wet cloth placed in the hive entrance. (Note that since the honey stomach contents were sampled early in the morning, the readings of sugar concentration for each day reflect mainly the level of foraging activity on the previous day.) Based on data in Kiechle 1961.

9.5. What Tells Water Collectors to Continue or Stop Their Activity?

Once a bee has begun collecting water, she must stay informed about her colony's need for more water and respond accordingly. If the need persists, she should continue collecting and perhaps even perform waggle dances to recruit others to the task; but if the need subsides, she should cease collecting. Clearly, the ability of the colony as a

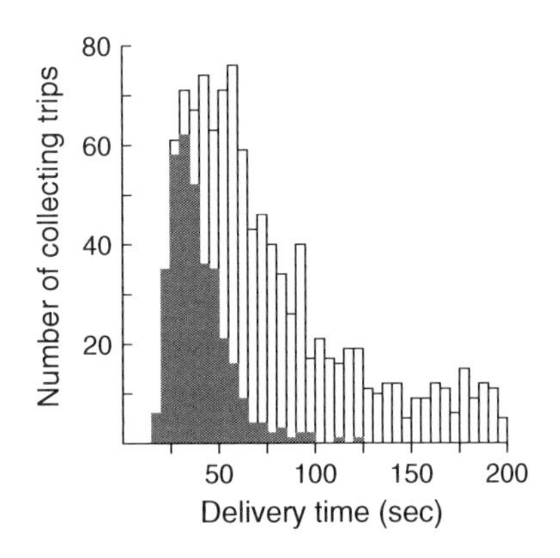

Figure 9.7 The relationship between delivery time and the tendency to perform waggle dances by water collectors. *Shaded bars:* collecting trips followed by a dance; *white bars:* collecting trips without a dance. The shorter the delivery time, the greater a bee's probability of performing a waggle dance, which will recruit new bees to help gather water. This relationship suggests that a short delivery time indicates to the water collectors their colony's need for a higher rate of water intake. After Lindauer 1954.

whole to adaptively modulate its water collection depends critically upon each water forager having ready access to current information about the colony's water need and adjusting her behavior in accordance with this information.

How does a bee engaged in water collection acquire information about her colony's need for additional water? In 1954, Lindauer suggested that a bee does so very easily each time she returns to the hive, simply by noting how quickly her load of water is taken from her by the bees inside the hive. This suggestion was based on Lindauer's observation that if there is an acute shortage of water, the hive bees take each water collector's load quickly and eagerly, and each water collector actively continues fetching water and may even perform a waggle dance. He also observed that if the colony's water needs have been met, the hive bees take each water load only slowly and reluctantly, and the water collectors cease collecting. Lindauer quantified these relations by measuring the delivery time (the time between the arrival of the water collector at the hive entrance and the completion of the transfer of her water load) for all water collectors and noting their subsequent behavior. As is shown in Figure 9.7, when the delivery time was less than about 40 sec, there was nearly always a dance; when the delivery time increased, the dances became rarer; and if the delivery took more than 2 min, there was no dancing at all. He also noted that as the delivery time became longer, a water collector's motivation to fetch additional water slackened, for she took longer and longer rest intervals in the hive and eventually stopped collecting.

Lindauer carefully pointed out that it may not be the delivery time per se which indicates the colony's need for water, for there are several other aspects of the unloading experience that also covary with a colony's water need. These include the search time (time between entering the hive and starting to unload), the number of bees simultaneously taking the fluid from a water collector, the liveliness of the antennal stroking during the unloading process, and the number of unloading rejections that a water collector experiences before finding a bee willing to take her load.[1] As water need increases, search times and unloading rejections decrease while the number of simultaneous unloaders and the liveliness of the antennal strokings increase. Figure 9.8 shows how one water collector experienced changes in sev-

1. A nectar or water collector experiences an unloading rejection when she encounters a hive bee who extends her tongue into the mouthparts of the forager and apparently tastes her load but then immediately withdraws her tongue and walks away.

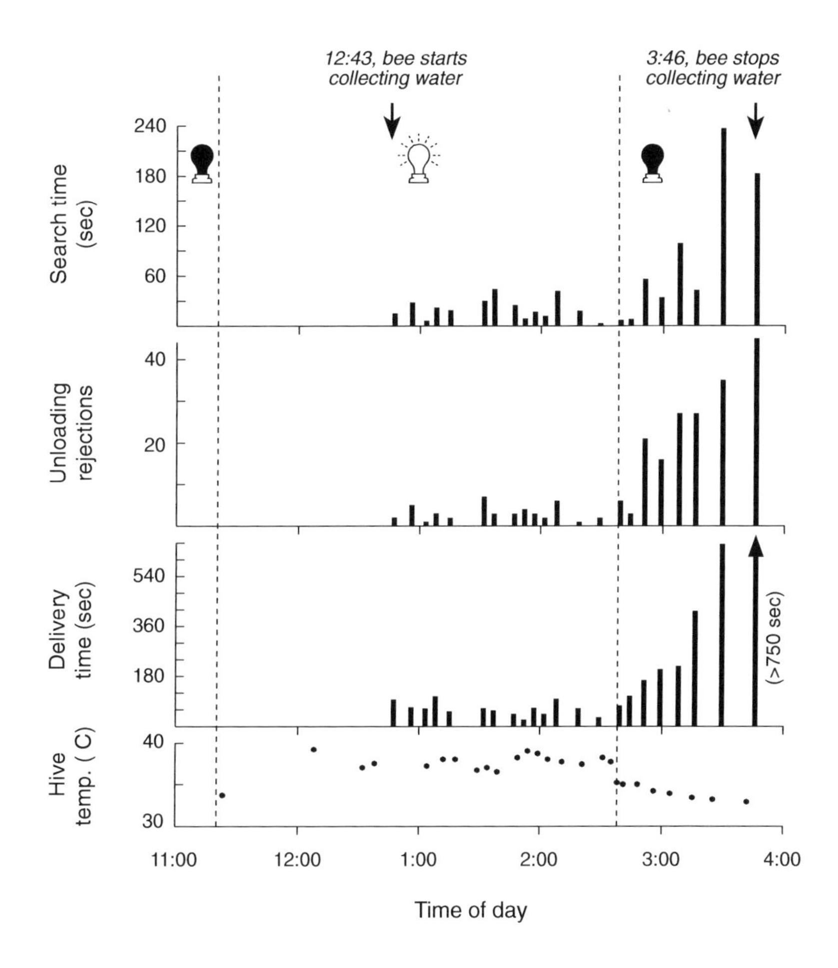

Figure 9.8 Changes in the multiple variables of a water collector's unloading experience as her colony's water need goes from high to low. As long as the hive was heated (light on at 11:20) and the colony's broodnest was threatened with overheating, each time the bee returned to the hive she experienced (1) a short search to find an unloader, (2) few encounters with hive bees that sampled but rejected her watery load, and (3) a short time to complete her delivery. But once the danger of overheating passed (light off at 2:38), the bee's search times, unloading rejections, and delivery times all increased. Ultimately, the bee stopped collecting water. Based on unpublished data of S. Kühnholz and T. D. Seeley.

eral of these variables as her colony's water need was experimentally switched from high to low, and how this bee eventually ceased bringing home water. For example, while the hive was heated, the bee experienced search and delivery times of 21 ± 12 sec and 64 ± 25 sec, but after the heat was turned off these times rose to 83 ± 85 sec and 326 ± 261 sec, values that are significantly higher ($x \pm$ SD, $P < 0.01$ for both comparisons). Likewise, the number of unloading rejections experienced by the water collector upon return to the hive was low (3.3 ± 1.8) when the hive was heated, and was dramatically higher (28.2 ± 10.1) when the heating lamps were switched off.

It remains to be investigated whether the indication of a colony's water need lies in the whole constellation of variables of the unloading experience, some subset of these variables, or perhaps just one special variable. To solve this puzzle, the different variables constituting the unloading experience must be uncoupled experimentally so that the effects of each on the behavior of water collectors can be assessed independently. This has yet to be accomplished. One can, however, compare the different variables in terms of how strongly they vary with changes in a colony's water need, and thereby determine which variable(s) might provide the strongest, most easily perceived cue or index of water need. Table 9.1 shows the changes in four variables—search time, delivery time, unloading rejections, and maximum number of unloaders—that were recorded when 5 different water collectors were followed trip by trip as their hive was first heated and then not heated with incandescent lamps (protocol shown in Figure 9.8). All four variables had a consistent pattern of change, either an increase or a decrease, when the colony's water need declined (lamps turned off), but they differed markedly in the magnitude of change. This is expressed for each variable by the ratio of the mean values observed with and without heat applied to the hive. The maximum number of simultaneous unloaders varied rather little (mean ratio = 1.5); hence that factor does not appear to be a decisive indicator of water need. Search time and delivery time varied much more strongly with water need (mean ratio = 6.6 and 4.3, respectively), and so they are likely to provide clearer information about the demand for water. What varied most strongly by far, however, is the number of unloading rejections (mean ratio = 14.5), which seems to suggest that unloading rejection may be the most important index of water need. But the value of an index depends not only on the quality of the information it gives but also on the precision with which it can be understood. The perception of *durations* (such as search times) is a well-known phenomenon for honey bees, but the capacity to count the *number* of discrete events (such as unloading rejections) is not (von Frisch 1967, p. 102). Thus one must question whether water collectors can count as many as 40 or more unloading rejections and can grasp differences in the number of rejections experienced on different returns to the hive. Taking all these facts into account, I come to the tentative conclusion that the search time to find an unloader is probably the most important indicator of a colony's water need.

Even though we do not yet have a large body of experimental evidence regarding the perception of a colony's water need, it seems highly likely that Lindauer was correct in proposing that water collectors sense their colony's need for a higher water intake by noting

Table 9.1. Comparison of the strength of indication of a colony's water need by several variables of a water collector's unloading experience. Based on unpublished data of S. Kühnholz and T. D. Seeley.

	Bee 1: 29 July 93			*Bee 2: 6 June 94*			*Bee 3: 8 June 94*			*Bee 4: 9 June 94*			*Bee 5: 30 June 94*			
	With heat	Without heat	Ratio	With heat	Without heat	Ratio	With heat	Without heat	Ratio	With heat	Without heat	Ratio	With heat	Without heat	Ratio	Mean of ratios
Search time (sec)	21	83	4.0	10	126	12.6	12	64	5.3	14	113	8.1	24	78	3.2	6.6
Delivery time (sec)	64	326	5.1	39	184	4.7	96	186	2.2	47	314	6.7	70	197	2.8	4.3
Unloading rejections	3.3	28.2	8.6	0.7	15.6	22.3	0.9	13.7	15.2	1.5	14.9	9.9	0.5	8.3	16.6	14.5
Max. no. of simultaneous unloaders	—	—	—	1.6	1.1	1.4	2.0	1.2	1.7	1.7	1.2	1.4	1.9	1.3	1.5	1.5

one or more variables of the unloading process. I say this partly because the correlation between a water collector's unloading experience and her subsequent collecting activity is so strong (Figure 9.8) and partly because I have performed experiments that indicate that *nectar* collectors sense their colony's need for a higher nectar intake by noting one or more variables of the unloading process (Seeley 1989a; see also Section 5.7.3). It is highly attractive to think that nectar collectors and water collectors sense the need for their respective materials in similar ways.

9.6. Why Does a Water Collector's Unloading Experience Change When Her Colony's Need for Water Changes?

The reason for a change in a water collector's unloading experience is the largest remaining question about the mechanisms whereby a honey bee colony regulates its water intake. As a first step toward answering it, it is important to note that what enables water collectors to unload faster—experience shorter search and delivery times—when their colony's water need rises, is almost certainly an increase in the fraction of the bees within the unloading area that accept water. There is no doubt that a rise in this proportion will shorten the average search time of the water collectors, because it will lower the expected number of bees that a water collector must sample to find one that accepts water (for a more detailed discussion of the relationship between search time and the composition of the bees in the unloading area, see the urn model in Section 5.7.4).

At any given time, an individual receiver bee evidently does not unload indiscriminately both water collectors and nectar collectors, but instead chooses between them, preferentially unloading one or the other type of collector. This exercise of choice is indicated by several pieces of evidence. First, the receivers can easily distinguish the two types of foragers by tasting the fluid that each returning forager offers. The fluids brought home by water collectors and nectar collectors have strongly disjunct distributions with respect to sugar concentration: water generally contains less than 3% sugar while nectar usually contains more than 12% sugar (Figures 2.12 and 9.3). In addition, what is known of the bees' ability to detect concentration differences in sugar solutions indicates that they can easily distinguish solutions as different as water and nectar (von Frisch 1967). A second piece of evidence, reported by several observers, is that foragers (both water and nectar collectors), upon return to the hive, often encounter bees who extend their tongues into the glossal groove of a forager bee as if they are ready to unload her, but then withdraw their tongues almost immediately and walk away (Lindauer 1954; Kiechle 1961; Seeley 1986; see also Figure 9.8). In such cases of unloading rejection, it looks very much as if the potential receiver bee had tasted the forager's load and decided to refuse it. This would happen if a water receiver happened to contact a nectar collector or if a nectar receiver contacted a water collector. A third line of evidence supporting the idea that a receiver bee preferentially accepts either nectar or water, but not both, is what Kiechle (1961, p. 169) observed while watching one amazing bee, number 221, which distinguished herself by repeatedly switching between collecting water and collecting a concentrated sucrose solution on the same day. This special bee belonged to a small colony occupying an observation hive that was housed in a flight room, and hence her behavior was easily monitored both inside and outside the hive. During a 2-hr observation period on 18 March 1960, Kiechle steadily followed bee 221 as she made 22 foraging trips, 13 to a watering place and 9 to a feeder containing a 2-mol/L sucrose solution. The two types of collecting trips were intermingled throughout the observation period, but only when the bee came home with sucrose solution was her forage quickly accepted by the receivers. Apparently, the receivers were discriminating between the water solution and the sugar solution, with most accepting only the latter.

The critical question can now be stated more precisely: What exactly is the process whereby the proportion of receiver bees accepting water changes whenever a colony's water need changes? In particular, why does the proportion of water receivers increase when the colony's water need increases? Consider two distinct hypotheses, shown graphically in Figure 9.9. The first, which seems to be implied in the writings of Lindauer (1954, 1971), is that there is one group of generalist receiver bees in a colony, and that these bees normally reject water and accept nectar but will *switch to accepting water* (and perhaps dilute nectar) and rejecting nectar (especially concentrated nectar) when they sense that their colony needs water. This hypothesis predicts a strong coupling, in the form of cross inhibition, between water collection and nectar collection because raising the number of receivers accepting water will lower the number of receivers accepting nectar, unless additional receivers are also activated when a water need arises. If this hypothesis is correct, it is likely that

Figure 9.9 Two hypotheses for how the proportion of the receiver bees accepting water changes when a colony's need for water changes. White and black arrows denote transfers of water and nectar, respectively. The oval in each diagram denotes the locus of control of water collection. According to the first hypothesis, the proportion changes because *generalist* receiver bees can switch back and forth between accepting nectar or water, depending on the colony's need for water. This hypothesis predicts inhibition of nectar foraging during times of intense water collection. According to the second hypothesis, the proportion of receiver bees accepting water changes because one group of *specialist* receivers, the water receivers, is present or absent, depending on the colony's need for water. On this hypothesis, the water receivers are nurse bees that sense a water shortage and come to unload water collectors so as to take water back to the broodnest. This hypothesis predicts little, if any, inhibition of nectar collection during periods of intense water collection.

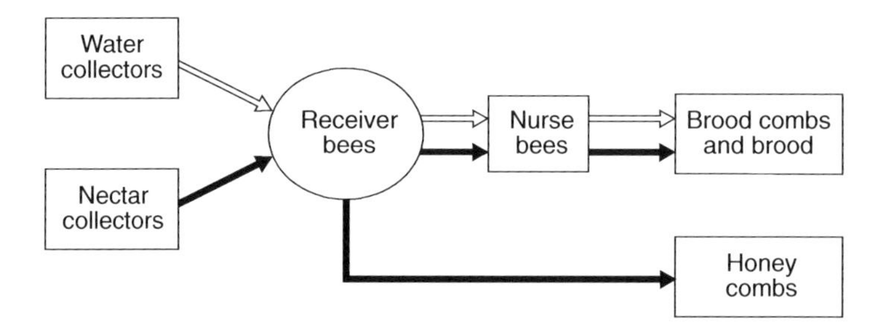

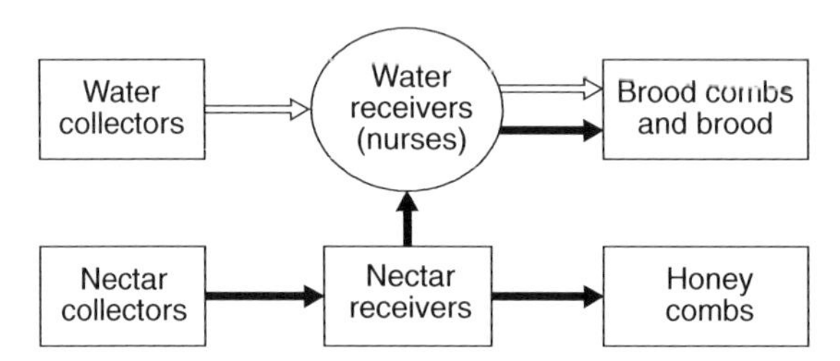

when a colony experiences an acute water need, its response will involve simultaneously increasing the intake of water and decreasing the intake of nectar, especially concentrated nectar.

The second hypothesis invokes the idea that there are two groups of specialist receiver bees in a colony, water receivers and nectar receivers, and that the water receivers are normally not receiving water but will begin to do so when they sense that their colony needs water. On this hypothesis, the water receivers are simply nurse bees that have sensed a water shortage in the broodnest and have moved to the unloading area to obtain water from returning water collectors, whereas the nectar receivers are the familiar food-storer bees that unload the returning nectar collectors (Figure 6.1). This hypothesis, in contrast to the first, predicts only a weak, or negligible, coupling between water collection and nectar collection, because by this hypothesis raising the number of receivers accepting water (nurses seeking water) does not automatically drive down the number of receivers accepting nectar (food-storer bees). So if this hypothesis is correct, when a colony experiences an acute water need, its response will involve increasing its intake of water but possibly not decreasing at all its intake of nectar, and certainly not selectively rejecting concentrated nectar. (The colony may slightly decrease its intake of nectar, because the migration of water receivers into the unloading area may make it more difficult for the nectar foragers to find nectar receivers, a situation which would slow the activity of the nectar foragers.)

Which hypothesis, if either, is correct? In the early 1950s, Lindauer (1954) conducted several experiments which provide indirect tests of these two hypotheses. The results of one such experiment are given in Figure 9.10, which shows the pattern of collection of water and 2 mol/L sucrose solution by a small colony in a flight room after being deprived of water for 2 days. For the first hour the foragers virtually ignored the rich sugar solution but avidly gathered the water, with nearly each collecting trip followed by a bout of dancing. This striking pattern suggests that at the start virtually all the colony's receiver bees were accepting water and rejecting the sugar solution, a phenomenon that is more easily explained by hypothesis 1 than by hypothesis 2. However, since the conditions which existed inside the colony at the start of this experiment may be extremely unusual, the results obtained may not be a good indicator of how water collection is usually controlled. For instance, it may be that when a colony is severely deprived of water, as was the study colony, every bee in the hive suffers acute thirst, so that even the bees that normally accept nectar and reject water (the nectar receivers) will temporarily accept water to quench their own thirst, but that after this emergency situation passes there will once again be separate groups of water receivers and nectar receivers, providing separate controls of these two materials. In this regard, it should be noted that between 10:00 and 12:00, after the water emergency had been dealt with, the colony's nectar collection increased while its water collection did not decrease. This fact suggests that the controls of the two collection processes were in fact not linked in a mutually inhibitory fashion; hence the results of this part of the experiment lend some support to hypothesis 2. All in all, it now seems clear that although this experiment clearly portrays the interesting foraging response of a water-deprived colony, it does not provide decisive evidence regarding the two hypotheses for how the water collection is controlled.

A second experiment is shown in Figure 9.11. In this study, Lindauer worked with a small colony occupying an observation hive inside a flight room, and presented it with four feeders containing sugar

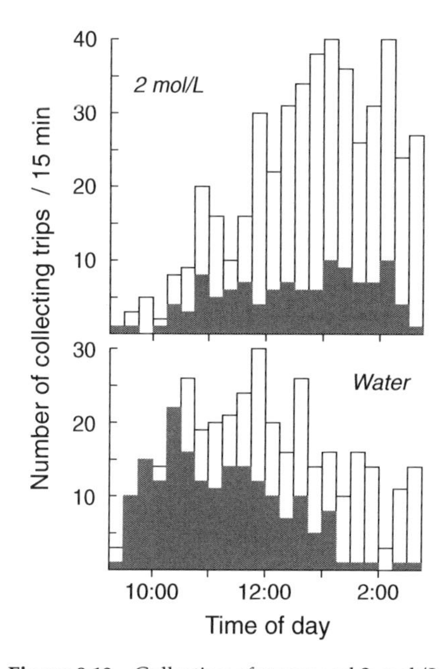

Figure 9.10 Collection of water and 2-mol/L sucrose solution by a colony confined in a flight room and deprived of water for the 2 previous days. *Shaded bars:* collecting trips followed by a dance; *white bars:* collecting trips without a dance. After Lindauer 1954.

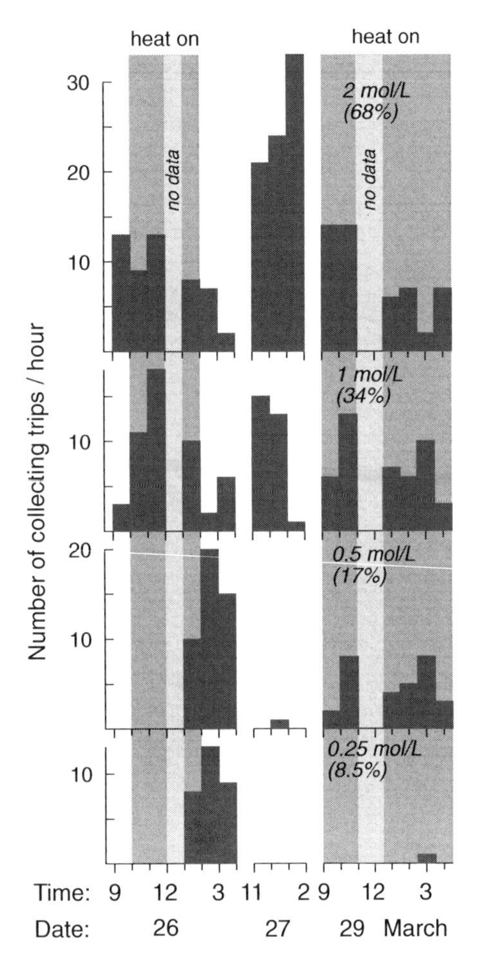

Figure 9.11 Foraging patterns of a colony housed in a flight room, offered four feeders with different concentration sugar solutions, and then heated with a bright lamp to induce a need for water. After Lindauer 1954.

solutions of different concentrations, ranging from rather dilute (0.25 mol/L) to highly concentrated (2 mol/L). No water source was provided. He monitored the total number of collecting trips to each feeder during hourly periods, and manipulated the colony's need for water by heating the hive interior with a lamp. In performing this experiment, Lindauer wanted to see if he could create a situation in which the receiver bees would accept only a dilute sugar solution and would reject a concentrated sugar solution. If this could be done, it would provide support for hypothesis 1. On the first day, 26 March, heating the hive clearly stimulated the collection of the two dilute sugar solutions: the mean number of collecting trips per hour to the 0.5- and 0.25-mol/L feeders rose from 0 and 0 during the first 3 hr of data collection (9:00 to 12:00) to 15 and 10 during the last 3 hr (1:00 to 4:00). For both feeders, the increase is statistically significant ($P < 0.01$). However, it is not so clear that the heating also inhibited the collection of the two concentrated sugar solutions: the mean number of collecting trips per hour to the 2.0- and 1.0-mol/L feeders did fall from 11.7 and 10.7 in the morning to 5.7 and 6.0 in the afternoon, but for neither feeder was the drop statistically significant ($P > 0.05$ and $P > 0.30$, respectively). Thus these results provide only equivocal support of hypothesis 1. The data gathered during the second trial of this experiment, on 29 March, are even less convincing because in this trial, for whatever reason, the heating seemed to have little effect on the colony's foraging. Thus this second experiment, like the first one, shows the pattern of a colony's response to increased need for water, but it does not clarify the underlying process.

In the summer of 1994, Susanne Kühnholz and I performed studies designed to test directly the two hypotheses presented above. Our basic plan was to look inside a colony and determine whether, when the colony starts collecting water, the bees receiving water are a new group, distinct from those previously observed receiving nectar. If hypothesis 2 is correct in all its details, the bees receiving water should be not only a different group, but also a younger group, relative to the bees receiving nectar, since this hypothesis predicts that water receivers are water-seeking nurse bees (see Figure 2.5 for a comparison of the typical ages of nurse bees and food-storer bees). We quickly learned that part of hypothesis 2 is incorrect: the water receivers are not thirsty nurse bees. This became clear when we took a colony of bees in an observation hive to the Cranberry Lake Biological Station, heated it with two 100-W lamps to stimulate water collection, located and labeled many of the colony's water collectors as they sucked up water from wet rocks along the lakeshore, and then determined the age distribution of the bees seen receiving water from these water collectors. We were able to determine the age distribution of the water receivers because we had added labeled, known-age bees to the colony (see Section 4.7) every third day for 3 weeks before attempting our experiment. The results from three trials of this experiment, conducted on 10–12 June 1994, are shown in Figure 9.12. The age distribution of the water receivers—they were middle-aged bees, some 12 to 23 days old—matched what one typically sees for food-storer bees, not nurse bees. Moreover, there was scarcely any overlap between the age distributions of the bees observed receiving water and those of the bees tending the queen, and it is the nurse bees in a colony who tend (feed and groom) the queen.

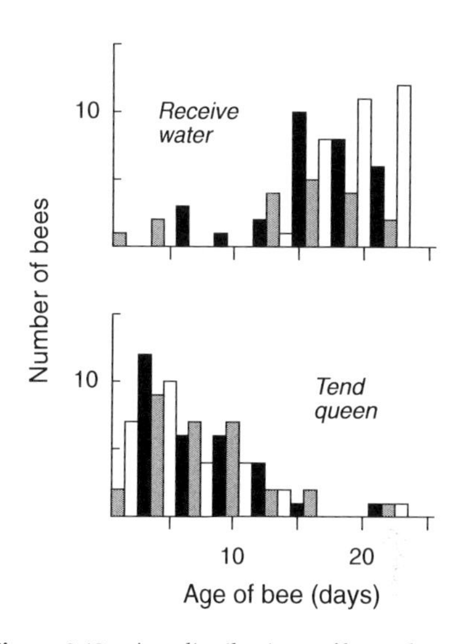

Figure 9.12 Age distributions of bees observed receiving water from water collectors or tending the queen. These data were gathered on three consecutive days (10–12 June 1994; each day's data are represented by a different pattern on the bars) by observing the water receivers and queen tenders in a colony occupying an observation hive. It was possible to determine the ages of approximately 10% of the bees that were observed because 50 0-day-old bees bearing identification labels had previously been introduced to the colony every 3 days, starting on 20 May 1994. Based on unpublished data of S. Kühnholz and T. D. Seeley.

This look at the age distribution of the water receivers revealed that they are drawn from the same age group as the nectar receivers, but it left open the question of whether water receivers are nectar receivers that have switched to accepting water during a water shortage (hypothesis 1). Alternatively, the water receivers might not be bees that normally are engaged in nectar reception, but instead might be bees that normally are engaged in some other task (or are inactive) and that start receiving water when the need arises (hypothesis 2, slightly modified). To distinguish between these two hypotheses, we performed two trials of the following experiment. First, in the morning, we labeled approximately half of the *nectar* receivers in our observation hive colony, by dotting paint on the thorax of each bee seen unloading any of 25 bees previously trained to forage at a sucrose solution feeder. Then, in the afternoon, we heated the observation hive to stimulate water collection, then identified and labeled 10–20 water collectors, and finally observed the unloadings of these water collectors to see if any of the water receivers were bees that had been labeled nectar receivers in the morning. If so, this would indicate that at least some of the water receivers derive from nectar receivers that have switched to accepting water. In the first trial, performed on 23 June 94, we witnessed 178 different bees receiving water in the afternoon, and observed that 32 (18%) of them were bees that had received nectar in the morning. In the second trial, performed on 3 July 94, we witnessed 126 different bees receiving water in the afternoon, and this time we found that 36 (29%) bore a paint mark identifying them as having been a nectar receiver earlier in the day.

These results reveal that *both* hypotheses described above contain part of the truth about how a colony increases the number of water receivers, and thus how it is that when a colony's water need increases its water collectors experience easier unloadings. On the one hand (hypothesis 1), it is now clear that some of the water receivers are former nectar receivers that have switched to water reception. But on the other hand (hypothesis 2), it also seems clear that not all the water receivers come from the pool of bees previously engaged in nectar reception. This is indicated by the fact that in both trials we labeled 45–55% of the colony's nectar receivers (specifically, in the two trials we labeled 307 and 378 nectar receivers in a colony that contained approximately 700 nectar receivers; see Section 6.3), yet only 18–29% (not 45–55%) of the water receivers were prior nectar receivers. Evidently, many of the water receivers, perhaps half or more, were middle-aged bees that were not previously involved in nectar reception. It seems, therefore, that our study colony had a reserve supply of labor among the middle-aged bees which it was able to draw on to cope with an emergency, in this case the threat of overheating.

The idea of a labor reserve within a colony, from which the colony can produce additional water receivers during an emergency need for water, explains another remarkable phenomenon that we observed repeatedly in the course of our experiments on the regulation of water collection: *a colony can massively increase its water collection without having to simultaneously decrease its nectar collection.* For example, on 20 June 1994, we heated our observation hive for 3 hr (from 12:30 to 3:30), monitored the traffic level of water collectors at the start and end of the heating, and throughout monitored the activities of a group of 15 labeled nectar foragers collecting a concentrated (2.5-mol/L) sucrose solution from a feeder. As a result of the heating, the colony's traffic level in water collectors rose by a factor of 45, from 0.4 to 18.2 bees/min into the hive. But at the same time, there was no drop in the collecting activity of the bees gathering the concentrated sugar solution. All 15 bees continued foraging at the feeder throughout the heating period, and their total traffic rate remained unchanged—2.4 bees/min into the hive—from start to end of the heating. The absence of any inhibition of nectar collection is perhaps most vividly indicated by the trip-by-trip records for one of the foragers exploiting the feeder (Figure 9.13). Despite the fact that this bee's colony was suffering a major heat stress (broodnest temperature nearly 40°C!) from our lamps and therefore had greatly boosted its water intake, this bee

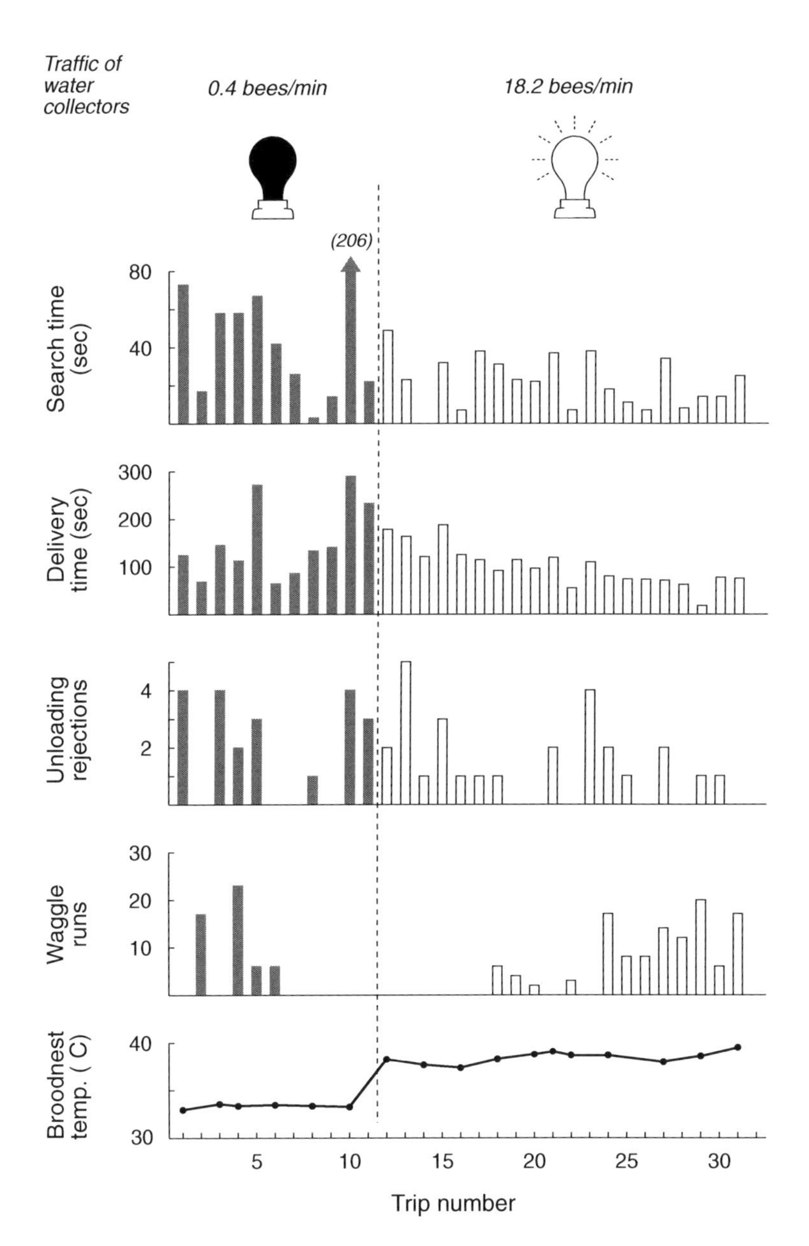

Figure 9.13 Trip-by-trip records of the unloading experiences of one nectar forager gathering a concentrated (2.5-mol/L) sugar solution throughout a time period during which her colony suffered a severe heat stress and so strongly raised its intake of water. This bee experienced neither increased difficulty in unloading her "nectar" nor decreased vigor in advertising her food source. This pattern, also observed on 2 other days with 2 other individuals, indicates that a colony can greatly boost its water collection without disrupting the activities of the nectar collectors. Based on unpublished data of S. Kühnholz and T. D. Seeley.

showed no signs of greater difficulty in unloading her highly concentrated food (no rise in search time, delivery time, or unloading rejections), and she showed no loss of enthusiasm to advertise the feeder by performing waggle dances (no decrease in the number of waggle runs performed per return to the hive).

In hindsight, it seems not at all surprising to have found evidence that the internal control of water collection by honey bee colonies involves a modulation of the number of bees functioning as fluid (nectar or water) receivers, rather than a fixed number of bees that accept different fluids under different conditions. The beauty of this feature of the control mechanism is that it enables a colony to modulate its water collection independently of its nectar collection, rather than having the two processes tightly linked in a mutually inhibitory relationship in which increased water collection would cause decreased nectar collection. Certainly, a colony is more likely to accumulate all the honey it needs for winter survival if its nectar collection is not disrupted every time its water collection must be boosted.

Summary

1. A honey bee colony gathers water for two purposes: lowering the broodnest temperature through evaporative cooling (Figure 9.1) and preparing food with proper water content for the larval brood. Thus a colony's water need rises both on hot days, when overheating threatens, and on cool days, when nectar foraging is hampered.

2. The water sector of a colony's economy involves a division of labor between the water collectors, who gather the water outside the hive, and the water consumers, who utilize it inside the hive. The hive bees' need for water varies widely and erratically, depending on the weather conditions, and hence a colony faces the problem of keeping its rates of water collection and consumption in balance. The same problem arises in the nectar sector, but whereas variation in activity in the nectar sector is *supply driven*, variation in the water sector is *demand driven* (Figure 9.2).

3. A colony can appropriately modulate its rate of water collection, for example by increasing it when the broodnest begins to overheat and decreasing it when the danger of overheating has passed (Figures 9.3 and 9.4). This raises the question of which bees collect water and thus adaptively adjust the colony's water intake.

4. The bees that collect water are a tiny subset (less than 1%) of the foragers. What determines whether a forager takes up water collection, as opposed to nectar or pollen collection, remains unknown. Water collectors often perform the task for many days in a row (Figure 9.5), and will sometimes specialize exclusively in water collection,

making many collecting trips per day so long as the water need persists.

5. Probably most future water collectors are prompted to begin fetching water by the recruitment dances of bees already engaged in collecting water. What activates the very first water collectors, though, may be an increase in the sugar concentration of the honey stomach contents of these bees. There is a strong correlation between the mean sugar concentration of the honey stomach contents of a colony's members and their motivation to drink water at the hive entrance (Figure 9.6).

6. A water collector probably acquires information about her colony's need for water, and thus whether she should continue or stop her collecting activity, by noting one or more variables of her unloading experience each time she returns to the hive. The greater the need, the quicker she is able to start her unloading, the sooner she can end her unloading, the fewer encounters she has with hive bees refusing her water load, and the more bees she has unloading her simultaneously (Figure 9.8). It remains to be investigated whether the indication of a colony's water need lies in the whole constellation of variables of the unloading experience, some subset of these variables, or perhaps one special variable. The fact that there is a strong correlation between a water collector's unloading experience and her subsequent behavior, including her dance activity and her continuation of water collection (Figure 9.7), suggests that the bee adjusts her behavior in response to her unloading experience. Thus it appears that water collectors and nectar collectors sense the colony's need for their respective materials through the same mechanism.

7. A water collector experiences faster, easier unloadings when her colony's water need is high. It seems clear that expedited unloading involves an increase in the number of bees accepting water. Two hypotheses have been proposed for how this might come about (Figure 9.9). In one, the additional water receivers are former nectar receivers that switch to water reception when they sense a need for water; in the second hypothesis, the additional water receivers are nurse bees that move to the unloading area and seek foragers bearing water when they need water in the broodnest. Two experiments were performed in the 1950s to test the first hypothesis (Figures 9.10 and 9.11), but neither one decisively proves or disproves this hypothesis. Recent experiments indicate that the additional water receivers come partly from the ranks of nectar receivers that switch to water recep-

tion and partly from the ranks of middle-aged bees (not young nurse bees; Figure 9.12) that are not nectar receivers. Evidently, a colony relies on a reserve supply of labor among the middle-aged bees that it can use as a source of additional water receivers. This ability apparently explains why a colony can massively increase its water collection without having to simultaneously decrease its nectar collection (Figure 9.13). The ability to modulate water collection independently of nectar collection no doubt helps a colony to accumulate all the honey it needs for winter survival.

OVERVIEW

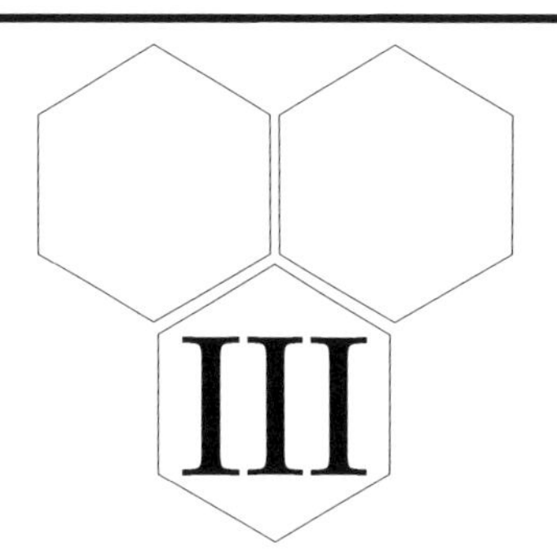

The Main Features of Colony Organization

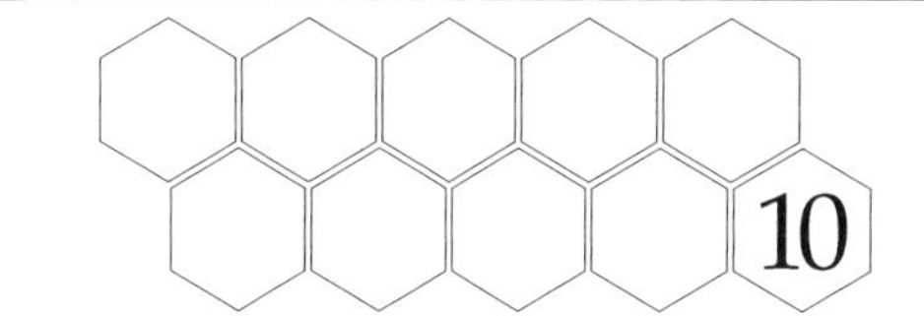

I began this book by posing a fundamental question about the evolution of biological organization: What are the devices of social coordination, built by natural selection, that have enabled certain species of social animals to make the transition from independent organism to integrated society? In the foregoing chapters, I reviewed the mechanisms of social physiology underlying the food-collection process of honey bee colonies, which is possibly the best-understood example of cooperative group functioning outside the realm of human society. In this chapter, I will summarize these findings by identifying the main features of honey bee colony organization. In addition, I will aim to place these features in a larger context by drawing comparisons between the inner workings of a bee colony and those of other kinds of functionally organized groups. These include multicellular organisms (groups of cells), colonies of marine invertebrates (groups of zooids), certain human organizations (groups of people), and multiprocessor computers (groups of electronic processors). All such highly cooperative groups share the basic problem of rationally allocating their members among various activities so that the more urgent needs of the ensemble are satisfied before the less urgent ones. They also share the problem of coordinating the actions of their members to achieve coherent patterns of activity. The solutions to these problems, however, vary greatly among the different types of integrated groups. By comparing these solutions, and reflecting on the functional significance of their similarities and differences, we can deepen our understanding of the mechanisms that make close cooperation a reality.

10.1. Division of Labor Based on Temporary Specializations

One virtually universal trait of functionally organized groups is division of labor, whereby each member of a group specializes in a subset of all the tasks required for successful group functioning. Sometimes the members' specializations are *permanent,* in which case individuals usually differ markedly in morphology in accordance with their different roles; and sometimes the specializations are *temporary,* in which case individuals remain basically uniform in structure. Examples of division of labor based on permanent specializations of morphologically differentiated individuals include the cells of vertebrate animals, where one can distinguish at least 200 cell types (Alberts et al. 1983); the zooids of siphonophore colonies, where one often sees a half dozen or so zooid types (Mackie 1963; Harvell 1994); and the workers of ant and termite colonies, where one sometimes finds two or more distinct physical castes (Wilson 1971; Oster and Wilson 1978). The honey bee colony exemplifies the second way of creating a division of labor, whereby physically similar individuals adopt only temporary specializations. Within a bee colony one can easily distinguish four main modes of labor specialization spread over the life of a bee—cleaner, nurse, food storer, and forager—and within each of these four general divisions one finds additional, more subtle specializations (Section 2.2). The foragers, for example, are plainly distinguishable as nectar collectors, pollen collectors, or water collectors. Each forager's specialty is apt to be only temporary, however, since a forager can readily switch from one substance to another (Sections 5.10 and 9.3).

The distinction between permanent and temporary specializations highlights an important question about biological organization: Why is the division of labor within cooperative ensembles sometimes organized one way, sometimes the other, and sometimes through a combination of permanent and temporary specializations? Part of the answer may be that the presence of permanently specialized members within a group somewhat limits the group's flexibility in responding to changes in the environment, since permanent specialists cannot switch tasks. Thus if a group consists entirely of permanent specialists, it can change its labor profile only slowly, by adjusting the birth (production) and death (elimination) rates of the different specialists. This reasoning suggests that future comparative studies of social organization will reveal that cooperative groups possess members with lifelong specializations only if the groups live in reasonably stable environments or have certain highly stable labor needs (such as defense) to which some of their members can be permanently devoted. In these situations, the benefits of permanent specialization, such as enhanced task performance resulting from the morphological adaptations which become possible when specialization is permanent, could outweigh the costs of permanent specialization, such as reduced flexibility in labor allocation.

It is generally thought that division of labor within living systems is favored by natural selection because the specialization that it makes possible enables individuals to increase their work efficiency, usually by enabling them to acquire special skills, based on special knowledge or special "equipment" (including morphology), or both (Miller 1978). We have seen, for example, that in nectar foraging by honey bee colonies the division of labor between nectar collectors and nectar processors raises the efficiency of nectar acquisition because it enables each nectar collector, which has acquired the special knowledge of the location of a rich flower patch, to gather nectar there intensively—before the flowers fade, competitors arrive, or darkness falls—without taking time out to process and store each nectar load she brings home (Chapter 6). High *efficiency* of task performance, however, is only half of what is required for a system of division of labor to yield benefits to the whole group. Equally essential is that the task an individual performs has high *importance,* where importance is a function of the group's labor needs (Simon 1976). This means that functionally organized groups with division of labor must possess mechanisms whereby the activity levels of their different specialists can be adjusted according to the group's needs. In the case of groups possessing temporary specialists, which can switch tasks, there must be mechanisms controlling the allocation of individuals among the different labor specializations.

It has traditionally been thought that one of the mechanisms whereby a honey bee colony achieves a proper labor allocation is the adoption by workers of different specializations at different ages. This practice will certainly divide the labor among the workers, and when one examines a colony's labor profile one does see a general pattern of individuals of different ages specializing in different tasks (Seeley 1982; Winston 1987; Figure 2.5 above). It is not clear, however, that the mechanism that produces this pattern involves age per se. Tofts and Franks (1992) suggests a mechanism distinct from physiological aging, one that moreover possesses the feature of generating an allocation of specialists which is closely matched to a colony's needs. The process Tofts and Frank propose for dividing labor—called "foraging for work"—is based on three assumptions: (1) each worker actively seeks work, and will switch to a nearby task when it fails to find work in its current task; (2) tasks are arranged approximately concentrically in the nest; and (3) workers are born into the first task, which is located in the nest center. Given these three assumptions, a colony will show the familiar pattern of age polyethism, which has been demonstrated by computer simulation (Tofts 1993). Basically, the birth of young workers increases the labor supply in the nest center, and this passively displaces (on average) the next older workers to the adjacent task zone, which in turn boosts the labor supply in this zone, causing the still older workers to move outward, and so on until the effects reach the outermost task zone, the outside environment, where the oldest workers labor and ultimately die (Figure 10.1, top).

Does the age polyethism pattern of honey bee colonies actually arise through the mechanisms expressed in the foraging-for-work hypothesis? There is no question that the second and third assumptions

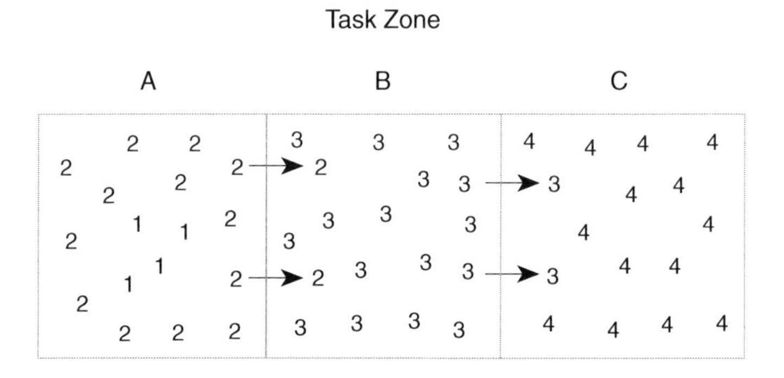

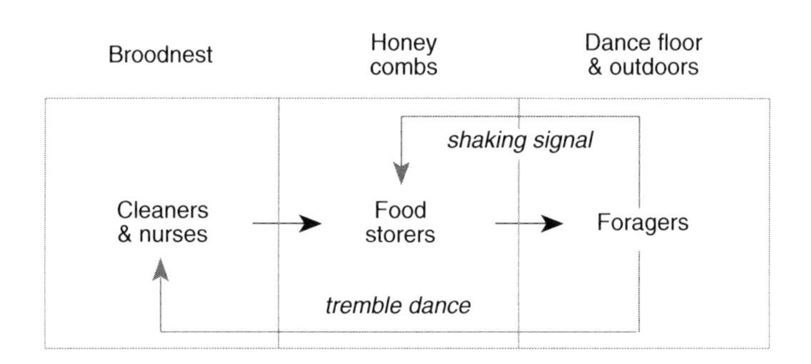

Figure 10.1 Division of labor. *Top:* The way a seemingly age-based division of labor can arise without any direct effect of age on behavior. Instead, a correlation between age and behavior can be generated through the spatial ordering of tasks into zones, coupled with a system in which individuals search ("forage") for work and shift from one task zone to the adjacent one when they fail to find adequate work. When young individuals arise in a task zone, or older individuals shift to a new task zone, they passively displace some of the current occupants by reducing the work there. The net direction of behavioral change is determined by individuals being born in one task zone (*A*), and dying in another task zone (*C*). Numbers represent individuals of different ages, from youngest (*1*) to oldest (*4*). *Bottom:* Application of the foraging-for-work hypothesis to honey bee colonies, with the addition of two special signals (the tremble dance and the shaking signal) by which information apparently flows from one task zone to another to stimulate bees to move to where the colony's labor needs are greatest.

are valid for honey bee colonies (see Figures 2.5 and 2.7), but it remains uncertain whether the first assumption is also correct, because the mechanisms of task switching are not fully understood for honey bees. The critical unknown is whether workers switch tasks entirely in response to external stimuli or whether these switches are also propelled by internal changes linked to aging. The latter possibility is plausible, for it is clear that as worker bees age they typically experience a rise in the level of juvenile hormone, and there is growing evidence that a change in the juvenile hormone level alters a bee's behavioral response thresholds, apparently by inducing changes in the central nervous system (reviewed in Robinson 1992). But it is also possible that the usual rise in juvenile hormone level is caused not by increases in a bee's age, but by changes in external stimuli, which could arise by the mechanisms described in the foraging-for-work hypothesis. One indication that age per se may have little, perhaps even no, effect on the juvenile hormone level of worker honey bees is an experiment performed by Robinson, Page, Strambi, and Strambi (1989) in which they created small colonies consisting of workers all of the same age, and found that regardless of the workers' age, there were always some with low hormone levels (the nurses) and always some with high hormone levels (the foragers). Thus it is clear that under certain experimental conditions the division of labor within honey bee colonies can be completely unrelated to age. This experiment suggests that the mechanisms underlying the division of labor within honey bee colonies may indeed be those of the foraging-for-work hypothesis, though certainly additional studies are needed to prove that age plays no role whatsoever in producing the division of labor.

Even if the foraging-for-work hypothesis proves basically correct, future studies are likely to reveal that natural selection has built elaborations on the basic mechanisms proposed by Tofts (1993). These additional devices could serve, for example, to strengthen a colony's ability to adaptively allocate its labor under changing conditions. The existence of such elaborations is suggested by certain recent discoveries about the organization of foraging in honey bee colonies. One is that the tremble dance evidently stimulates nurses to switch to food storing when additional food storers are needed (Section 6.4), and another is that the shaking signal may stimulate food storers to switch to foraging when additional foragers can be profitably deployed (Section 6.2) (Figure 10.1, bottom). If these communication signals prove to have these effects, then it will be clear that worker bees switch tasks not only when they experience difficulty finding work in their current task zone, but also when they receive a signal indicating a need for more workers in a nearby task zone. In short, we may find that workers can be "pulled" as well as "pushed" from one zone to another, via special signals that allow information to flow from one task zone to another. Clearly, many intriguing mysteries remain regarding the mechanisms by which honey bee colonies and other functionally organized groups achieve an effective division of labor. Their elucidation is one of the fundamental problems of biology.

10.2. Absence of Physical Connections between Workers

A second feature of the organization of honey bee colonies is a lack of tight structural links between the bees inside a hive. The absence of tissue connections between colony members is characteristic of social insect colonies in general (Wilson 1971), but stands in marked contrast to the internal architecture of functional units at other levels of biological organization. The molecules in cells and the cells in multicellular organisms generally function within an anatomically fixed network of interactions. Indeed, the structural relations of cells within organisms are often so stable that the patterns of cellular interconnections provide an invaluable guide to how an organ system—such as the visual cortex (Van Essen and Maunsell 1983; Livingstone and Hubel 1988)—works as a unit. It should be noted, however, that not all group-level units are built without strong connections between their subunits. In the colonies of marine invertebrates—including bryozoans, ascidians, corals, and siphonophores—the component zooids (which originally were single unitary organisms) are organically linked. In these colonies, as in multicellular organisms, we find fixed spatial relations of the subunits and the elaboration of colony-level structures built of the bonded subunits, such as common vascular systems, nervous systems, and jet propulsion systems (Mackie 1986).

The absence of physical connections between the members of a honey bee colony has numerous consequences for how they communicate with one another, share resources, and collaborate in common tasks. It means, for example, that there are no nervous system connections between the bees in a hive. Within multicellular organisms, the tight links between cells capable of propagating electrical signals (neurons) makes possible the construction of a nervous system, an organism-wide network for the rapid transmission of information to specific targets (Horrobin 1964). The absence of analogous links between the organisms in a social insect colony means that a colony of bees, for instance, lacks an analogous colony-wide network for high-speed communication. Interestingly, the colonial marine invertebrates, presumably by virtue of the connectedness of their zooids, have evolved group-level nervous systems based on electrical impulses traveling interzooidally via nerves and excitable epithelia. In some species, the colonial nervous system even includes such sophisticated features as giant axons for coordinating rapid escape responses by the colony as a whole (Mackie 1984, 1986).

Although the absence of tissue connections between the members of honey bee (and other social insect) colonies evidently prevents the rapid propagation of electrical signals between workers, it does not hinder the colony-wide spread of information through chemical signals. As we have seen, chemical signals can pass between the physically separate worker bees either via the atmosphere inside the hive, as with the volatile alarm pheromones that coordinate defense responses, or via the food that the workers exchange through trophallaxis, as with the proteins that appear to regulate pollen collection (Section 8.4). The bees have also evolved a special pathway for speedy and widespread chemical signal conduction: messenger bees that pick up the queen's pheromones and then travel about the broodnest actively dispersing this olfactory indicator of the queen's presence (Section 1.2). This messenger mechanism, which takes advantage of the physical independence of workers, can be viewed as a primitive means of achieving relatively rapid telecommunications from one sender to many receivers. Multicellular organisms, whose subunits are generally immobile, employ an entirely different solution to the problem of speeding up long-range chemical signaling within the plant or animal body: letting the fluid flowing within the vascular system quickly transport the signal molecules (hormones) to all parts of the organism (Bowles 1990; Snyder 1985).

The mobility of the bees inside a hive has numerous consequences for colony design besides its effects on the colony's communication systems. For instance, it means that certain organizational problems must be solved that do not arise in entities built of interlocked subunits, such as minimizing the time and energy workers spend moving about inside the hive looking for work. The honey bee colony's solution to this travel-cost problem appears to involve its division of labor for the within-hive tasks. We have seen that this is a rather coarse division of labor, with just two main groupings of tasks for bees that have completed their initial job of cleaning cells: broodnest work and food-storage area work. In each location there are many different jobs. Hence the behavioral repertoire of a bee working in either of these two locations usually includes many different job skills. (Section 2.2). By generalizing in several tasks (though still a subset of all the tasks occurring inside a hive), rather than specializing in just one task, workers no doubt reduce their search times and travel costs between tasks (Seeley 1982). This idea that spatial efficiency influences the division of within-nest labor also applies to certain ant species (Wilson 1976; Sendova-Franks and Franks 1993) and may prove true for social insects in general.

The bee's capacity for independent travel has evidently posed not only problems but also opportunities for organizational design. For example, in principle, it enables the members of a colony to personally gather information over a wide area inside the hive, thereby enabling them to base their actions on broad knowledge of the colony's internal state. Lindauer (1952) aptly named this process "patrolling." It remains unclear, however, exactly how important this patrolling is to bees. So far, the analysis of the foraging process indicates that foragers do not conduct wide-ranging reconnaissance trips while inside the hive. Unemployed foragers, for instance, do not survey the dance floor to locate the dancer advertising the best food source (Section 5.10). Likewise, there is neither evidence that pollen foragers travel about the combs and size up their colony's pollen reserve to assess the need for further pollen collection (Section 8.3), nor any indication that nectar and water foragers conduct broad surveillance of the honeycombs or broodnest to judge the need for greater nectar or water collection (Sections 5.7 and 9.5). Rather, all three types of foragers evidently rely on indirect indicators of their colony's forage needs: protein hunger and search time to find a receiver bee. And monitoring these indicators does not require widespread patrolling. At the same time, however, it is likely that the bees working inside the hive—the cell cleaners, brood tenders, food storers, and so forth—do travel about a good deal; so their behavior patterns probably reflect some integration of information gleaned from different locations within the hive. To cite just one example, a food-storer bee may decide to build additional storage comb when she senses difficulty finding places to store the fresh nectar (Section 7.2). This sense of difficulty probably arises only after the bee has searched across her colony's honeycombs and encountered cell after cell filled with honey. Obviously, such direct monitoring of conditions throughout the group is not possible for individuals with fixed locations in their group, such as the cells within a worker bee or the zooids within a bryozoan colony. They must rely instead on information that is gathered either directly from their immediate neighborhood or indirectly from distant locations through sophisticated communication systems (neural and hormonal). Thus it seems that a worker bee's ability to change locations inside the hive, and so conduct wide-ranging reconnaissance, means that she will have comparatively little need for direct communication with her nestmates, a subject discussed further in Section 10.4.

In summary, there are profound differences in internal design between organizations whose subunits are mobile and those whose subunits are immobile. The members of a honey bee colony, which have retained their physical autonomy and thus can move throughout the hive, still possess the broad behavioral abilities of their solitary ancestors (the one conspicuous loss being the full power of reproduction). Also, their communication pathways function only as labile links between largely independent individuals. By contrast, the cells of a multicellular organism, which function in fixed positions within the organism, perform only a narrow set of biochemical activities, the ones appropriate to each cell's particular anatomical location. Moreover, many cells, such as muscle cells, possess stable lines of communication to receive the steady flow of information from distant sites which they need to function properly. In short, the coordination mechanisms that one finds inside a organization depend greatly on whether the topology of the organization's members is fluid or fixed.

10.3. Diverse Pathways of Information Flow

Many authors have rightly observed that the formation of a higher-level unit by integrating lower-level units will succeed only if the emerging organization acquires the appropriate "technologies" for passing information between its members (see, for example, Wiener 1961; Boulding 1978; and Wright 1988). Coherence entails communication. Thus it is clear that to understand how any thoroughly integrated entity works, we must know how information passes between its subunits. In the case of a honey bee colony, we now know that information flows between its members via mechanisms that are remarkably diverse and often curiously subtle. Natural selection has shaped the workers so that they are sensitive to virtually all stimulus variables that contain useful information: the recruitment dances of nestmates, the temperature of the nest interior, their hunger for protein, the shape of a beeswax cell, the difficulty of unloading forage, the scent of the queen, and countless others. Evidently, evolution has been highly opportunistic in building pathways of information flow inside a beehive.

To gain perspective on this subject, it is useful to recognize Lloyd's (1983) distinction between signals and cues. Signals are information-bearing actions or structures that have been shaped by natural selection specifically to convey information. Cues are variables that likewise convey information, but have not been molded by natural selection to express this information. Both kinds of information-rich variables provide reliable information to the individuals noticing them, but signals do so expressly whereas cues do so only incidentally. One of the more important lessons to emerge from the analysis of foraging by honey bee colonies is that much, perhaps even most, of the communication within a colony occurs via cues rather than signals.[1] Indeed, the story of honey bee foraging contains so far just three signals—the waggle dance (Section 5.2), the tremble dance (Section 6.3), and the shaking signal (Section 6.2)—but numerous cues, including search time to locate a nectar receiver (Section 5.7), nectar influx (Section 7.2), fullness of the honeycombs (Section 7.2), protein level in the shared food (Section 8.4), perhaps the odor of brood (Section 8.3) and the presence of empty pollen cells (Section 8.5), and search time to locate a water receiver (Section 9.5).

The predominance of cues over signals within highly integrated groups may reflect two basic facts regarding the evolution of communication systems. The first is that information transfer will evolve more readily when it occurs via cues than via signals because the evolution of cuing involves only the formation of an adaptive response to a pre-existing stimulus (the cue) whereas the evolution of signaling involves the adaptive modification of both a stimulus (the signal) and a response. It may even be that natural selection builds a signal-

1. Biologists generally restrict the term "communication" to information transfer via signals (Otte 1974; Dawkins 1986). But because so much of the information flow inside honey bee colonies occurs via cues, I find it more practical to have the term "communication" denote information transfer via cues as well as signals.

based mechanism of communication only when there are no cues that adequately express the information to be conveyed. Thus, for example, bees possess their highly evolved waggle dance signal because there are no cues in a successful forager's behavior which incidentally specify the location of her food source. The second, and perhaps more important, fact is that the process of group integration is largely a matter of information flow from group to individual, so that each individual can tune its activity in accordance with the activities of the other group members (Simon 1976). This group-to-individual information flow will occur mainly via cues because the group-level indicators to which individuals respond for coordinating their actions are most likely to be by-products of the combined activities of a colony's members (cues) rather than group-level phenomena that evolved specifically for information transfer (signals). Consider, for instance, the impressive example of coordination among a colony's foragers as they adjust their colony's choosiness among nectar sources in relation to the colony's foraging status (Section 5.7). This coordination occurs because each nectar forager responds to the difficulty of unloading her nectar upon return to the hive, an action which reflects the actions of all the food storers and nectar foragers in the colony. Unloading difficulty provides a reliable flow of information from colony to individual, and clearly it is a cue, not a signal.

Another important feature of the diverse pathways of information flow inside honey bee colonies is that not all are conveniently marked by a prominent communication display. When considering information flow via signals, one almost always finds a conspicuous display by the sender, since most signals within cooperative groups will be molded by natural selection so that they provide strong, clear messages to the receivers. But when one considers information flow via cues, one finds that the variables expressing information are often exceedingly subtle, since in this situation receivers respond to stimuli that are just intrinsic components of their social environment. The contrast between signals and cues in apparency of information transfer is illustrated by the waggle dance and the search time to find an unloader, a signal and a cue used by foragers. The former is an acoustically and visually striking behavior which attracts the attention of other bees, while the latter is a silent and invisible variable which causes foragers to quietly adjust their internal thresholds of dance response (Section 5.7). Both stimuli are highly informative, but only with the former is there an obvious flow of information. A sim-

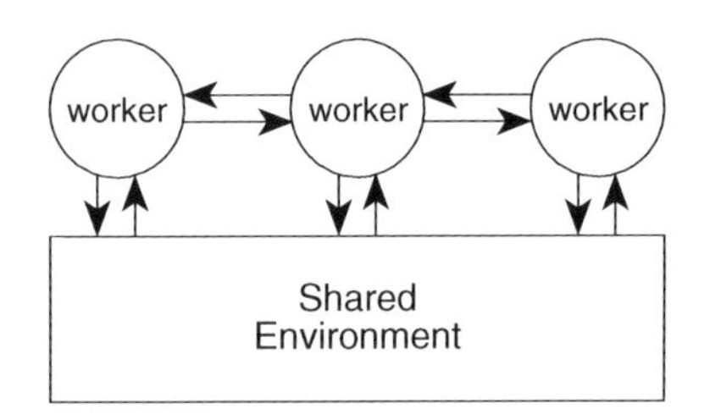

Figure 10.2 Direct *(horizontal arrows)* and indirect *(vertical arrows)* information flow between the workers in a colony. Indirect flow is possible because each colony member can both influence and be influenced by the multicomponent shared environment inside the hive (the combs and their contents, the shared food, and the hive atmosphere). For example, each nurse bee within the colony affects the broodnest temperature with her heating and cooling activities, and in turn responds to the broodnest temperature by adjusting her thermoregulation activities.

ilar contrast in apparency of information transfer occurs in the social organization of foraging by colonies of fire ants *(Solenopsis saevissima)* (Wilson 1962). Here again there is a conspicuous production of a signal—the laying of a chemical trail to a rich food source—together with a less striking cue, the degree of crowding at the food source. This crowding is an automatic by-product of recruitment to the food, and hence it is not at all apparent that it is an important source of information to the ants. Experimental analysis indicates, however, that this cue provides negative feedback to regulate the recruitment produced by the signal.

The signal-cue dichotomy helps us appreciate another significant fact about the mechanisms of information exchange in honey bee colonies, which is that much of the information flow between bees occurs indirectly, through some component of the shared environment, rather than directly (Figure 10.2). Virtually all such indirect interactions are mediated by cues rather than signals. An example is the transfer of information for the reduction of a colony's pollen collection after several days of successful foraging. The reduction of pollen foraging involves hundreds or thousands of bees (Section 8.2), yet these bees never need to come together and exchange information directly. Their foraging activities can be completely and efficiently coordinated by information contained in the liquid food exchanged among the bees in a hive. Thus as each pollen forager brings in more and more pollen, the level of protein in the shared food rises, and the foragers receive the information that their colony's pollen need has been met (Section 8.4). Another example of information flow through the shared environment is the thermoregulation of the hive. A colony maintains the central broodnest at 33 to 36°C in the face of ambient temperatures that may range from –20 to 45°C (Section 9.2). The coordinated heating and cooling of the broodnest occurs automatically: each nurse bee responds to the temperature of her immediate environment by appropriately heating it (by making intense isometric contractions of her flight muscles) or cooling it (by fanning her wings to draw cooler air into the area and, when overheating is extreme, by spreading water for evaporative cooling) (Heinrich 1985). In effect the temperature of the air and comb inside a hive provides a communication network regarding a honey bee colony's heating and cooling needs. Studies of other social insects have also revealed the importance for colony organization of indirect information flow through the shared environment. One is Grassé's (1959) work on nest building in termites, which shows that the nest building activities of termites can be completely and efficiently coordinated by information embodied in the structure of a partially completed nest. Thus one termite might begin building a column by depositing a fecal pellet; a second termite receives guidance about where to place her pellet by the position of the pellet left by the first termite.

Studies of information flow within honey bee and other social insect colonies ultimately may reveal that more information is exchanged indirectly rather than directly. The use of the shared environment as a communication pathway has certain advantages, such as providing easy passage of information from a group to an individual whenever an individual responds to the environmental effects of the group. And as mentioned previously, group-to-individual information flow is central to the process of group integration. Communication through the shared environment also has the major advantage that it allows easy *asynchronous* transfer of information between individuals. The importance of this feature is underscored by the fact that designers of multiprocessor computers, faced with the problem of creating an efficient way for processors to communicate, have discovered that sometimes it is more effective to link each processor to a shared memory rather than to construct an intricate set of communication paths among the processors (an "interprocessor communication bus") and complex protocols by which two processors can communicate. The principal advantage of communicating via shared memory is that asynchronous communication is easy because the shared memory provides a buffer between any two processors (Baskett and Hennessy 1986; Gelernter 1989). Each processor writes the results of its computations in the shared memory, where they are available to all the other processors whenever they are needed, and it reads from the shared memory the information it needs to perform its computations. This indirect data transfer among processors via a shared memory is closely analogous to the indirect communication among bees via their shared environment.

Traditionally, studies of communication in colonies of honey bees and other social insects have emphasized the conspicuous communication processes that involve signals honed by natural selection, such as the waggle dance. There is no question that these processes are extremely important. Nevertheless, close analysis of the food collection process in honey bee colonies has revealed many subtle communication processes that involve cues, many of which are variables of the shared environment whose informational importance is not at all obvious. One important lesson of such studies is that we must think creatively and watch closely when tracing the pathways of information flow inside colonies. Given that natural selection always builds biological machinery in an opportunistic fashion, this lesson probably applies to the communication networks inside living systems at all levels of biological organization.

10.4. High Economy of Communication

While it is true that a honey bee colony contains an astonishingly intricate web of information pathways, built of diverse signals and cues, evidently it is also true that the total amount of information shared within a colony is smaller than one might expect for such a well-integrated entity. Certainly, one of the more surprising truths to emerge from the analysis of a colony's food collection process is the remarkable economy of communication among the bees involved in foraging. The picture that has emerged of a colony's foraging operation is one of an ensemble of largely independent individuals that rather infrequently exchange information (directly or indirectly) with one another and of a flow of information that is smaller than it initially appears.

This pattern of highly economical communication is seen most clearly in the honey bee's best-understood communication process: recruitment to food sources through the waggle dance. Consider first the fact that when a bee returns from a highly profitable patch of flowers and advertises her food source with a dance, she does not provide a detailed description of the properties of the source. Her dance signal contains only information about the distance, the direction, and the *overall goodness* of the flowers. (A dancer also conveys information about the scent of her flowers, but she does this through the passive release of odor molecules from her body and the food she carries, not by an active encoding of the scent information in the dance signal.) Thus she withholds the full details of what makes her flower patch an attractive forage site—proximity to the hive, abundance of nectar or pollen, high sugar concentration of the nectar, and so forth—and instead shares only her personal summary of these facts. This summarizing greatly simplifies the message she shares with her nestmates. What is equally noteworthy, though, is that even the simplified message is not fully received by the dance followers. The bees following a waggle dance acquire information about the distance and direction of the dancer's food source, but not about the profitability of this source (Section 5.10). The information about profitability, coded in the total number of waggle runs performed in a dance, is "received" by the colony as a whole, for it influences the number of bees recruited by each dancer, but evidently it is never communicated from dancer to dance follower. Thus through the senders' summarizing of information, and the receivers' ignoring of information, the bees show a surprising economy of information transfer in the waggle dance.

Not only are the communication interactions of foragers less informative than expected, but also their frequency appears to be remarkably low. This results partly from the fact that much of the time foragers are outside the hive, where they are separated from their nestmates, but even when foragers are inside the hive, they send and receive messages only rather rarely. Each time a nectar forager comes home, she receives one message about the colony's need for more nectar (via the cue of time spent searching for unloaders; Section 5.7), but almost never receives information about alternative food sources (it is exceedingly rare for an employed forager to follow dances; (Section 5.9). Moreover, she rarely sends a message about her food source, as indicated by the fact that only a small minority of the nectar foragers—those returning from highly profitable nectar sources—perform waggle dances (Section 5.2). Evidently, most nectar foragers send just two messages during each return to the hive: requests for help in unloading shortly after arriving and requests for food shortly before departing. Pollen foragers and water collectors apparently show a similar pattern of infrequent information exchange with nestmates (Sections 8.2 and 9.5). The unemployed foragers, however, may have a somewhat higher frequency of communication, since they typically follow one dancer after another until a food source is found (Section 5.10). But even for the unemployed foragers, the multiple bouts of dance following are spread over a period lasting several hours, if not a couple of days. Also, each bout of dance following by an unemployed forager normally involves following just one dance thoroughly, whereupon the forager leaves the hive to search for the advertised forage site. Furthermore, there is no evidence of a two-way exchange of information between dancer and follower. The follower does not send a request for information; she simply receives whatever information is presented to her. Thus it appears that for unemployed foragers, as for employed foragers, communication occurs only infrequently.

It now seems clear that even though the foragers in a honey bee colony possess the splendid waggle dance behavior for sharing information about forage sites, and even though they are sensitive to multiple cues indicating their colony's forage needs, the individual forager experiences surprisingly meager and infrequent exchanges of information with her hivemates. As for the nonforager bees in a colony, which work entirely inside the hive and so experience frequent contacts with nestmates, for them the amount of communication is probably higher. But how much higher remains unknown, and I suspect that it is less than we currently imagine.

To place the honey bee's economy of communication in a larger context, I suggest that the intensity of communication within colonies falls well below that within units at lower levels of biological organization: organisms and cells. This suggestion reflects the idea that the members of an organization, regardless of its specific composition, will experience a trade-off between *action* and *interaction.* With forager honey bees, for example, time spent communicating—producing or following waggle dances—is time that cannot be spent gathering food. It is crystal clear that this trade-off applies strongly to one special kind of organization, the multiprocessor computer, where an important design constraint is the incompatibility of computation and communication (Denning and Tichy 1990). At any given moment, a processor is either computing (performing instructions) or communicating (sending, receiving, or waiting for messages from other processors). The architects of multiprocessor computers have learned that one way to keep the costly communication time limited to a small fraction of the computation time is to build coarse-grained systems, ones in which the processors are relatively complex and capable of mostly independent computations. Fine-grained systems, built of many simple processors, are instead characterized by a high ratio of communication to computation (see, for example, *The Connection Machine* [Hillis 1986]). If we assume that there is an analogous "computation-communication" trade-off in living systems, it seems reasonable to suppose that as the highest level of functional integration has increased (from cell to organism to society), and the capabilities of the subunits forming the integrates have expanded, the fraction of time that each subunit spends communicating has dwindled. If so, a honey bee colony, with its economy of communication, should be seen as representing one extreme within a spectrum of biological organizations whose granularity—the ability of the components to function independently—ranges widely, from the interdependency of molecules to the self-sufficiency of organisms.

10.5. Numerous Mechanisms of Negative Feedback

The organization of a honey bee colony includes multiple pathways of negative feedback for the control of key variables of a colony's physiology. Negative feedback plays a particularly important role in a colony's foraging process because a colony experiences strong fluctuations in the external supply or the internal demand (or both) for all three of its forage commodities: nectar, pollen, and water. These fluctuations induce discrepancies between a colony's collection rate and its consumption rate for each commodity, and if these were to persist for long the colony would acquire either a burdensome surplus or a dangerous deficit for each of the materials. Mechanisms of negative feedback provide a colony's foragers with the information they need to adjust their colony's collection rate and consumption rate to eliminate the mismatches between supply and demand. In essence, a negative feedback mechanism consists of a set of processes coupled in cyclic manner (forming a feedback loop), and among these processes there is an odd number of inhibitory couplings (Jones 1973). The basic challenge in studies of negative feedback, at all levels of biological organization, is to trace the cyclic pathway through which the feedback operates and to understand the excitatory and inhibitory links between the processes within this pathway.

Figure 10.3 summarizes our current understanding of the looped pathways of feedback control involved in regulating the collection of nectar, pollen, and water. Two general features of these feedback mechanisms stand out. One is that these are *highly distributed* systems of feedback control. Thus they differ markedly from the control systems found in multicellular organisms, which typically consist of specialized components located in discrete anatomical sites. These usually include sensors that measure the variable to be controlled, a controlling device that reads the information from the sensors, and effectors that respond to signals from the controlling device to restore the variable to its desired level (Grodins 1963). For example, in mammals the regulation of carbon dioxide level in the blood arises through the joint action of a set of CO_2 receptors in the arteries, a respiratory

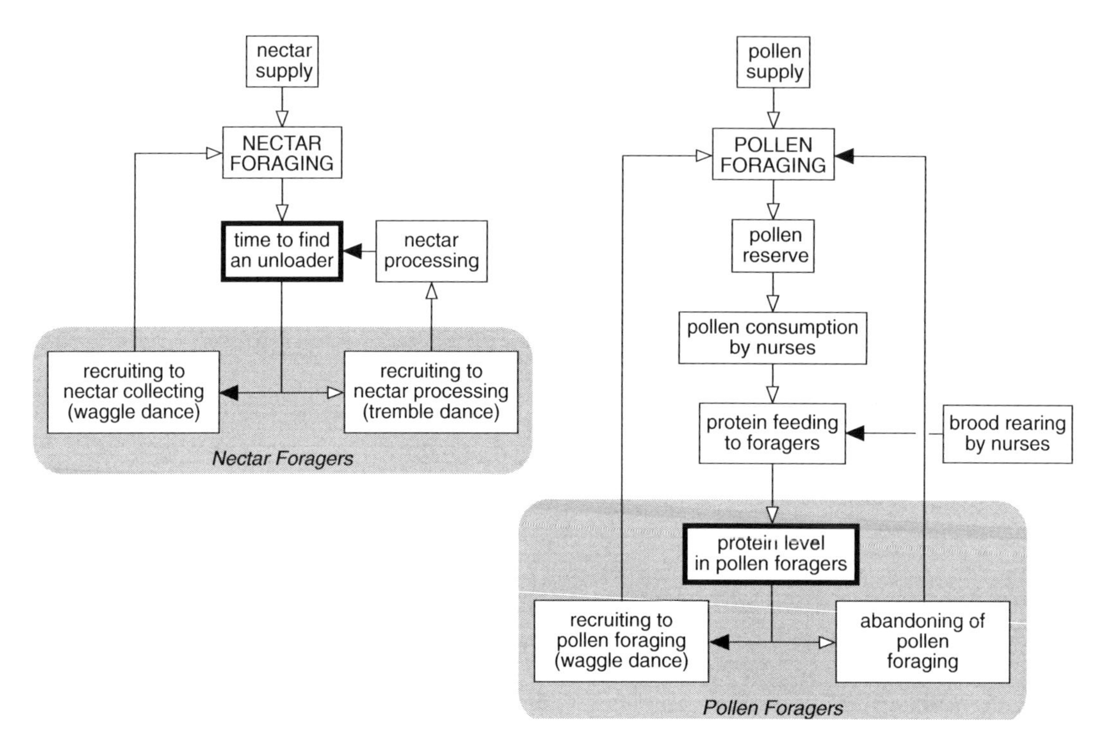

center in the brain which receives neural signals from the receptors, and a set of respiratory muscles associated with the lungs which adjust the animal's breathing rate in response to signals from the brain (Horrobin 1970). Within honey bee colonies, in contrast, we find that sensor, controller, and effector are not distinct components within a colony, but that they all occur together inside each of the bees that is involved in the regulation of a given variable of colony physiology. For example, a colony regulates its pollen reserve through the actions of its many pollen foragers, each of which senses the need for additional pollen, decides what set of foraging behaviors is most appropriate, and then acts on this decision. Even though the control of a colony's pollen intake is broadly distributed among its members, coherent regulation nevertheless occurs because the behavior of each pollen forager is ultimately linked to a single shared variable, the size of the colony's pollen reserve. The distributed nature of this control system actually confers the important advantage of high reliability, since the whole system will continue to function rather well even if some of its parts begin to function improperly.

The second general feature of the feedback mechanisms within a

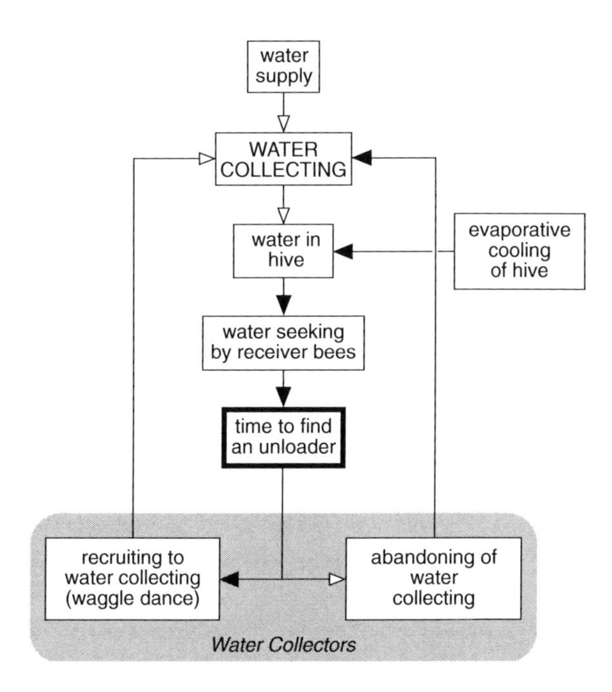

Figure 10.3 Pathways of negative feedback whereby a colony maintains a balance between collection and consumption (or processing) for each of its three forage commodities. *White arrows:* excitation; *black arrows:* inhibition. For each type of forager, the box outlined in bold denotes what is currently understood to be the key control variable for each commodity, that is, the variable that the foragers actually sense for negative feedback on their actions. The gray areas demarcate the activities of the foragers, and show how the feedback functions of sensor, controller, and effector are merged within individual bees. Notice an important difference in the control of nectar collection versus pollen or water collection: the former lacks a pathway of inhibition. A colony's response to a high level of nectar collection is a rise in processing activity, whereas its response to a high level of pollen or water collection is ultimately a reduction in collecting activity. This difference reflects the fact that a colony's demand for nectar is effectively infinite, while its needs for pollen and water are strictly finite.

colony's foraging operation is that the variables to which the foragers respond for feedback are only indirect indicators of discrepancy between supply and demand. This means that the bees monitor and regulate variables whose stability is not directly important to a colony's well-being. Thus water collectors adjust their behavior to keep the water supply matched to the water demand, but they do so not by measuring water collection and water consumption, but by detecting changes in the time spent searching for unloaders. Likewise pollen foragers, which apparently know nothing about their colony's internal supply of pollen or its demand for pollen, are able to modulate their behavior properly by detecting changes in their protein level. The same idea applies to nectar foragers, which perform waggle dances and tremble dances to keep the colony's rates of nectar collection and processing in balance, but do so without any knowledge of these two variables of a colony's foraging operation. Instead, like the water collectors, they monitor the seemingly inconsequential variable of search time during unloading. Why have bees evolved feedback loops based on control variables with little intrinsic importance? Almost certainly because these variables are far more easily sensed than the underlying variables of supply and demand, yet they provide all the feedback information that is needed to correct a mismatch between supply and demand. Hence in the design of the negative feedback loops within a honey bee colony, once again natural selection has devised mechanisms for colony functioning that require only minimal information collection by a colony's members.

10.6. Coordination without Central Planning

The most thought-provoking feature of a honey bee colony is its ability to achieve coordinated activity among tens of thousands of bees without a central authority. There is no evidence whatsoever of a control hierarchy in colonies of bees, with certain individuals acquiring information, deciding what needs to be done, and issuing instructions to other individuals that then perform the necessary tasks. As the biblical King Solomon observed for colonies of ants, there is "neither guide, overseer, nor ruler." In particular, it is clear that the queen bee does not supervise the activities of the worker bees. She does emit a chemical signal, the blend of "queen substance" pheromones, which provides negative feedback regulating the colony's production of queens (Section 1.2), but neither this nor any other signal from the queen can provide comprehensive instructions to the thousands of workers within a hive.

Coherence in honey bee colonies depends instead upon mechanisms of decentralized control which give rise to natural selection processes—that is, the differential persistence and proliferation of independent agents (see Section 5.12)—analogous to those that create order in the natural world and in the competitive market economies of humans. How this control operates is now best understood for the forager allocation process, wherein a "friendly competition" among a colony's foragers enables them to distribute themselves among an array of sources of nectar, pollen, and water in a way that reflects the desirability of each food source for the colony (Figure 10.4). The main arena for this competition is the dance floor inside the hive, a kind of labor clearinghouse where employed foragers advertise their work sites and unemployed foragers listen to these announcements to find suitable jobs. Employed foragers visiting highly desirable forage sites persist at these sites and produce vigorous dances which arouse unemployed foragers to join them, whereas employed foragers visiting

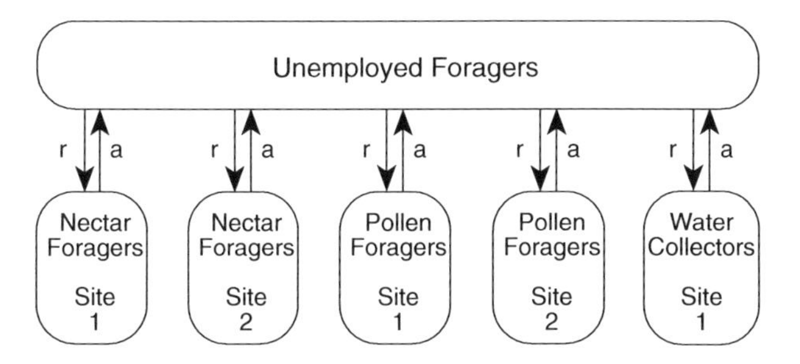

Figure 10.4 Distribution of a colony's foragers among forage sites without central supervision. The number of foragers employed at each site is determined simply by the history of recruitment (*r*) and abandonment (*a*) for each site. The employed foragers compete with one another for recruits from the pool of unemployed foragers. This competition is, however, "friendly" in that each forager honestly adjusts the strength of her waggle dance signal in relation to the desirability of her food source. Desirability reflects both the profitability of the source and the colony's need for its commodity (nectar, pollen, or water). Because the unemployed foragers respond essentially at random to the employed foragers' recruitment signals, recruitment is strongest to the most desirable sites. And because the employed foragers decide honestly whether or not to abandon a forage site, abandonment is strongest from the least desirable sites. The net result is an increase in the foragers at the most desirable sites and a decrease at the least desirable sites. Thus a competitive system pools the knowledge and actions of thousands of individuals to achieve a rational allocation of a colony's foragers without a central intelligence.

less desirable sites will tend to refrain from announcing them to the unemployed foragers and may even cease working at these sites. In this way, a huge mass of information which is dispersed among the employed foragers is neatly exploited without the need to convey it to a central authority. Indeed, no member of the colony ever acquires anything like a synoptic view of its colony's foraging operation.

To fully appreciate the beauty of the colony's method of forager allocation, one must make special note of two details of bee biology which enable bees to take full advantage of the power of a competitive market process. The first is the high quality of the information presented on the dance floor by the employed foragers. In particular, recall that the duration of each dance is a function not only of the profitability of the forage site being advertised, but also of the colony's need for the commodity being advertised (Section 5.8). Thus, for example, a nectar forager will perform a super-stimulating, 100 waggle run dance only if her nectar source is extremely rich *and* her colony's need for more nectar is extremely high. Because the production of each dance involves an integration of information about conditions both outside and inside the hive, a colony's overall allocation pattern is appropriate to both the colony's opportunities and its needs. One might wonder, however, whether the employed foragers' advertisements are always honest and hence whether they always provide accurate information for the unemployed foragers. It seems reasonable to believe that dancing bees present only reliable information on the dance floor, for any dishonesty in dance production should be eliminated by natural selection. Imagine, for example, that a mutation were to arise which causes bees to present false information on the dance floor, such as performing a lengthy dance for a forage site with low desirability. This would surely diminish the economic—and ultimately the reproductive—success of the colony. Hence the mutation would be eliminated. We can be confident, therefore, that the array of recruitment dances within a beehive provides an excellent guide to the labor market for the unemployed foragers.

The second detail of honey bee biology that favors the competitive mechanism of forager allocation is the way that unemployed foragers sample with minimal bias the information presented on the dance floor. Each unemployed forager follows just one dance, chosen more or less at random, before she leaves the hive to search for a forage site (Sections 5.10 and 8.6). This pattern, repeated by hundreds or thousands of unemployed foragers each day, means that each dance advertisement receives the audience that it deserves. Thus not only is the information on the dance floor accurately presented, but also it is fairly assessed. Moreover, the unemployed foragers are always open to additional reports of fresh discoveries, so that no one group of dancers ever monopolizes the information exchange. This "open market" policy toward sources of information means that a colony can flexibly respond to changing conditions outside the hive, allocating foraging effort toward new opportunities whenever they arise. This openness and flexibility is illustrated by the fact that even though every food source is advertised at first by just one forager—the enterprising scout bee that made the discovery—if a source offers a plentiful supply of highly desirable forage, eventually it will become massively advertised in the hive and then it will become a major focus of the colony's foraging.

One feature of the unemployed foragers' behavior deserves special mention, for at first glance it would seem to diminish the effectiveness of the competitive process of labor allocation. This is the failure of unemployed foragers to survey broadly the information displayed on the dance floor. This may seem distinctly odd, since well-informed consumers are crucial to the proper functioning of competitive markets in human society (Samuelson 1973), but apparently they are not in the bee society. This difference stems from the fact that whereas the individual human being seeks to maximize his or her personal profits, the individual honey bee seeks to maximize her colony's profits. And the latter goal is best achieved if each unemployed forager refrains from broadly monitoring the dance floor to identify the single most desirable food source. Both empirical and theoretical studies (Section 5.14) indicate that over time a colony achieves a distribution of its foragers among food sources such that the marginal rate of value accumulation from each source is equal, and that this labor distribution, though not always providing the highest possible rate of value accumulation, is generally close to the optimum. If, instead, each unemployed forager identified the one best food source being advertised in the hive, the allocation process would produce an all-or-none response, and it seems virtually certain that the colony's labor allocation would be far less efficient. Time lags in the recruitment process would lead to overinvestment of labor in the one source that was initially most profitable and underinvestment in all alternatives that initially were less profitable but eventually would become more profitable.

A second conspicuous discrepancy between the market mechanisms found in human and bee societies also deserves mention: the absence of prices in honey bee colonies. This lack may make the smooth running of a honey bee colony's economy seem especially puzzling to the economist, for prices are crucial to coordination in human economic systems. Specifically, prices enable people to wisely adjust their production activities in accordance with changes in the supply-demand ratios for different good and services without having broad knowledge of these changes. Or as Friedrich Hayek (1945) so vividly expressed it, the price system is "a kind of machinery for registering change, a system of telecommunications which enables individual producers to watch merely the movement of a few pointers to adjust their activities to changes of which they may never know more than is reflected in the price movement." Although forager bees have no prices for nectar, pollen, and water, they do possess other indicators which register changes in the supply-demand ratio for each of their forage commodities. For nectar and water, the functional analog of price is evidently the time each forager spends searching in the hive for bees willing to receive her nectar or water load (Sections 5.7 and 9.5). The effect on a forager bee of experiencing a short search is like that on a human worker of seeing a high price: production (collection) of the material is stimulated. For pollen, the analogue is evidently the protein hunger that each forager feels (Section 8.4), such that sensing a deep hunger is like noting a high price, and this triggers a rise in pollen production (collection). The essential thing is that each of these three indicators reliably rises and falls in accordance with changes in the supply-demand ratio of its corresponding material, hence each one provides the same information as does price. By responding to these indicators, each forager adjusts her activity to the collection (supply) and consumption (demand) activities of her hivemates, with the overall result that the labors of a colony's foragers stay closely matched to their colony's needs.

Why is there no central planning within a honey bee colony? This question is best answered, I believe, by noting the essence of the problem a colony must solve in allocating its foragers among food sources: *adaptation to countless changes in the particular circumstances of time and place.* Some of these changes occur inside the hive as a colony's resource needs vary from hour to hour, but most occur outside the hive as the colony's many, widely scattered food sources vary in profitability from minute to minute. A central authority could, in principle, best coordinate a colony's foraging efforts in light of these changes, for an organization blessed with an omniscient and clever central planner can indeed coordinate its members' activities better than can one with a decentralized administration (Chandler 1977). Such centralized control requires, however, that a tremendous amount of information—usually dispersed among all the individuals within an organization—be communicated to a central decision-making unit, that this unit then integrate all this knowledge, and finally that it issue instructions to all the participants (Simon 1981). But we cannot expect that this is how the bees solve their problem, for their communication systems are rather rudimentary. They have not evolved anything like a colonial nervous system which would allow information to flow rapidly and efficiently to a central decision-maker. Moreover, there is no individual or organized subset of individuals within a colony capable of processing the huge mass of information from the foragers, the food storers, and the nurse bees that must be integrated in devising a rational allocation of a colony's foraging efforts. Therefore, it is not surprising that the bees solve this allocation problem by means of decentralized control. The decisions regarding how a colony should deploy its foragers are left to the foragers themselves, especially the employed foragers, each of which possesses intimate knowledge of the facts of her particular food source. Such limited knowledge is, of course, not sufficient. Each employed forager must acquire additional information about the colony's forage needs, so that she can adjust her actions—such as advertising or abandoning her source—to fit with the full pattern of changes experienced by the colony. This is accomplished by means of easily perceived cues that indicate to each bee in summary manner which of a colony's forage needs are most urgent. Thus natural selection has devised elegant mechanisms whereby the members of a honey bee colony can solve the daunting forager allocation problem without needing to assemble in one mind all the relevant information that is in fact dispersed among all the bees in a hive.

Enduring Lessons from the Hive

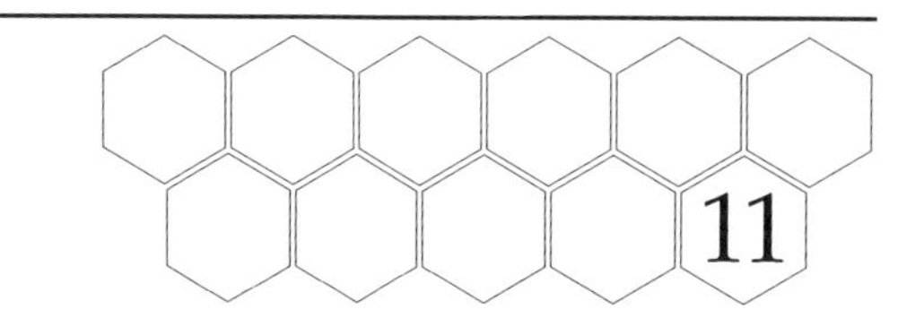

During the past 15 summers, I have spent most of my time trying to unravel the mechanisms which enable the members of a honey bee colony to work together in coordinated fashion as they gather their food from the flower patches dotting the countryside around their hive. This book reviews the results of these efforts. It necessarily presents a multitude of details, but I hope that the summary at the end of each "results" chapter, together with the "main features" chapter, has helped the reader achieve a general view of the mechanisms of social coordination inside a beehive. In this final chapter, I wish to stress two overarching insights that have emerged from these studies of the bees.

The first is that the system of control devices found in a honey bee colony is extremely sophisticated and endows a colony with exquisite powers of adaptive response, both to internal changes and to external contingencies. Certainly, all biologists are keenly aware of the amazing adaptive responses of cells and organisms, and are awed by the complexity of the underlying mechanisms of cellular and organismal physiology. But probably few biologists recognize that evolution has likewise endowed certain animal societies with impressive abilities and has fashioned elaborate mechanisms of communication and control inside these societies to produce their remarkable group-level skills. I suspect this fact is little appreciated even among sociobiologists, and for good reason: it is only rather recently that we have had both the *conceptual framework* (especially the ideas of Hull and Dawkins on replicators and interactors) and the *empirical findings* (such as those presented in this book) to demonstrate the reality of extremely rich functional organization at the group level.

With respect to the empirical evidence regarding colonial foraging in honey bees, for example, I think it is fair to say that 15 years ago we had only a vague sense of the foraging abilities of a bee colony and even less awareness of the complicated variety of mechanisms underlying these abilities. In 1978, no one knew that a colony can thoroughly monitor a vast region around the hive for rich food sources, nimbly redistribute its foragers within an afternoon, fine-tune its nectar processing to match its nectar collecting, effect cross inhibition between different forager groups to boost its response differential between food sources, precisely regulate its pollen intake in relation to its ratio of internal supply and demand, and limit the expensive process of comb building to times of critical need for additional storage space. Moreover, we lacked precise knowledge regarding many of the bee-level actions underlying such colony-level abilities, including the performance of tremble dances to reallocate labor, the transfer of proteinaceous food from nurses to foragers to regulate pollen collection, the use of search time to assess supply/demand relations, and the use of friendly competition on the dance floor to allocate foragers among flower patches. Today, in contrast, we easily see a honey bee colony as a sophisticated, group-level vehicle of gene survival, and we know that natural selection has equipped it with a wealth of ingenious mechanisms for successful group functioning. It seems likely that future studies of social physiology, in honey bees and other highly social animals, will further substantiate the view that animal societies, like cells and organisms, can possess a high level of functional organization.

The second general insight concerns the devices which nature uses in making a honey bee colony function as a unit. I think it is now clear that we can think of a bee colony as a bag of tricks, indeed, an almost bewildering bundle of special-purpose, tailor-made tricks, evolved through natural selection to solve the various problems faced by a colony. For this reason, it is probably an exercise in futility to seek grand principles of colony functioning, and the challenge of social physiologists is instead to uncover the diverse gadgets used to implement colony functioning.

In the course of this work there will surely be many surprises. The analysis of the honey bee colony has revealed, for example, that one cannot even embrace the seductive generalization that the bee colony is a complicated combination of rather *simple* tricks. To be sure, some tricks are wonderfully elegant—simple but effective—and involve bees operating nicely with surprisingly limited information, such as when they acquire information about the colony's need for more nectar simply by noting the search time to find unloaders, or when they allocate themselves among food sources simply by means of competition among foragers, not central planning. But some other tricks are impressively complex, such as when a bee integrates multivariable information about a flower patch to assess its overall goodness, combines this with information about the colony's need for her forage commodity, and then produces a waggle dance of appropriate duration. Equally complicated is the ability of bees to learn the location of a rich food source and then express this information in a waggle dance. Thus it seems clear that in order to lay bare how a colony actually works, we must be ever mindful of the full range of complexity in the social behavior of animals. Indeed, recognition of the ingenuity of natural selection in devising mechanisms of social coordination strikes me as the quintessence of the knowledge generated by these studies of the bees.

Glossary / Bibliography / Index

Glossary

This list contains terms used in this book which have special meanings in biology, entomology, and beekeeping. It is designed to make it possible to read this book without having to refer to textbooks in these subjects.

Adaptation. In biology, a particular structure, physiological process, or behavior that has been favored by natural selection, hence a trait that helps an organism survive and reproduce. Also, the evolutionary process that leads to the formation of such a trait.

Age polyethism. The regular changing of labor activities by members of a colony as they age.

Allele. A particular form of a gene, distinguishable from other forms or alleles of the same gene.

Antennation. Touching with the antennae, either as a sensory probe or as a tactile signal to another insect.

Apiary. A place where a group of hives of honey bees are kept.

Ascidians. The members of the class Ascidiacea of the phylum Chordata. They are also known as sea squirts. Colonial organization has arisen in several lines. In the socially most advanced species, colonies are organized such that the member zooids are physically united under a common tunic and their incurrent siphons ("mouths") open to the outside but the excurrent siphons empty into a common cloaca.

Brood. The immature members of a colony collectively, including eggs, larvae, and pupae.

Broodnest. The central, roughly spherical region of a colony's nest where the brood is reared.

Bryozoans. The members of the phylum Bryozoa. They are also known as moss animals, though most live in the sea. Nearly all of the 4000 known species form sessile colonies, which grow by budding. In many species there is polymorphism among the zooids within a colony, with different individuals reduced or otherwise modified to serve in defense, cleaning, or reproduction.

Building cluster. The loose cluster of bees that assembles itself where comb is being built. Nearly all the bees hang quietly in the cluster, secreting wax and perhaps also warming the building site.

Burr comb. The bits and pieces of comb which bridge the main curtains of comb in a hive.

Comb. A double layer of cells crowded together in a regular array.

Crossing-over. The process whereby homologous chromosomes, while engaged in meiosis, perform reciprocal swaps of genetic material. The result is the production of almost infinite variety in the genetic constitution of gametes.

Cue. An information-bearing action or structure that expresses information only incidentally, not because it has been shaped by natural selection to do so.

Dance floor. The region of the combs in a beehive, usually located just inside the entrance opening, where most of the waggle dances are performed.

Diploid. Having a chromosome set consisting of two copies (homologues) of each chromosome. A diploid individual generally arises through the fusion of two gametes. (Contrast with haploid.)

Drone. A male honey bee.

Drone cell. A special type of beeswax cell built to hold the immature stages of drones. Larger in diameter than a worker cell, it is also used for honey storage.

Eclosion. The emergence of the adult insect from the pupa.

Employed forager. A worker bee that is engaged in exploiting a patch of flowers.

Eukaryotes. Single-celled and multicellular organisms whose cells contain their genetic material inside a membrane-bound nucleus and that possess other membrane-bound cell organelles, such as mitochondria and chloroplasts. Includes all life above the level of bacteria and blue-green algae: protozoa, fungi, plants, and animals. (Contrast with prokaryotes.)

Feedback control. The control of an early stage of a multipart process in relation to the level of the product of a later stage. In negative feedback control, there is a negative correlation between the level of activity in the early stage and the level of product of the later stage.

Food-storer bee. A member of the labor group that is specialized to receive and store nectar and perform other tasks arising in the peripheral, food-storage region of the nest. Usually one of the middle-aged individuals in a honey bee colony.

Frame. In beekeeping, the wooden structure that surrounds and supports each beeswax comb inside a manmade beehive.

Functional organization. Organization that has been favored by natural selection, hence that contributes to the survival and reproduction of the cell, organism, or society in which the organization is found.

Gamete. The mature sexual reproductive cell: the egg or the sperm.

Genotype. The genetic constitution of an organism. (Contrast with phenotype.)

Germ cell. A sexual reproductive cell: an egg or sperm.

Grooming. The cleaning of the body surfaces of one's self or of nestmates by licking with the tongue and stroking with the legs.

Haplodiploidy. The mode of sex determination in which males arise from haploid (unfertilized) eggs and females from diploid (fertilized) eggs.

Haploid. Having a chromosome set consisting of just one copy of each chromosome, as in most gametes. (Contrast with diploid.)

Heterozygous. Referring to a diploid organism having different alleles of a given gene on both (homologous) chromosomes. (Contrast with homozygous.)

Homeostasis. The maintenance of a steady state, by means of self-regulation through mechanisms of negative feedback control.

Homozygous. Referring to a diploid organism having identical alleles of a given gene on both (homologous) chromosomes. (Contrast with heterozygous.)

Honey stomach. The expandable portion of the alimentary canal, located in the abdomen just before the midgut, that serves as a receptacle for nectar and water in honey bees.

Hymenoptera. The order of insects that comprises the wasps, bees, and ants.

Hypopharyngeal glands. Glandular structures located in the head of an adult worker bee that produce proteinaceous secretions which are fed to the brood and also various enzymes which serve in the conversion of nectar to honey.

Ideal free distribution. The distribution of individuals over the environment that arises if each individual freely chooses where to be based on an assessment of the relative attractiveness of all the alternative locations (territories, forage patches, and so on). More individuals will settle in the more attractive sites, and fewer in the less attractive sites. Ultimately, as the more attractive sites become crowded and individuals move into previously less attractive sites, all sites become equally attractive.

Instar. Any stage between molts (casting off of the outgrown skin) during the course of development in insects.

Interactor. Any discrete entity, such as an individual organism or a colony of bees, that houses replicators and functions as a unit to preserve and propagate the replicators inside it. (Synonymous with vehicle.)

Juvenile hormone. In insects in general, a hormone that maintains larval development. In honey bees in particular, this hormone also mediates changes in the physiology of the adults.

Larva. The stage between egg and pupa, found in certain insects, including the honey bee. An intensively feeding and growing stage.

Life cycle. The complete span of the life of an organism, from the moment of fertilization to the time it dies.

Mandibles. The jaws, or anterior pair of mouthpart structures, of an insect.

Meiosis. The cellular processes that lead to the formation of germ cells (gametes) for sexual reproduction. In particular, a diploid cell divides twice to form four daughter cells, but the original cell's chromosomes are replicated only once, so that each daughter cell receives only half the chromosomes present in the original cell. The fusion of two gametes restores the original chromosome number. See also Rules of Meiosis.

Metazoan organism. Any of the multicellular animals with the exception of the sponges.

Midgut. The middle portion of the alimentary tract of an insect, where the products of digestion are absorbed.

Mutation. A change in the genetic material: the source of the genetic variants among which natural selection chooses.

Nasanov's gland. A gland, located on the apical segment of a bee's abdomen, that secretes a pheromone that serves to attract other honey bees.

Natural selection. The differential survival and reproduction by individuals of different genetic types but belonging to the same genetic population. This is the main guiding force in evolution.

Nectar flow. A period of intense nectar secretion by plants during which a colony of bees is able to produce much honey. Also called a honeyflow.

Nurse bee. A member of the labor group that is specialized to care for the brood and perform other tasks arising in the central, broodnest region of a beehive. Usually one of the younger individuals in a honey bee colony.

Ommatidium. One of the basic visual units of the insect compound eye. Each ommatidium consists of a lens system and several light-sensitive cells, and so functions as a small eye, registering the intensity and the color of the light coming from a small portion of an insect's visual field.

Organelle. Any of the organized structures that are found in cells. Examples include mitochondria, chloroplasts, nuclei, ribosomes, and contractile vacuoles.

Organism. Any living animal or plant, unicellular or multicellular.

Patriline. The members of a social insect colony who share the same father.

Patrolling. The act of walking about and investigating the nest interior.

Phenotype. The observable properties of an individual as they have developed under the combined influences of the genetic constitution of the individual and the environmental factors it has experienced. (Contrast with genotype.)

Pheromone. A chemical substance that is used in communication between organisms of the same species.

Pollen basket. A smooth area, bordered on each side with a fringe of long curved hairs, on the outer surface of each hind leg of a worker bee. A foraging bee packs pollen here for transport back to the hive.

Polymorphism. In colonies of invertebrates, the coexistence of two or more morphologically and functionally distinct types of colony members of the same sex.

Proboscis. The extensible, tubular mouthparts of a bee.

Prokaryotes. Single-celled organisms whose genetic material is distributed throughout the cell, rather than being contained in a membrane-bound nucleus. Includes bacteria and blue-green algae. An earlier stage in the evolution of life than the eukaryotes, since mitochondria, chloroplasts, and certain other organelles in eukaryotic cells are, in origin, symbiotic prokaryotic cells. (Contrast with eukaryotes.)

Propolis. A collective term for the plant resins collected by bees and brought into their hive to seal cracks in the walls, reinforce the wax combs, and create a clean, smooth coating over the interior surfaces.

Pupa. The nonfeeding stage between the larva and adult in certain insects, including the honey bee, during which development to the final adult form is completed.

Queen. The reproductive female in a social insect colony.

Queen cell. A special type of beeswax cell built to house immature queen honey bees.

Queen excluder. A piece of beekeeping equipment; specifically, a screen whose apertures allow the passage of worker honey bees but not of the larger queen. It allows a beekeeper to confine a colony's queen to a certain region of the hive and thereby to segregate the brood and honey.

Queenright. Referring to a honey bee colony that contains a fully functioning queen.

Queen substance. The set of pheromones by which the queen honey bee attracts worker bees and signals her presence to them.

Recombination. The formation of new combinations of alleles on chromosomes during meiosis, through crossing-over (the reciprocal exchange of corresponding segments between two homologous chromosomes). The result is the production of almost infinite variety in the genetic constitution of gametes.

Recruit. A forager honey bee that locates a new food source by following one or more waggle dances of her nestmates rather than searching independently.

Relatedness. The fraction of genes identical between two individuals by virtue of common descent. Also known as the degree of relatedness.

Replicator. Any entity of which copies can be made, for example, genes and ideas.

Rules of meiosis. The processes during the formation of gametes which normally guarantee that the haploid gametes contain an unbiased sample of the genes in the diploid cells that divide to form the gametes. Each allele in an organism's genetic constitution has, therefore, a precisely 50% chance of finding itself in any particular gamete of the organism.

Scout. A forager honey bee that looks for a new food source by independent searching rather than by following the waggle dances of her nestmates.

Segregation. The separation of alleles, on homologous chromosomes, from one another during meiosis. The result is that each of the haploid daughter cells produced by meiosis contains one or the other member of each pair of alleles found in the diploid mother cell, never both.

Shaking signal. The signal whereby successful forager bees evidently stimulate nonforagers to move onto the dance floor and become foragers. Signal production consists basically of a bee's dorsoventral vibration of the body, especially the abdomen, while holding another bee. Also called the vibration dance and the dorso-ventral abdominal vibration (D-VAV).

Signal. An information-bearing action or structure that has been shaped by natural selection specifically to convey information.

Siphonophores. Colonial invertebrates in the order Siphonophora of the phylum Cnidaria. Their colonies are characterized by possessing at least two types of members (zooids)—gastrozooids for prey capture and digestion and gonozooids for reproduction—and by the ability to swim freely in the open sea.

Social physiology. The scientific study of the functioning of highly integrated societies and of the individual organisms of which they are composed.

Somatic cell. Any of the cells of the body except the germ cells.

Spermatheca. The sac-like structure in a female insect in which sperm are stored.

Swarming. The method of colony reproduction in honey bees in which the queen and approximately half the workers of a colony quickly leave the parental hive and fly to an exposed site nearby. There they cluster while scout bees search for a suitable nest cavity. Finally, they all fly to this nesting site and establish a new colony.

Tongue-lashing. The behavior in which a worker bee repeatedly extends and contracts her proboscis, thereby drawing a droplet of water or nectar expressed from the mouth into a thin film that can evaporate quickly. This activity is done to cool the hive interior and to ripen nectar into honey.

Tremble dance. The dance whereby worker bees serving as nectar foragers stimulate additional bees to function as food storers when the colony's rate of nectar collection has risen above its capacity for nectar processing. This dance also inhibits the production of waggle dances and hence prevents a further rise in the nectar influx.

Trophallaxis. The transfer of liquid substances from one member of an insect colony to another.

Unemployed forager. A worker bee that has previously foraged or is otherwise prepared to forage but is not actually engaged in exploiting a patch of flowers.

Unloading rejection. The experience of a nectar forager or water collector upon return to the hive when another bee places her tongue into the mouthparts of the forager and apparently tastes her load but then immediately withdraws her tongue and walks away.

Vehicle. Any discrete entity, such as an individual organism or a colony of bees, that houses replicators and functions as a unit to preserve and propagate the replicators inside it. (Synonymous with interactor.)

Waggle dance. The dance whereby honey bee workers communicate the location of desirable food sources.

Wax glands. The glands, located on the underside of a worker bee's abdomen, that secrete beeswax.

Worker. One of the nonreproductive females in a social insect colony.

Worker cells. The standard-sized cells built of beeswax, used to hold the immature stages of worker honey bees and the stores of honey and pollen in a honey bee colony.

Zooid. One of the asexually produced members of a marine invertebrate colony, such as a coral, bryozoan, or siphonophore colony.

Bibliography

Alberts, B., D. Bray, J. Lewis, M. Raff, K. Roberts, and J. D. Watson. 1983. *Molecular Biology of the Cell.* Garland: New York.

Alexander, R. D. 1974. The evolution of social behavior. *Annual Review of Ecology and Systematics* 5:325–383.

Allen, D. M. 1959. The "shaking" of worker honeybees by other workers. *Animal Behaviour* 7:233–240.

Al-Tikrity, W. S., A. W. Benton, R. C. Hillman, and W. W. Clarke, Jr. 1972. The relationship between the amount of unsealed brood in honeybee colonies and their pollen collection. *Journal of Apicultural Research* 11:9–12.

Avitabile, A. 1978. Brood rearing in honey bee colonies from late autumn to early spring. *Journal of Apicultural Research* 17:69–73.

Badertscher, S., C. Gerber, and R. H. Leuthold. 1983. Polyethism in food supply and processing in termite colonies of *Macrotermes subhyalinus* (Isoptera). *Behavioral Ecology and Sociobiology* 12:115–119.

Barker, R. J. 1971. The influence of food inside the hive on pollen collection by a honeybee colony. *Journal of Apicultural Research* 10:23–26.

Bartholdi, J. J., III, T. D. Seeley, C. A. Tovey, and J. H. Vande Vate. 1993. The pattern and effectiveness of forager allocation among flower patches by honey bee colonies. *Journal of Theoretical Biology* 160:23–40.

Baskett, F., and J. L. Hennessy. 1986. Small shared-memory multiprocessors. *Science* 231:963–966.

Bates, D. E. B., and N. H. Kirk. 1985. Graptolites, a fossil case-history of evolution from sessile, colonial animals to automobile superindividuals. *Proceedings of the Royal Society of London B* 228:207–224.

Bernard, C. 1865. *Introduction à l'étude de la médecine expérimentale.* Baillière: Paris.

———1927. *An Introduction to the Study of Experimental Medicine,* trans. H. C. Greene. Macmillan: New York. (Reprinted 1957; Dover: New York.)

Boch, R. 1956. Die Tänze der Bienen bei nahen und fernen Trachtquellen. *Zeitschrift für vergleichende Physiologie* 38:136–167.

Bonner, J. T. 1974. *On Development: The Biology of Form.* Harvard University Press: Cambridge.

Boulding, K. E. 1978. *Ecodynamics: A New Theory of Societal Evolution.* Sage: Beverly Hills.

Bowles, D. 1990. Signals in the wounded plant. *Nature* 343:314–315.

Breed, M. D., H. H. W. Velthuis, and G. E. Robinson. 1984. Do worker honey bees discriminate among unrelated and related larval phenotypes? *Annals of the Entomological Society of America* 77:737–739.

Bronowski, J. 1974. New concepts in the evolution of complexity. In *Philosophical Foundations of Science,* ed. R. J. Seeger and R. S. Cohen, pp. 133–151. Reidel: Dordrecht.

Buss, L. W. 1987. *The Evolution of Individuality.* Princeton University Press: Princeton.

Butler, C. G. 1945. The influence of various physical and biological factors of the environment on honeybee activity. An examination of the relationship between activity and nectar concentration and abundance. *Journal of Experimental Biology* 21:5–12.

Camazine, S. 1993. The regulation of pollen foraging by honey bees: how foragers assess the colony's need for pollen. *Behavioral Ecology and Sociobiology* 32:265–272.

Camazine, S., and J. Sneyd. 1991. A model of collective nectar source selection by honey bees: self-organization through simple rules. *Journal of Theoretical Biology* 149:547–571.

Chandler, A. D., Jr. 1977. *The Visible Hand: The Managerial Revolution in American Business.* Harvard University Press: Cambridge.

Crailsheim, K. 1986. Dependence of protein metabolism on age and season in the honeybee (*Apis mellifica carnica* Pollm.). *Journal of Insect Physiology* 32:629–634.

———1990. The protein balance of the honey bee worker. *Apidologie* 21: 417–429.

———1991. Interadult feeding of jelly in honeybee (*Apis mellifera* L.) colonies. *Journal of Comparative Physiology B* 161:55–60.

Crailsheim, K., L. H. W. Schneider, N. Hrassnigg, G. Bühlmann, U. Brosch, R. Gmeinbauer, and B. Schöffmann. 1992. Pollen consumption and utilization in worker honeybees *(Apis mellifera carnica):* dependence on individual age and function. *Journal of Insect Physiology* 38:409–419.

Crane, E. 1990. *Bees and Beekeeping: Science, Practice and World Resources.* Cornell University Press: Ithaca.

Crick, F. C. 1988. *What Mad Pursuit: A Personal View of Scientific Discovery.* Basic Books: New York.

Crozier, R. H., and R. E. Page. 1985. On being the right size: male contributions and multiple mating in social Hymenoptera. *Behavioral Ecology and Sociobiology* 18:105-115.

Dade, H. A. 1977. *Anatomy and Dissection of the Honeybee,* 2nd ed. International Bee Research Association: London.
Darwin, C. R. 1878. *The Effects of Cross and Self Fertilization in the Vegetable Kingdom,* 2nd ed. Murray: London. (Reprinted 1989; New York University Press: New York.)
Dawkins, M. S. 1986. *Unravelling Animal Behaviour.* Longman: Essex, Eng.
Dawkins, R. 1982. *The Extended Phenotype: The Gene as the Unit of Selection.* Oxford University Press: Oxford.
———1989. *The Selfish Gene: New Edition.* Oxford University Press: Oxford.
Denning, P. J., and W. F. Tichy. 1990. Highly parallel computation. *Science* 250:1217–1222.
Dreller, C., and W. H. Kirchner. 1993. Hearing in honeybees: localization of the auditory sense organ. *Journal of Comparative Physiology A* 173:275–279.
Dukas, R., and P. K. Visscher. 1994. Lifetime learning by foraging honey bees. *Animal Behaviour* 48:1007–1012.
Eberhard, W. G. 1980. Evolutionary consequences of intracellular organelle competition. *Quarterly Review of Biology* 55:231–249.
———1990. Evolution in bacterial plasmids and levels of selection. *Quarterly Review of Biology* 65:3–22.
Eckert, J. E. 1933. The flight range of the honeybee. *Journal of Agricultural Research* 47:257–285.
Erickson, E. H., Jr., S. D. Carlson, and M. B. Garment. 1986. *A Scanning Electron Microscope Atlas of the Honey Bee.* Iowa State University Press: Ames.
Esch, H. 1963. Über die Auswirkung der Futterplatzqualität auf die Schallerzeugung im Werbetanz der Honigbiene *(Apis mellifica). Zoologischer Anzeiger* 26(suppl.):302–309.
———1964. Beiträge zum Problem der Entfernungsweisung in den Schwänzeltänzen der Honigbienen. *Zeitschrift für vergleichende Physiologie* 48:534–546.
Fewell, J. H., and M. L. Winston. 1992. Colony state and regulation of pollen foraging in the honey bee, *Apis mellifera* L. *Behavioral Ecology and Sociobiology* 30:387–393.
Fewell, J. H., R. C. Ydenberg, and M. L. Winston. 1991. Individual foraging effort as a function of colony population in the honey bee, *Apis mellifera* L. *Animal Behaviour* 42:153–155.
Franks, N. R. 1989. Army ants: a collective intelligence. *American Scientist* 77:138–145.
Free, J. B. 1967. Factors determining the collection of pollen by honeybee foragers. *Animal Behaviour* 15:134–144.
Free, J. B., and I. H. Williams. 1971. The effect of giving pollen and pollen supplement to honeybee colonies on the amount of pollen collected. *Journal of Apicultural Research* 10:87–90.
Fretwell, S. D. 1972. *Populations in a Seasonal Environment.* Princeton University Press: Princeton.

Fretwell, S. D., and H. L. Lucas, Jr. 1970. On territorial behavior and other factors influencing habitat distribution in birds. *Acta Biotheoretica* 19:16–36.

Frisch, K. von. 1923. Über die "Sprache" der Bienen, eine tierpsychologische Untersuchung. *Zoologische Jahrbücher. Abteilung für allgemeine Zoologie und Physiologie der Tiere* 40:1–186.

———1940. Die Tänze und das Zeitgedächtnis der Bienen im Widerspruch. *Naturwissenschaften* 28:65–69.

———1967. *The Dance Language and Orientation of Bees.* Harvard University Press: Cambridge.

Frumhoff, P. C. 1991. The effects of the cordovan marker on apparent kin discrimination among nestmate honey bees. *Animal Behaviour* 42:854–856.

Gahl, R. A. 1975. The shaking dance of honey bee workers: evidence for age discrimination. *Animal Behaviour* 23:230–232.

Gary, N. E. 1971. Magnetic retrieval of ferrous labels in a capture-recapture system for honey bees and other insects. *Journal of Economic Entomology* 64:961–965.

Gary, N. E., P. C. Witherell, and K. Lorenzen. 1978. The distribution and foraging activities of common Italian and "Hy-Queen" honey bees during alfalfa pollination. *Environmental Entomology* 7:233–240.

Gelernter, D. 1989. The metamorphosis of information management. *Scientific American* 261(Aug.):66–73.

Getz, W. M. 1991. The honey bee as a model kin recognition system. In *Kin Recognition,* ed. P. G. Hepper, pp. 358–412. Cambridge University Press: Cambridge.

Gould, J. L. 1976. The dance-language controversy. *Quarterly Review of Biology* 51:211–244.

Gould, J. L., and C. G. Gould. 1988. *The Honey Bee.* Freeman: New York.

Gould, J. L., M. Henerey, and M. C. MacLeod. 1970. Communication of direction by the honey bee. *Science* 169:544–554.

Grassé, P.-P. 1959. La reconstruction du nid et les coordinations interindividuelles chez *Bellicositermes natalensis* et *Cubitermes sp.* La théorie de la stigmergie: essai d'interprétation du comportement des termites constructeurs. *Insectes Sociaux* 6:41–83.

Grodins, F. S. 1963. *Control Theory and Biological Systems.* Columbia University Press: New York.

Hamilton, W. J., III, and K. E. F. Watt. 1970. Refuging. *Annual Review of Ecology and Systematics* 1:263–286.

Harvell, C. D. 1994. The evolution of polymorphism in colonial invertebrates and social insects. *Quarterly Review of Biology* 69:155–185.

Hayek, F. A. 1945. The use of knowledge in society. *The American Economic Review* 35:519–530.

Heinrich, B. 1980. Mechanisms of body-temperature regulation in honeybees, *Apis mellifera. Journal of Experimental Biology* 85:61–87.

———1985. The social physiology of temperature regulation in honeybees. In *Experimental Behavioral Ecology and Sociobiology,* ed. B. Hölldobler and M. Lindauer, pp. 393–406. Sinauer: Sunderland, Mass.

———1993. *The Hot-Blooded Insects: Strategies and Mechanisms of Thermoregulation.* Harvard University Press: Cambridge.
Hellmich, R. L., and W. C. Rothenbuhler. 1986. Relationship between different amounts of brood and the collection and use of pollen by the honey bee *(Apis mellifera). Apidologie* 17:13–20.
Hepburn, H. R. 1986. *Honeybees and Wax: An Experimental Natural History.* Springer-Verlag: Berlin.
Hillis, W. D. 1986. *The Connection Machine.* MIT Press: Cambridge.
Horrobin, D. F. 1964. *The Communication Systems of the Body.* Basic Books: New York.
———1970. *Principles of Biological Control.* Medical and Technical Publications: Aylesbury, Eng.
Houston, A., P. Schmid-Hempel, and A. Kacelnik. 1988. Foraging strategy, worker mortality, and the growth of the colony in social insects. *American Naturalist* 131:107–114.
Hull, D. L. 1980. Individuality and selection. *Annual Review of Ecology and Systematics* 11:311–332.
———1988. Interactors versus vehicles. In *The Role of Behavior in Evolution,* ed. H. C. Plotkin, pp. 19–50. MIT Press: Cambridge.
Jeffree, E. P. 1955. Observations on the decline and growth of honey bee colonies. *Journal of Economic Entomology* 48:723–726.
Jeffree, E. P., and M. D. Allen. 1957. The annual cycle of pollen storage by honey bees. *Journal of Economic Entomology* 50:211–212.
Jones, R. W. 1973. *Principles of Biological Regulation: An Introduction to Feedback Systems.* Academic Press: New York.
Keller, L., and P. Nonacs. 1992. The role of queen pheromones in social insects: queen control or queen signal? *Animal Behaviour* 45:787–794.
Kelley, S. 1991. The regulation of comb building in honey bee colonies. Senior honors thesis, Division of Biological Sciences, Cornell University.
Kiechle, H. 1961. Die soziale Regulation der Wassersammeltätigkeit im Bienenstaat und deren physiologische Grundlage. *Zeitschrift für vergleichende Physiologie* 45:154–192.
Killion, E. E. 1992. The production of comb and bulk comb honey. In *The Hive and the Honey Bee,* ed. J. M. Graham, pp. 705–722. Dadant: Hamilton, Ill.
Kirchner, W. H. 1993. Vibrational signals in the tremble dance of the honeybee, *Apis mellifera. Behavioral Ecology and Sociobiology* 33:169–172.
Kirchner, W. H., C. Dreller, and W. F. Towne. 1991. Hearing in honeybees: operant conditioning and spontaneous reactions to airborne sound. *Journal of Comparative Physiology A* 168:85–89.
Kirchner, W. H., and M. Lindauer. The causes of the tremble dance of the honeybee, *Apis mellifera. Behavioral Ecology and Sociobiology* 35:303–308.
Kirchner, W. H., M. Lindauer, and A. Michelsen. 1988. Honeybee dance communication: acoustical indication of direction in round dances. *Naturwissenschaften* 75:629–630.
Knaffl, H. 1953. Über die Flugweite und Entfernungsmeldung der Bienen. *Zeitschrift für Bienenforschung* 2:131–140.

Körner, I. 1939. Zeitgedächtnis und Alarmierung bei den Bienen. *Zeitschrift für vergleichende Physiologie* 27:445–459.

Leigh, E. G., Jr. 1991. Genes, bees and ecosystems: the evolution of a common interest among individuals. *Trends in Ecology and Evolution* 6:257–262.

Lensky, Y. 1964. Comportement d'une colonie d'abeilles à des températures extrêmes. *Journal of Insect Physiology* 10:1–12.

Levin, M. D. 1961. Distribution of foragers from honey bee colonies placed in the middle of a large field of alfalfa. *Journal of Economic Entomology* 54:431–434.

Lindauer, M. 1948. Über die Einwirkung von Duft- und Geschmacksstoffen sowie anderer Faktoren auf die Tänze der Bienen. *Zeitschrift für vergleichende Physiologie* 31:348–412.

———1952. Ein Beitrag zur Frage der Arbeitsteilung im Bienenstaat. *Zeitschrift für vergleichende Physiologie* 34:299–345.

———1954. Temperaturregulierung und Wasserhaushalt im Bienenstaat. *Zeitschrift für vergleichende Physiologie* 36:391–432.

———1961. *Communication among Social Bees.* Harvard University Press: Cambridge.

———1971. *Communication among Social Bees,* 2nd ed. Harvard University Press: Cambridge.

———1975. *Verständigung im Bienenstaat.* Fischer: Stuttgart.

Livingstone, M., and D. Hubel. 1988. Segregation of form, color, movement, and depth: anatomy, physiology, and perception. *Science* 240:740–749.

Lloyd, J. E. 1983. Bioluminescence and communication in insects. *Annual Review of Entomology* 28:131–160.

Mackie, G. O. 1963. Siphonophores, bud colonies, and superorganisms. In *The Lower Metazoa: Comparative Biology and Phylogeny,* ed. E. C. Dougherty, pp. 329–337. University of California Press: Berkeley.

———1984. Fast pathways and escape behavior in Cnidaria. In *Neural Basis of Startle Behavior,* ed. R. C. Eaton, pp. 15–42. Plenum: New York.

———1986. From aggregates to integrates: physiological aspects of modularity in colonial animals. *Philosophical Transactions of the Royal Society of London B* 313:175–196.

Maynard Smith, J. 1986. *The Problems of Biology.* Oxford University Press: Oxford.

———1988. Evolutionary progress and levels of selection. In *Evolutionary Progress,* ed. M. H. Nitecki, pp. 219–230. University of Chicago Press: Chicago.

Michener, C. D. 1974. *The Social Behavior of the Bees.* Harvard University Press: Cambridge.

Miller, J. G. 1978. *Living Systems.* McGraw-Hill: New York.

Milum, V. G. 1955. Honey bee communication. *American Bee Journal* 95:97–104.

Moritz, B., and K. Crailsheim. 1987. Physiology of protein digestion in the midgut of the honeybee (*Apis mellifera* L.). *Journal of Insect Physiology* 33:923–931.

Moritz, R. F. A., and E. E. Southwick. 1992. *Bees as Superorganisms: An Evolutionary Reality.* Springer-Verlag: Berlin.

Muller, W. J., and H. R. Hepburn. 1992. Temporal and spatial patterns of wax secretion and related behavior in the division of labour of the honeybee *(Apis mellifera capensis). Journal of Comparative Physiology A* 171:111–115.

Naumann, K., M. L. Winston, K. N. Slessor, G. D. Prestwich, and F. X. Webster. 1991. Production and transmission of honey bee queen (*Apis mellifera* L.) mandibular gland pheromone. *Behavioral Ecology and Sociobiology* 29:321–332.

Neukirch, A. 1982. Dependence of the life span of the honeybee *(Apis mellifica)* upon flight performance and energy consumption. *Journal of Comparative Physiology* 146:35–40.

Nieh, J. C. 1993. The stop signal of honey bees: reconsidering its message. *Behavioral Ecology and Sociobiology* 33:51–56.

Noonan, K. C. 1986. Recognition of queen larvae by worker honey bees *(Apis mellifera). Ethology* 73:295–306.

Núñez, J. A. 1966. Quantitative Beziehungen zwischen den Eigenschaften von Futterquellen und dem Verhalten von Sammelbienen. *Zeitschrift für vergleichende Physiologie* 53:142–164.

———1970. The relationship between sugar flow and foraging and recruiting behaviour of honey bees (*Apis mellifera* L.). *Animal Behaviour* 18:527–538.

———1982. Honeybee foraging strategies at a food source in relation to its distance from the hive and the rate of sugar flow. *Journal of Apicultural Research* 21:139–150.

Oldroyd, B. P., T. E. Rinderer, and S. M. Buco. 1990. Nepotism in the honey bee. *Nature* 346:707–708.

———1991. Honey bees dance with their super-sisters. *Animal Behaviour* 42:121–129.

Oldroyd, B. P., T. E. Rinderer, S. M. Buco, and L. D. Beaman. 1993. Genetic variance in honey bees for preferred foraging distance. *Animal Behaviour* 45:323–332.

Oster, G. F., and E. O. Wilson. 1978. *Caste and Ecology in the Social Insects.* Princeton University Press: Princeton.

Otte, D. 1974. Effects and functions in the evolution of signaling systems. *Annual Review of Ecology and Systematics* 5:385–417.

Page, R. E., Jr. 1986. Sperm utilization in social insects. *Annual Review of Entomology* 31:297–320.

Page, R. E., Jr., and M. D. Breed. 1987. Kin recognition in social bees. *Trends in Ecology and Evolution* 2:272–275.

Page, R. E., Jr., and E. H. Erickson, Jr. 1984. Selective rearing of queens by worker honey bees: kin or nestmate recognition. *Annals of the Entomological Society of America* 77:578–580.

———1988. Reproduction by worker honey bees (*Apis mellifera* L.). *Behavioral Ecology and Sociobiology* 23:117–126.

Page, R. E., Jr., and G. E. Robinson. 1991. The genetics of division of labour in honey bee colonies. *Advances in Insect Physiology* 23:117–169.

Page, R. E., Jr., G. E. Robinson, and M. K. Fondrk. 1989. Genetic specialists, kin recognition and nepotism in honey-bee colonies. *Nature* 338:576–579.

Park, O. W. 1923. Water stored by bees. *American Bee Journal* 63:348–349.

———1925. The storing and ripening of honey by honeybees. *Journal of Economic Entomology* 18:405–410.

———1929. Time factors in relation to the acquisition of food by the honeybee. *Research Bulletin of the Iowa Agricultural Experiment Station* 108:184–226.

Ratnieks, F. L. W. 1988. Reproductive harmony via mutual policing by workers in eusocial Hymenoptera. *American Naturalist* 132:217–236.

———1993. Egg-laying, egg-removal, and ovary development by workers in queenright honey bee colonies. *Behavioral Ecology and Sociobiology* 32:191–198.

——— The role of the queen Dufour gland in egg removal in the honey bee. *Journal of Apicultural Research* (in press).

Ratnieks, F. L. W., and H. K. Reeve. 1992. Conflict in single-queen hymenopteran societies: the structure of conflict and processes that reduce conflict in advanced eusocial species. *Journal of Theoretical Biology* 158:33–65.

Ratnieks, F. L. W., and P. K. Visscher. 1989. Worker policing in the honeybee. *Nature* 342:796–798.

Raveret Richter, M., and K. D. Waddington. 1993. Past foraging experience influences honey bee dance behaviour. *Animal Behaviour* 46:123–128.

Rhein, W. von. 1951. Über die Ernährung der Drohnenmaden. *Zeitschrift für Bienenforschung* 1:63–66.

Ribbands, C. R. 1952. Division of labour in the honeybee community. *Proceedings of the Royal Society of London B* 140:32–42.

———1953. *The Behaviour and Social Life of Honeybees.* Bee Research Association: London. (Reprinted 1964; Dover: New York.)

Rinderer, T. E. 1982a. Regulated nectar harvesting by the honeybee. *Journal of Apicultural Research* 21:74–87.

———1982b. Volatiles from empty comb increase hoarding by the honey bee. *Animal Behaviour* 29:1275–1276.

———1983. Regulation of honey bee hoarding efficiency. *Apidologie* 14:87–92.

Rinderer, T. E., and J. R. Baxter. 1978. Effect of empty comb on hoarding behavior and honey production of the honey bee. *Journal of Economic Entomology* 71:757–759.

———1979. Honey bee hoarding behaviour: effects of previous stimulation by empty comb. *Animal Behaviour* 27:426–428.

———1980. Hoarding behavior of the honey bee: effects of empty comb, comb color, and genotype. *Environmental Entomology* 9:104–105.

Robinson, G. E. 1992. Regulation of division of labor in insect societies. *Annual Review of Entomology* 37:637–665.

Robinson, G. E., and R. E. Page, Jr. 1989. Genetic determination of nectar foraging, pollen foraging, and nest-site scouting in honey bee colonies. *Behavioral Ecology and Sociobiology* 24:317–323.

Robinson, G. E., R. E. Page, Jr., C. Strambi, and A. Strambi. 1989. Hormonal and genetic control of behavioral integration in honey bee colonies. *Science* 246:109–112.

Robinson, G. E., B. A. Underwood, and C. E. Henderson. 1984. A highly specialized water-collecting honey bee. *Apidologie* 15:355–358.

Rood, J. P. 1983. The social system of the dwarf mongoose. In *Advances in the Study of Mammalian Behavior,* ed. J. F. Eisenberg and D. G. Kleiman, pp. 454–488. American Society of Mammalogists: Shippensburg, Pa.

Rösch, G. A. 1927. Über die Bautätigkeit im Bienenvolk und das Alter der Baubienen. Weiterer Beitrag zur Frage nach der Arbeitsteilung im Bienenstaat. *Zeitschrift für vergleichende Physiologie* 6:264–298.

Roubik, D. W. 1989. *Ecology and Natural History of Tropical Bees.* Cambridge University Press: New York.

Ruttner, F. 1988. *Biogeography and Taxonomy of Honeybees.* Springer-Verlag: Berlin.

Samuelson, P. A. 1973. *Economics,* 9th ed. McGraw-Hill: New York.

Schick, W. 1953. Über die Wirkung von Giftstoffen auf die Tänze der Bienen. *Zeitschrift für vergleichende Physiologie* 35:105–128.

Schmid, J. 1964. Zur Frage der Störung des Bienengedächtnisses durch Narkosemittel, zugleich ein Beitrag zur Störung der sozialen Bindung durch Narkose. *Zeitschrift für vergleichende Physiologie* 47:559–595.

Schmid-Hempel, P., and T. Wolf. 1988. Foraging effort and life span of workers in a social insect. *Journal of Animal Ecology* 57:500–521.

Schmid-Hempel, P., A. Kacelnik, and A. I. Houston. 1985. Honeybees maximize efficiency by not filling their crop. *Behavioral Ecology and Sociobiology* 17:61–66.

Schmid-Hempel, P., M. L. Winston, and R. C. Ydenberg. 1993. Foraging of individual workers in relation to colony state in the social Hymenoptera. *Canadian Entomologist* 125:129–160.

Schneider, S. S., J. A. Stamps, and N. E. Gary. 1986a. The vibration dance of the honey bee. I. Communication regulating foraging on two time scales. *Animal Behaviour* 34:377–385.

———1986b. The vibration dance of the honey bee. II. The effects of foraging success on daily patterns of vibration activity. *Animal Behaviour* 34:386–391.

Schoener, T. W. 1971. Theory of foraging strategies. *Annual Review of Ecology and Systematics* 2:369–404.

Seeley, T. D. 1979. Queen substance dispersal by messenger workers in honeybee colonies. *Behavioral Ecology and Sociobiology* 5:391–415.

———1982. Adaptive significance of the age polyethism schedule in honeybee colonies. *Behavioral Ecology and Sociobiology* 11:287–293.

———1983. Division of labor between scouts and recruits in honeybee foraging. *Behavioral Ecology and Sociobiology* 12:253–259.
———1985. *Honeybee Ecology: A Study of Adaptation in Social Life.* Princeton University Press: Princeton.
———1986. Social foraging by honeybees: how colonies allocate foragers among patches of flowers. *Behavioral Ecology and Sociobiology* 19:343–354.
———1987. The effectiveness of information collection about food sources by honey bee colonies. *Animal Behaviour* 35:1572–1575.
———1989a. Social foraging in honey bees: how nectar foragers assess their colony's nutritional status. *Behavioral Ecology and Sociobiology* 24:181–199.
———1989b. The honey bee colony as a superorganism. *American Scientist* 77:546–553.
———1992. The tremble dance of the honey bee: message and meanings. *Behavioral Ecology and Sociobiology* 31:375–383.
———1994. Honey bee foragers as sensory units of their colonies. *Behavioral Ecology and Sociobiology* 34:51–62.
Seeley, T. D., and R. A. Morse. 1976. The nest of the honey bee (*Apis mellifera* L.). *Insectes Sociaux* 23:495–512.
Seeley, T. D., and C. A. Tovey. 1994. Why search time to find a food-storer bee accurately indicates the relative rates of nectar collecting and nectar processing in honey bee colonies. *Animal Behaviour* 47:311–316.
Seeley, T. D., and W. F. Towne. 1992. Tactics of dance choice in honey bees: do foragers compare dances? *Behavioral Ecology and Sociobiology* 30:59–69.
Seeley, T. D., and P. K. Visscher. 1988. Assessing the benefits of cooperation in honeybee foraging: search costs, forage quality, and competitive ability. *Behavioral Ecology and Sociobiology* 22:229–237.
Seeley, T. D., S. Camazine, and J. Sneyd. 1991. Collective decision-making in honey bees: how colonies choose among nectar sources. *Behavioral Ecology and Sociobiology* 28:277–290.
Sendova-Franks, A., and N. R. Franks. 1993. Task allocation in ant colonies within variable environments (a study of temporal polyethism: experimental). *Bulletin of Mathematical Biology* 55:75–96.
Sherman, P. W., J. U. M. Jarvis, and R. D. Alexander. 1991. *The Biology of the Naked Mole-Rat.* Princeton University Press: Princeton.
Sherman, P. W., T. D. Seeley, and H. K. Reeve. 1988. Parasites, pathogens, and polyandry in social Hymenoptera. *American Naturalist* 131:602–610.
Shuel, R. W. 1992. The production of nectar and pollen. In *The Hive and the Honey Bee,* ed. J. M. Graham, pp. 401–436. Dadant: Hamilton, Ill.
Simon, H. A. 1962. The architecture of complexity. *Proceedings of the American Philosophical Society* 106:467–482.
———1976. *Administrative Behavior: A Study of Decision-making Processes in Administrative Organization,* 3rd ed. Free Press: New York.
———1981. *The Sciences of the Artificial,* 2nd ed. MIT Press: Cambridge.
Snodgrass, R. E. 1956. *Anatomy of the Honey Bee.* Cornell University Press: Ithaca.

Snyder, S. H. 1985. The molecular basis of communication between cells. *Scientific American* 253(Oct.):132–141.

Southwick, E. E. 1983. Nectar biology and nectar feeders of common milkweed, *Asclepias syriaca* L. *Bulletin of the Torrey Botanical Club* 110:324–334.

Stabentheiner, A., and K. Hagmüller. 1991. Sweet food means "hot dancing" in honeybees. *Naturwissenschaften* 78:471–473.

Starr, C. K. 1984. Sperm competition, kinship, and sociality in aculeate Hymenoptera. In *Sperm Competition and the Evolution of Animal Mating Systems,* ed. R. L. Smith, pp. 428–459. Academic Press: Orlando.

Stephens, D. W., and J. R. Krebs. 1986. *Foraging Theory.* Princeton University Press: Princeton.

Todd, F. E., and C. B. Reed. 1970. Brood measurement as a valid index to the value of honey bees as pollinators. *Journal of Economic Entomology* 63:148–149.

Tofts, C. 1993. Algorithms for task allocation in ants (a study of temporal polyethism: theory). *Bulletin of Mathematical Biology* 55:891–918.

Tofts, C., and N. R. Franks. 1992. Doing the right thing: ants, honeybees, and naked mole-rats. *Trends in Ecology and Evolution* 7:346–349.

Turrell, M. J. 1972. An investigation of the factors influencing the development of the wax glands in the honey bee, *Apis mellifera* L. M.Sc. thesis, Entomology Department, Cornell University.

Valentine, J. W. 1978. The evolution of multicellular plants and animals. *Scientific American* 239(Aug.):66–78.

Van Essen, D. C., and J. H. R. Maunsell. 1983. Hierarchical organization and functional streams in the visual cortex. *Trends in Neuroscience* 6:370–375.

Visscher, P. K. 1982. Foraging strategy of honey bee colonies in a temperate deciduous forest. M.Sc. thesis, Entomology Department, Cornell University.

———1986. Kinship discrimination in queen rearing by honey bees *(Apis mellifera). Behavioral Ecology and Sociobiology* 18:453–460.

———1989. A quantitative study of worker reproduction in honey bee colonies. *Behavioral Ecology and Sociobiology* 25:247–254.

———1995a. Reproductive conflict in honey bees: a stalemate of worker egg laying and policing. Unpublished manuscript.

———1995b. Nepotism in queen rearing by honey bees: a reexamination of evidence. Unpublished manuscript.

Visscher, P. K., and R. Dukas. 1995. Honey bees recognize development of nestmates' ovaries. *Animal Behaviour* 49:542–544.

Visscher, P. K., and T. D. Seeley. 1982. Foraging strategy of honeybee colonies in a temperate deciduous forest. *Ecology* 63:1790–1801.

Waddington, K. D. 1982. Honey bee foraging profitability and round dance correlates. *Journal of Comparative Physiology* 148:297–301.

———1985. Cost-intake information used in foraging. *Journal of Insect Physiology* 31:891–897.

———1990. Foraging profits and thoracic temperature of honey bees. *Journal of Comparative Physiology B* 160:325–329.

Waddington, K. D., and W. H. Kirchner. 1992. Acoustical and behavioral correlates of profitability of food sources in honey bee round dances. *Ethology* 92:1–6.

Waldbauer, G. P., and S. Friedman. 1991. Self-selection of optimal diets by insects. *Annual Review of Entomology* 36:43–63.

Weaver, N. 1979. Possible recruitment of foraging honeybees to high-reward areas of the same plant species. *Journal of Apicultural Research* 18:179–183.

Weaver, W. 1948. Science and complexity. *American Scientist* 36:536–544.

Wehner, R., and M. V. Srinivasan. 1984. The world as the insect sees it. In *Insect Communication*, ed. T. Lewis, pp. 29–47. Academic Press: London.

Wenner, A. M., P. H. Wells, and F. J. Rohlf. 1967. An analysis of the waggle dance and recruitment in honey bees. *Physiological Zoology* 40:317–344.

Werren, J. H., U. Nur, and C. Wu. 1988. Selfish genetic elements. *Trends in Ecology and Evolution* 3:297–302.

Wiener, N. 1961. *Cybernetics; or Control and Communication in the Animal and the Machine*, 2nd ed. MIT Press: Cambridge.

Williams, G. C. 1992. *Natural Selection: Domains, Levels, and Challenges.* Oxford University Press: New York.

Willis, L. G., M. L. Winston, and K. N. Slessor. 1990. Queen honey bee mandibular pheromone does not affect worker ovary development. *Canadian Entomologist* 122:1093–1099.

Wilson, D. S., and E. Sober. 1989. Reviving the superorganism. *Journal of Theoretical Biology* 136:336–356.

Wilson, E. O. 1962. Chemical communication among workers of the fire ant *Solenopsis saevissima* (Fr. Smith): 1. The organization of mass-foraging. *Animal Behaviour* 10:134–147.

———1971. *The Insect Societies.* Harvard University Press: Cambridge.

———1975. *Sociobiology: The New Synthesis.* Harvard University Press: Cambridge.

———1976. Behavioral discretization and the number of castes in an ant species. *Behavioral Ecology and Sociobiology* 1:141–154.

Winston, M. L. 1987. *The Biology of the Honey Bee.* Harvard University Press: Cambridge.

Winston, M. L., and K. N. Slessor. 1992. The essence of royalty: honey bee queen pheromone. *American Scientist* 80:374–385.

Winston, W. L. 1991. *Operations Research: Applications and Algorithms*, 2nd ed. PWS-Kent: Boston.

Wolf, T. J., and P. Schmid-Hempel. 1989. Extra loads and foraging life span in honeybee workers. *Journal of Animal Ecology* 58:943–954.

———1990. On the integration of individual foraging strategies with colony ergonomics in social insects: nectar collection in honeybees. *Behavioral Ecology and Sociobiology* 27:103–111.

Wolf, T. J., P. Schmid-Hempel, C. P. Ellington, and R. D. Stevenson. 1989. Physiological correlates of foraging efforts in honey-bees: oxygen consumption and nectar load. *Functional Ecology* 3:417–424.

Wright, R. 1988. *Three Scientists and Their Gods: Looking for Meaning in an Age of Information.* Times Books: New York.

Woyciechowski, M. 1990. Do honey bee, *Apis mellifera* L., workers favour sibling eggs and larvae in queen rearing? *Animal Behaviour* 39:1220–1222.

Woyciechowski, M., and A. Lomnicki. 1987. Multiple mating of queens and the sterility of workers among eusocial Hymenoptera. *Journal of Theoretical Biology* 128:317–327.

———1989. Worker reproduction in social insects. *Trends in Ecology and Evolution* 4:146.

Young, D. 1989. *Nerve Cells and Animal Behaviour.* Cambridge University Press: Cambridge.

Index

abandonment: of food sources, 133–136, 259; of tasks, 195–198, 204, 209
adaptation in the dance response, 99–101
age polyethism, 29–31, 155, 177–181, 202–203, 231, 240–244. *See also* division of labor
allocation: of bees among tasks, 167–168, 173–174, 241–244; of foragers among flower patches, 54–59, 134–142, 145–151, 258–261; of foragers between nectar and pollen, 195–196, 207–209
annual cycle of colonies, 34–36, 43–44
ants, 5, 240, 250
Apis mellifera: *carnica*, 75–76; *caucasica*, 75; *ligustica*, 75–76; *mellifera*, 75
ascidians, 244
auditory communication. *See* sound communication

beekeeping, 22, 60, 62, 77, 108, 157
beeswax. *See* wax
Bernard, Claude, 21
Bronowski, Jacob, 4
brood care. *See* nurse bees
bryozoans, 244

capturing bees, 78
caste. *See* division of labor
cells: drone, 33–34, 177–178; queen, 8, 34; worker, 33–34
central planning, 258–262, 265
chemical communication, 11–12, 13, 25–26, 245
choosing among flower patches, 54–59, 134–142, 145–151
cleaner bees, 29–31, 242, 246
coding information in waggle dances, 37–39, 90–94, 98–102
colonial marine invertebrates, 5, 240, 244
colony: economics, 39–45, 193; foundation, 34–36; life cycle, 34–36
colony population, measurement of, 83
comb: builders of, 177–181; construction of, 61–63; drone, 33–34, 177–178; odor of, 190; significance of empty, 62–63, 73, 181–190
communication: economy of, 252–255; honesty in, 259–260; of the location of food sources, 36–39, 98–102; of the need for more food storers, 163–173, 242–245; of the need for more foragers, 158–162, 242–245; of the need for pollen, 198–207; of the need for water, 220–226; of the presence of the queen, 11, 245. *See also* cues; pheromones; signals
complexity, 17–19
computers, multiprocessor, 251, 254
computer simulations, 17, 21
conflict: between colonies, 60; within colonies, 7, 11–13, 36
control: hierarchy of, 258–262; system of feedback, 255–258
cooling of the hive. *See* thermoregulation
coordination problems, 155–156, 213–214, 239, 249, 258–262

corals, 244
Crick, Francis C., 17, 19
cross inhibition, 114–115, 142–145, 227–234
cues, 184, 220, 224–226, 248–252

dance duration: definition of, 82; significance of, 90–92; variation in, 91–92, 98–102, 132–134
dance floor, 40–49, 88–89, 122–123, 258–259
dance response, adaptation in, 99–101
dances, communicative. *See* round dance; tremble dance; waggle dance
dance threshold. *See* threshold for waggle dances
Darwin, Charles R., 53
data collection techniques, 81–83
Dawkins, Richard, 4, 13, 263
decentralized control, 258–262
decision making: at the colony level, 54–66, 134–142; at the individual level, 88, 94–99, 117–119, 196–198, 226–227
delivery time: definition of, 82; of nectar, 107–108; of water, 222–225
digestive tract, 25–26, 202
division of labor: among foragers, 85–87; among workers in general, 29–31, 155–156, 231, 240–244; permanent vs. temporary, 240–241
dorso-ventral abdominal vibration (D-VAV). *See* shaking signal
drone comb. *See* comb
drones, 10–11, 14

economic systems of humans, 258–262
economy of the colony. *See* colony
employed foragers. *See* forager bees
empty comb. *See* comb
evolution of biological organization, 3–7, 239, 263–265
excitation processes, 144, 172, 188–190, 202–204, 220–221, 255–258. *See also* shaking signal;, tremble dance; waggle dance

fanning, 26, 30, 212, 250
feedback processes, 198–207, 255–258
feeders, sugar water, 77–79
filtering of information, 88
flight muscles: as a heat source, 26, 43, 250; as a sound source, 26, 37, 166
flower patches, allocation of foragers among, 54–59, 84–154, 258–261
flowers: availability of, 43–45, 61; foragers' search for, 50–52; scent of, 36, 39, 108, 252
food: adjustment in the bees' selectivity among sources of, 59–61, 104–113; bees' search for, 50–52; processing of, 39–41; requirements of the colony, 42–43; scent of, 39; types of, 39–41. *See also* nectar; pollen
food-storer bees, 26, 29–31, 40–42, 82, 107–113, 155–156, 162–174, 231–234, 246
forager bees: age of, 26, 30–31; allocation among flower patches, 54–59, 134–142, 145–151, 258–261; allocation between nectar and pollen, 195–196, 207–209; employed vs. unemployed, 85, 122–124, 134, 207–209, 258–261; knowledge possessed by, 85, 106–110, 122–132; novice, 86, 113, 124; number of in the colony, 85, 87, 156–162; as sensory units, 98–104
foraging: behavior of workers, 87–88, 132–134; colonial patterns of, 47–50, 54–59, 63–65; colony's range of, 47–50; tempo, 117, 133–134, 197–198; theory of, 94–98; trip time, 95, 156–157, 195–198. *See also* colony economics
foraging for work, hypothesis of, 242–244
foraging strategy: of the colony, 46, 50, 60, 102, 149–151; of the individual bee, 94–98, 187, 260–261

genetic variation among workers, effects of, 7, 31, 219
glands: Dufour, 13; hypopharyngeal, 25–26, 29, 39, 202–204; Nasanov's, 25–26, 39; poison, 24–25; salivary, 25–26, 41; wax, 25–26, 179–181

Hayek, Friedrich A., 261
heating the hive. *See* thermoregulation
hoarding assay, 190
homeostasis, 194–195, 205, 212
honey: consumption of by the colony, 43; making of, 26, 30, 41; storage of, 33–34, 41, 214–215
honey stomach, 25, 184–187, 213, 220–221

hormones, 243, 245, 247
Hull, David L., 4, 263
human society, 258–262
hunger, for protein, 206–207, 209, 248, 261
hypopharyngeal glands, 25–26, 29, 39, 202–204

ideal free distribution, 151
inactive workers, importance of, 173–174
individual variation in behavior, 98–102, 219–220
information: acquisition by colonies, 85–121; acquisition by individuals, 94–98, 107–110, 113–119, 122–132, 181–187, 206–207, 220–226, 246–247, 258; flow of, 247–252, 262 (*see also* communication); flow via the shared environment, 250–252; integration of, about conditions inside the hive, 223–224, 246–247; integration of, about the properties of a food source, 94–98, 100–101, 118, 252, 265; integration of, about foraging conditions, 119, 258
inhibition processes, 142–145, 172, 188–190, 195–196, 202–204, 216–217, 221–226, 255–258
interactor concept, 4, 263
inventory policy 63–64, 193

jerking dance. *See* shaking signal
juvenile hormone, 243

kin recognition, 14–16, 130–131

labeling, of bees, 18, 47, 73, 79–80
labor reserves. *See* inactive workers
levels of analysis, 16–19
levels of selection, 4–5

market economies, 258–262
mathematical modeling, 17, 20–21, 111–113, 137–142, 145–151
meiosis, rules of, 14
messenger bees, 11, 245
multicellular organisms, 4, 14, 240, 244–247, 255–256

Nasanov's gland, 25–26
natural selection, 4, 17–18, 136, 142, 248–249, 252, 258–260, 263–264
nectar: composition of, 40, 107, 217; consumption of, 43; foraging for, 94–98; load size for a worker, 42, 134, 156–157; processing of, 26; storage of, 41, 185–187
nectar collection rate of the colony: acquisition of information about, 107–110; adjustment of, 52–54, 156–162, 256; daily variation in, 44–45, 61; importance of, 62–63, 66, 104–106
nectar flow, 43–45, 103–104, 156, 183
nectar processing rate of the colony, adjustment of, 113, 162–174, 256
nectar-source profitability: assessment of, 94–98, 113–119; coding of in waggle dances, 90–94, 98–102; criterion for, 94–98; importance of, 90–102; variation in, 91–92
negative feedback, 198–207, 255–258
nepotism, 14–16
nervous system, 114, 119, 243, 244–245, 247, 255–256, 262
nest: architecture of, 31–34; cleaning of, 29–31; construction of, 61–63; site selection for, 34; thermoregulation of, 43, 65–66, 212–213, 215–218, 250
nurse bees, 26, 29–31, 41, 173, 201–205, 213, 221, 228–231, 242–243, 250

observation hive, 7, 18, 48–49, 71–75
odor. *See* flowers, scent of; pheromones
optimal foraging theory, 94–98, 149–152
ovarian development, in workers, 9, 13

paint, for labeling bees, 79–80
patriline recognition. *See* kin recognition
patrolling, 29, 246–247
pheromones: alarm, 245; brood, 11; queen, 11–13, 245, 258
poison gland, 24–25
pollen: composition of, 39; consumption of, 43; foraging for, 24; load size for a worker, 42, 196–198; odor in hive, 201; storage of, 33–34, 39, 63–65, 193–195; supply and demand, 193–195
pollen basket, 23–24
pollen collection rate of colony, adjustment of, 63–65, 193–207, 256
pollen reserve: importance of, 63–65, 193; influence of, 194–207, 256
price, role of, 261

profitability of nectar sources. *See* nectar-source profitability
propolis, 24, 34, 42–43
protein: flux in colony, 39–43, 202–204; hunger, 206–207, 209, 248, 261; and the nutrition of bees, 43, 202–203, 206–207

quality of food sources. *See* nectar-source profitability
queen: cells of, 8, 34; court of, 11–12; mating behavior of, 7, 12, 31, 101–102; pheromones and, 11–13, 245, 258; role of, 8, 258; virgin, 14–16, 36

receiver bees, 40, 227–234. *See also* food-storer bees
recruitment: communication process, 36–39; rate of, 52–54, 135–136, 157–158, 209
regulation: of comb building, 61–63, 177–187; of nest temperature, 65, 212–213; of pollen collection, 63–65, 193–207, 256; of water collection, 65–66, 212–234, 257
reserves: of bees, 43, 173–174, 232–234 (*see also* inactive workers); of honey, 43, 214–215; of pollen, 43, 193–195; of water, 184, 214–215
round dance, 96, 100–101

salivary glands, 25–26, 41
sampling: of bees for observation, 82; of dances on the dance floor, 123–132; 260–261; of potential receiver bees in the unloading area, 111–113
scale hive, 43, 82–83
scent of food, 39, 77–78, 252
scout bees, 34–36, 54, 85–88
scouting. *See* food, bees' search for
searching for food: by colonies, 50–52; by individual bees, 85–88
search time, to find a food-storer bee: definition of, 82; reason for variation in, 111–113, 222–234; significance of, 107–110, 115–119, 169–170, 249–250, 256–257
seasonal effects, 97–98, 103–104, 183–184
selectivity among food sources, adjustment of, 59–61, 104–113
senses of worker bees, 26–28
shaking signal, 158–162, 242–244, 248
shared environment, information flow via the, 250–252
shifting between tasks. *See* tasks, switching between
signals, 248–252. *See also* pheromones; shaking signal; tremble dance; waggle dance
Simon, Herbert A., 3, 20, 241, 249
siphonophores, 5, 240, 244
Smith, John Maynard, 19
social physiology, definition of, 6
sound communication, 26, 27, 37–39, 92–93
spatial efficiency, 245–246
specialization on tasks. *See* tasks, specialization of bees among
storage space for honey, importance of, 189
supervision, absence of, 114–118, 136, 258–262
supply and demand relationships, 163, 193, 213–214, 255–258, 261
swarming, 34–36, 44, 179, 184
switching between tasks. *See* tasks, switching between

tasks: skill in performance of, 155, 241; spatial arrangement of in hive, 29–34, 88–89, 242–243; specialization of bees among, 31, 219–220, 227–228, 240–244, 246; switching between, 159, 207–209, 219, 227–234, 242–244; variety of in a colony, 29–31
task switching. *See* allocation, of bees among tasks
taste, 27, 201, 220, 226
telecommunications, 245, 261
temperature control. *See* thermoregulation
tempo of work, 117, 133–134
termites, 5, 240, 251
thermoregulation: in colonies, 43, 65–66, 212–213, 215–218, 250; in individuals, 93, 134, 218
thirst, 65–66, 220–221, 229
threshold for waggle dances: individual variation in, 98–102; mechanisms controlling adjustment of, 104–113; tuning of, 102–107
tongue-lashing, 41, 212–213
topology of a group's members, 244–247

touch, 201
traffic, at the hive entrance, 107–108, 157, 232–233
training, of bees, 78–79
travel-cost problem, 89, 245–246
tremble dance: behavior pattern of, 165–167; cause of, 167–170; effect of, 163–165, 170–173, 215, 242–244; location of in the hive, 167, 242–244; meaning of, 163; as a signal, 248; sounds of, 166–167
trophallaxis, 245

unemployed foragers. *See* forager bees
unloader bees. *See* food-storer bees
unloading: experience of, 108–113, 222–234; location of, 108; rejections during, 222–227; time required (*see* delivery time)
urn model, 111–113

vehicle concept, 4
ventilation, of the hive. *See* fanning; thermoregulation
vibration dance. *See* shaking signal

waggle dance: behavior pattern of, 36–39, 89–90; diurnal pattern of, 86; following of, 124–132, 219; liveliness of, 92–94, 128–130; location of in the hive, 48–49, 88–89; as a signal, 248; sounds of, 37–39, 92–93; threshold of, 92–93, 99, 102–107. *See also* dance duration
water: collection of, 40–41, 65–66, 212, 215–220; composition of, 40–41, 217; consumption of, 43, 212–213; receivers, 227–234; supply and demand, 213–214; use in cooling, 41, 65, 212–213
water-collection rate of colony, adjustment of, 65–66, 212–234, 257
water collectors, 40–41, 115, 218–220, 227
wax: glands, 25–26, 179–181; handling of, 26; production of, 29–30, 61–63, 177–187
weather, effects of, 118, 183, 190, 220–221
workers: anatomy of, 23–28; egg laying by, 5–13; life history of, 28–31; policing, 12–13; reproduction of, 5–13